Entropy-Enthalpy Compensation

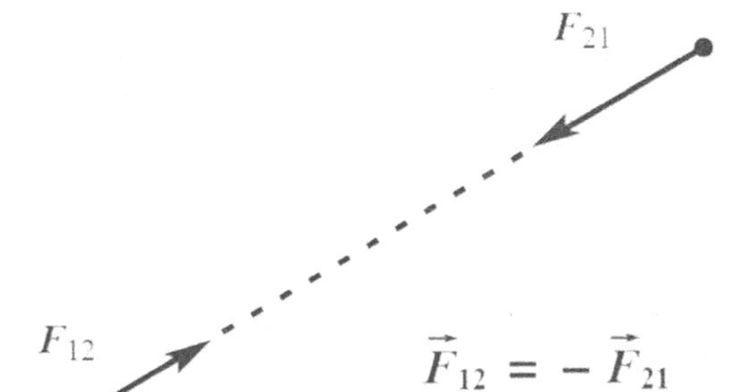

陰

Entropy–Enthalpy Compensation

Finding a Methodological Common Denominator through Probability, Statistics, and Physics

edited by

Evgeni Starikov
Bengt Nordén
Shigenori Tanaka

JENNY STANFORD
PUBLISHING

Published by

Jenny Stanford Publishing Pte. Ltd.
Level 34, Centennial Tower
3 Temasek Avenue
Singapore 039190

Email: editorial@jennystanford.com
Web: www.jennystanford.com

British Library Cataloguing-in-Publication Data
A catalogue record for this book is available from the British Library.

Entropy-Enthalpy Compensation: Finding a Methodological Common Denominator through Probability, Statistics, and Physics

ISBN 978-981-4877-30-5 (Hardcover)
ISBN 978-1-003-05625-6 (eBook)

In memory of

my teacher Prof. Dr. V. Ya. Maleev (1930–2018)

and my colleague Prof. Dr. Heinz Gamsjäger (1932–2016).

Contents

Preface xi

1. Entropy-Enthalpy Compensation and Exploratory Factor Analysis of Correlations: Are There Common Points? **1**

B. Nordén and E. B. Starikov

1.1 Introduction 2
1.2 Results and Discussion 4
1.2.1 Macroscopic Thermodynamics Considered from the Standpoint of van der Waals Equation of State 4
1.2.2 Correctness of Our Macroscopic-Thermodynamic Approach 10
1.2.3 What Is the Actual Difference between Gibbs and Helmholtz Functions? 11
1.2.4 The Actual Physical Sense of the EEC 15
1.2.5 Statistical-Mechanical Standpoint 18
1.2.6 What Is the Actual Probability Distribution behind the Statistical Mechanics? 20
1.2.7 Bayesian Statistical Thermodynamics of Real Gases 25
1.2.8 Applicability of Linhart's Approach to Real Gases 29
1.2.9 Is There Some Physical Connection between Boltzmann's and Gibbs' Entropy Formulae? 33
1.2.10 Can Our Approach Be Really Productive? 37
1.2.11 A Methodological Perspective 39
1.2.12 What Is the Actual Zest of Our Approach? 40
1.3 Conclusions 42
1.4 Outlook 43

Appendix 1 to Chapter 1 **49**

Appendix 2 to Chapter 1: Methodological Roots and Significance of Energetics **55**
A2.1 Introduction 55
A2.2 Energetics Is a Generally Applicable Concept 59
A2.2.1 Foreword 59
A2.2.2 The First Definition of Entropy 61
A2.2.3 Introduction and Preliminary Concepts 63
A2.2.4 Succinct Presentation of Thermodynamic Principles 76
A2.2.4.1 Joule–Mayer principle 77
A2.2.4.2 Principle of Carnot–Clausius 80
A2.2.5 Energy and the Forms of Sensitivity 87
A2.2.6 Third Part 102
A2.2.6.1 The muscle system and energetics 102
A2.2.6.2 Analogy between the muscle system and the nervous system 108
A2.2.6.3 Energetics and the nervous system 116
A2.2.6.4 Energetics and the nervous system (*Continued*) 122
A2.2.7 Thermodynamic Design of Some Mental Situations 129
A2.2.8 Summary and Conclusions 141
A2.3 Our General Conclusion 149
A2.3.1 The Balance of Bodies: Types of Body Balance 151
A2.3.2 Our Immediate Comment 153
A2.4 How to Employ the Ideas of Energetics: A Methodological Reiteration 159
A2.4.1 How to Make a Mechanical Theory of Mental Phenomena 159
A2.4.2 167
A2.4.3 173
A2.4.4 181

A2.4.5 The Senses: Theory of the Consecutive Images 184
A2.4.6 Demential Law by Paul Janet 188
A2.4.7 Psychoses 191
A2.4.8 Mechanical Representation of Psychic Phenomena 193
A2.4.8.1 Mechanism of dementia 195
A2.4.8.2 Mechanism of sensations 197
A2.4.8.3 Mechanism of psychoses 198
A2.4.8.4 Consequences 200
A2.4.8.5 Influence of the cerebral inertia coefficient 201
A2.4.9 Conclusion 210

Appendix 3 to Chapter 1: A Methodological Outlook **213**

2. Polynomial Exploratory Factor Analysis on Molecular Dynamics Trajectory of the Ras-GAP System: A Possible Theoretical Approach to Enzyme Engineering **239**

E. B. Starikov, Kohei Shimamura, Shota Matsunaga, and Shigenori Tanaka

2.1 Introduction 240
2.2 Results and Discussion 243
2.2.1 Linear Exploratory Factor Analysis Results 243
2.2.2 Nonlinear Exploratory Factor Analysis Results 245
2.3 Detailed Description of the Method 249
2.3.1 Difference between Confirmatory and Exploratory Factor Analysis 250
2.3.2 Difficulty Factors in Factor Analysis 253
2.3.3 Difference between Linear and Nonlinear Factor Analysis 256
2.3.4 The System under Study: Choosing the Proper Variables to Analyze the Macromolecular Dynamics 257
2.3.5 Technical Details of the MD Simulation and Data Processing 259

2.3.5.1 The system setup 259
2.3.5.2 MD simulation procedure 259
2.3.5.3 Analyses of MD trajectories 261
2.4 Conclusion 261

Supplementary Material to Chapter 2 **268**

Index 379

Preface

From our school years, we know that when discussing physics, we cannot avoid mentioning the mathematical notions of probability and statistics. Statistics is the conventional product of the probability theory. The mathematical foundations of statistics are very well developed to enable its efficient usage as a skillful tool in processing experimental findings (see, for example, Ref. [1] and the references therein).

With the aid of statistics, we may get useful, physically relevant information from experiments, with the ultimate correctness criterion being the appropriate number of experimental attempts dictated by statistics. To properly express the obtainable information, we first need to know the pertinent functional form that must be fitted to the measurement findings. We may somehow choose functional forms, parameterize them, and fit the constructs chosen to the measured data, but statistics help us estimate the goodness of fit—thus validating or helping us discard our parameters' estimates—and, hence, our functional form choice.

After having correctly estimated the necessary parameters, we may go in for diverse inferences, by duly testing our hypotheses using a number of the relevant statistical tools. Finally, all these efforts boil down to some credible picture of the physical phenomenon/process under study. Another efficient way of arriving at some credible physical picture is by numerical modeling, nowadays performed on computers, which is why we dub it "computer simulation."

In any kind of modeling, we start with physically plausible axioms or even just hypotheses and then mathematically arrive at some formulae, which must be compared to the relevant experimental data in order to prove or disprove the correctness of the hypotheses/axioms. However, at this point, physics does get into conceptual trouble. We take for granted that microscopically seen, all the macroscopically observable objects are consistent with some discrete distribution of microscopic particles (atoms and molecules). Hence, we build up general microscopic models explaining the diversity of macroscopically observable phenomena from the microscopic standpoint.

One important and successful step is well known since more than 100 years ago and is connected with the advancement of statistical and quantum mechanics. However, some general difficulties still remain. Indeed, as soon as we adopt the atomistic picture of matter and deal therefore with atoms and molecules, the problem of interatomic and intermolecular coupling arises. This coupling will obviously be manifested in the experimental data, but how should we theoretically interpret what we have measured?

Purely mechanically seen, if we treat atoms as small balls, the first idea would be to consider both repulsion and attraction among them. When treating repulsion, it is important to know whether our "balls" are rigid. Analytically treating the problem (writing down and solving the resulting equations of motion) after duly considering both latter points together is practically impossible. We must introduce some reasonable approximations.

This is just what the great peers Ludwig Boltzmann and Josiah Willard Gibbs were ingeniously able to do: initially considering the particles as rigid (hard balls) to treat the interactions among them just as rigid collisions (Boltzmann) or initially neglecting any coupling as a whole (Gibbs). The results of these simple approaches were both very encouraging, and one managed to draw a number of seminal conclusions.

Meanwhile, further detailed research (such as the spectroscopy of the "black-body radiation") posed the next problem: The atoms/molecules turn out to be far from rigid. Hence, to describe the coupling among the "soft" balls is a several-orders-of-magnitude more complicated mathematical task than to treat the "rigid"/statistically independent ones. Black-body radiation is where thermodynamics is of crucial significance. Even if the system under study is not involved in any kind of transformation process, black-body radiation is present and detectable. This is just the observable result of thermal motion, that is, the conventional dynamics the "soft" atomic/molecular constituents of the matter do take part in. To write down the anticipated equations of motion is throughout possible, but to solve them analytically is exceedingly difficult. Such was the quest for further approximations.

Most importantly, thermal motion looks empirically like a kind of chaotic *perpetuum mobile* (as clearly visualized by Brownian motion, for instance), without being so in fact: it represents the result of competition (and, eventually, equilibration/compensation) among some relevant driving forces (let us call them "enthalpic factors") on the one hand and some pertinent obstacles (let us call them "entropic factors") on the other hand, just as any realistic dynamical process.

In fact, Clausius, Boltzmann, and Gibbs suggested studying a largely simplified version of the above rather complicated full story. Their genius helped us to grasp one very important thing: as we do not really know and hence cannot foretell who might be the actual winner in the above-mentioned eternal competition, let us avoid considering the latter *explicitly* and instead take help from the probability theory and mathematical statistics to get an *implicit* description, which would still be correct and valid, nonetheless.[a]

Howbeit, the proper solution finally came up and turned out to be successful: quantum mechanics as developed by Max Planck, Louis de Broglie, Erwin Schrödinger, Werner Heisenberg, and numerous colleagues all around the world. Indeed, atoms/molecules are throughout "soft" because they consist of relatively hard nuclei and much softer clouds of electrons. These softer parts should obey some different mechanical laws in comparison to our conventional macroscopically based representations. What could the proper model for such dynamics look like?

Here comes the important turning point: a "desperate move" [2] by Planck to introduce the seminal idea of energy quantum. Moreover, Planck had simply taken over the expression for entropy as a logarithmic function of some magic probability, as conjectured by Boltzmann. Neither Boltzmann nor Planck was somehow analyzing the physical sense of this probability. Planck could just demonstrate that both his "desperate move" and Boltzmann's formula are correct and physically valid.

[a]Here is the ingenious and seminal sense of the famous Boltzmann/Gibbs formulae to connect entropy and probability: it is throughout possible to build up valid physical theories *without* taking care of *what* the probability and, consequently, the entropy really is. Remarkably, Carnot could arrive at the very eve of *building-up the true explicit description* of the whole story. Undoubtedly, he could also manage going along with this, and Carnot's ideas were definitely in the wind, for there were a number of colleagues all around the world who were duly picking up the banner Carnot had dropped. Do we know anything about them and their results?

Consequently, the further tremendous success of Planck's idea, which resulted in some properly and successfully working theoretical models, has two natural sides. In what follows, we do not discuss the apparent side, the success of the model involved, for that is already undoubted. The points we would like to discuss here are instead the difficulties brought about by fetishizing the model and how to overcome them. *Nos autem non judicas victores*... we are discussing *our own* dealing with the seminal research results.

Philosophically seen, the idea of the energy quantum is throughout materialistic: roughly speaking, we treat the energy notion as a kind of sausage, which might be cut into extremely thin slices. Does this, in itself, introduce some novel physical insight? The true answer: unfortunately not.

We shall next discuss the life and work of Simon Ratnowsky (1884–1945), who could prove that the very idea of energy quantum stems from considering the zero of absolute temperature, which is physically unreachable. Moreover, Ratnowsky could rigorously show that all the formulae derivable starting from the energy quantum hypothesis are well obtainable from Gibbs' results, without taking the latter hypothesis into account.[b]

The main aim of this communication has been to trigger a discussion of how to fruitfully use the powerful tools of mathematical statistics without declining into metaphysical discourses and addressing the question whether it is possible and useful to *explicitly* treat what we might very well describe *implicitly*?

Our answer is: Yes, sure!

Chapter 1 of this book demonstrates that an explicit picture of the ubiquitous interatomic/intermolecular coupling as a dialectic competition of actions (enthalpic factors) and counteractions

---

[b]Is quantum physics dispensable then? Do we need any philosophical yard work when dealing with such a model? Nice posers, which are not to be discussed here in detail, though they are definitely worth raising. Interestingly, the straightforward link between quantum mechanics and equilibrium statistical mechanics/ thermodynamics, just as Ratnowsky could clearly reveal, is currently being talked of [3–11]. To our mind, this discussion arises from nothing more than sheer difficulties with the rational interpretation of the former [12–19] and the logical inconsistencies of the latter [20]. The true reason of the difficulties just mentioned is the ignorance about the actual physical sense of the notion of entropy. Meanwhile, there are colleagues who could duly clarify the latter point. This is why, here we shall continue the discussion started in Ref. [20].

(entropic factors) is physically meaningful: the very fact of enthalpic-entropic compensation/equilibration might be viewed as a kind of microscopic phase transition.

This opens the door to studying mechanisms of diverse microscopic processes in detail. Of extreme importance here should be computer simulations (in particular, molecular dynamics). Nowadays, we might obtain credible results this way, for in a Cartesian space, we may generate the full account of trajectories and velocities of all the atoms in the molecules under study.

This way, we arrive at the true picture of the thermal motion plus any possible dynamics beyond the latter. The poser is, how do we distinguish between both? Of mechanistic interest is the definitive atomic/molecular dynamics beyond the overwhelming thermal motion, with the latter properly hiding, but not cancelling, the former.

Of much help in this situation is the explorative factor analysis (EFA) of correlations. Indeed, interatomic/intermolecular coupling might mathematically be considered a kind of correlation. Again, mathematically, we may transform the dynamics of the myriad particles coupled to each other into a proper picture of myriad correlated dynamical normal modes. We might now perform EFA, which is a unique method of multivariate statistics.

This is just the point where actual physics begins: EFA enables us to detect dynamical factors hidden by the noise. The former delivers the desired mechanistic picture, whereas the latter corresponds to the ubiquitous thermal motion.

While Chapter 1 discusses the general methodological modalities of the approach outlined above, Chapter 2 presents its detailed description and demonstrates how to employ it for studying the mechanisms of enzymatic reactions. This might help solve difficult problems of the rational protein/enzyme design, aside from opening other interesting strategic perspectives in the general theoretical physical chemistry field.

Evgeni Starikov

2021

References

1. F. James (2006). *Statistical Methods in Experimental Physics*, 2nd ed., World Scientific, New Jersey, London, Singapore, Beijing, Shanghai, Hong Kong, Taipei, Chennai.
2. H. Kragh (2000). Max Planck: the reluctant revolutionary, *Phys. World*, **13**, pp. 31–35.
3. M. O. Scully (2002). Extracting work from a single heat bath via vanishing quantum coherence II: microscopic model, *AIP Conference Proceedings*, **643**, pp. 83–91.
4. H. Linke (2003). Coherent power booster, *Science*, **299**, pp. 841–842.
5. M. O. Scully, M. S. Zubairy, G. S. Agarwal, H. Walther (2003). Extracting work from a single heat bath via vanishing quantum coherence, *Science*, **299**, pp. 862–864.
6. G. Sturm (2003). Warum die Tasse nicht nach oben fällt: Thermodynamik, Entropie und Quantenmechanik, *Quanten.de Newsletter*, Juli/August 2003, pp. 1-6
7. A. Cabello, M. Gu, O. Gühne, J.-Å. Larsson, K. Wiesner (2016). Thermodynamical cost of some interpretations of quantum theory, *Phys. Rev. A*, **94**, p. 052127.
8. C. E. A. Prunkl, C. G. Timpson (2018). On the thermodynamical cost of some interpretations of quantum theory, *Stud. Hist. Philos. Sci. B*, **63**, pp. 114–122.
9. D. Castelvecchi (2017). Battle between quantum and thermodynamic laws heats up, *Nature*, **543**(7647), pp. 597–598.
10. A. Cabello, M. Gu, O. Gühne, J.-Å. Larsson, K. Wiesner (2019). The thermodynamical cost of some interpretations of quantum theory. Reply to Prunkl and Timpson, and Davidsson, arXiv.org > quant-ph > arXiv:1901.00925.
11. F. Binder, L. A. Correa, C. Gogolin, J. Anders, G. Adesso (2019). *Thermodynamics in the Quantum Regime: Recent Progress and Outlook*, Springer Nature Switzerland AG.
12. J.-Å. Larsson (2005). The quantum and the random: similarities, differences, and "contradictions," *AIP Conference Proceedings*, **810**, pp. 353–359.
13. R. F. Streater (2007). *Lost Causes in and Beyond Physics*, Springer, Berlin, Heidelberg, New York.
14. C. A. Fuchs (2011). *Coming of Age with Quantum Information: Notes on a Paulian Idea*, Cambridge University Press, Cambridge, New York, Melbourne, Madrid, Cape Town, Singapore, Sao Paulo, Delhi, Dubai, Tokyo, Mexico City.

15. C. A. Fuchs, N. D. Mermin, R. Schack (2014). An introduction to QBism with an application to the locality of quantum mechanics, *Am. J. Phys.*, **82**, pp. 749–754.
16. H. Atmanspacher, C. A. Fuchs (2014). *The Pauli-Jung Conjecture and Its Impact Today*, Imprint Academic, Great Britain.
17. H. C. von Baeyer (2016). *QBism: The Future of Quantum Physics*, Harvard University Press, Harvard, USA.
18. B. Nordeén (2016). Quantum entanglement: facts and fiction; how wrong was Einstein after all? *Q. Rev. Biophys.*, **49**, p. e17.
19. J. Bricmont (2018). *Quantum Sense and Nonsense*, Springer International Publishing AG.
20. E. B. Starikov (2019). *A Different Thermodynamics and Its True Heroes*, Pan Stanford Publishing, Singapore.

Acknowledgments

Evgeni B. Starikov would like to express his sincere gratitude to Dr. Ramandeep Singh Johal, associate professor (physics), Indian Institute of Science Education & Research, Mohali, Punjab, India, for detailed and thorough discussions on the Bayesian approach to statistical mechanics, and to Dr. Michael Eckert (Deutsches Museum, Munich, Germany) for providing us with the copy of an important historical document cited in Appendix 3.

The authors would especially like to acknowledge editors Archana Ziradkar and Bhavana Singh for careful, critical, thoughtful editing, which has significantly increased the quality of the publication at hand.

Chapter 1

Entropy-Enthalpy Compensation and Exploratory Factor Analysis of Correlations: Are There Common Points?

B. Nordén[a,*] and E. B. Starikov[a,b]
[a]*Chemistry and Chemical Engineering, Chalmers University of Technology, Göteborg, Sweden*
[b]*Graduate School of System Informatics, Kobe University, 1-1 Rokkodai, Nada, Kobe 657-8501, Japan*
starikow@port.kobe-u.ac.jp

Here we analytically demonstrate that entropy-enthalpy compensation (EEC) may be physically understood as a general expression of microscopic phase transition. This idea can be fruitful when studying in detail the microscopic mechanisms of diverse physical-chemical processes. However, it clearly goes beyond conventional equilibrium thermodynamics and statistical physics. Does this mean that we must completely discard the latter or thoroughly modify the former or both? Here we discuss how to approach this important methodological problem. Aside from

**BN: In memory of my colleague Ingmar Grenthe (1933–2020).*

Entropy-Enthalpy Compensation: Finding a Methodological Common Denominator through Probability, Statistics, and Physics
Edited by Evgeni B. Starikov, Bengt Nordén, and Shigenori Tanaka

ISBN 978-981-4877-30-5 (Hardcover), 978-1-003-05625-6 (eBook)
www.jennystanford.com

biophysical chemistry, as EEC is frequently observed in biological contexts, the theme, as it is, should be of general interest. An aim of this communication is to trigger a discussion of whether it will be possible to *explicitly* treat what we might very well describe *implicitly*.

1.1 Introduction

We continue discussing a different approach to thermodynamics and statistical mechanics (energetics) triggered long ago by numerous authors—and summarized by us (see Refs. [1–17] and the references therein). We suggest reinferring the general functional dependence of thermodynamic potentials (both Helmholtz and Gibbs functions A and G, respectively) on the intensive variables of state P (pressure) and T (absolute temperature). Indeed, already some 150 years ago Gibbs had published such formal inferences, but a number of novel results calling for our urgent attention have appeared in the meantime.

Before that, Engelbrektsson had used energetics to derive a universal thermodynamic equation of state and Franzén could experimentally verify it (see Ref. [1] for details). Of ultimate importance would be to try building up the statistical-mechanical foundations for their results.

Here we suggest a way to tackle this project. We extend the approach by Linhart [1]. He was considering ideal gas models solely, like Boltzmann and Gibbs. Following Gibbs' approach, Linhart could not only infer the Boltzmann–Planck formula $S = k{\cdot}\ln W$ but also clarify its actual physical sense. Now we must extend Linhart's approach to dealing with realistic non-ideal aggregate states of matter. Hence, we must consider both repulsive and attractive forces among atoms—unlike in an ideal gas, where rigid collisions are the only way of interatomic coupling.

We suggest employing the approach used in Ref. [18], where the authors infer a thermodynamic equation of state. We pick up the idea from Ref. [18] that van der Waals approximation should be the starting point to integrate the Clapeyron–Clausius equation of state, which is thermodynamically rigorous. Thus, van der Waals equation reveals the actual functional dependence of the system's volume on P and T. Indeed, the work in Ref. [18] is not the sole approach to the problem, and we shall discuss the pros and cons of the approaches presently available.

We use the statistical-mechanical approach to go beyond itself, so we need to reanalyze its foundations. The general poser is the actual physical meaning of the probability notion. We do encounter serious problems when trying to find the answer. Herewith, we just trigger the discussion. Equilibrium thermodynamics forms the conceptual basis of statistical mechanics. The former is not logically closed, for its entropy notion has an unclear physical sense. Instead of clarifying this point itself, equilibrium thermodynamics calls for the probabilistic interpretation of entropy and thus needs statistical mechanics. In turn, the latter cannot clarify the physical sense of the entropy notion itself as well. As a result, we arrive at a clear-cut methodological problem: Do we really need the statistics? If so, to what extent and how should we use it? In fact, the actual sense of entropy, while well known, is not widely recognized [1]. Energetics reveals the clear physical sense of entropy: it is correspondent to ubiquitous hindrances/obstacles/resistances (HOR) on the way to success of any realistic process.

Any process requires some driving force to start with. Kinetic energy is the source of the latter. When speaking of thermal motion, it is logical to relate the kinetic energy to the heat content, that is, enthalpy. Starting a process immediately triggers the relevant HOR. Newtonian mechanics states that the source of kinetic energy is potential energy. In turn, the latter results from coupling among the systems' parts. Hence, correlations should bear a profound physical sense. Further, Newton's third basic law reads: "Any action must encounter counteraction." Hence, any driving force must encounter counteraction due to ubiquitous HOR. There is more to the story: the higher the driving force, the more the HOR. The good news, according to Carnot, Clausius, and Lord Kelvin, is that HOR are never infinite. They must reach their maximum level. If we have a driving force enough to *equilibrate/compensate* the maximum of HOR, we achieve the aim of the process under study. Neither Carnot nor Clausius nor Thomson had time to formulate the story properly. It seems obvious due to the ingenious efforts of several colleagues [1].

It is the entropy-enthalpy compensation (EEC) that describes the interplay between driving forces and the relevant HOR summary. Nevertheless, we should never discard probability theory and statistics. The point is how to use them properly. Linhart proves no contradiction between the "energetic" and "probabilistic" entropy pictures [1]. Moreover, powerful computer-based simulation approaches relieve us nowadays of the pain when treating complicated mathematical/numerical problems.

Here we suggest a combination of analytical and computer-based approaches resulting in exploratory factor analysis of correlations (EFAC). The latter is a multivariate statistical method, so we continue using statistics. We discuss the physical sense of EEC and how EFAC may help describe the impact of the latter.

1.2 Results and Discussion

1.2.1 Macroscopic Thermodynamics Considered from the Standpoint of van der Waals Equation of State

To start with, the authors [18] suggest considering the conventional van der Waals equation

$$\left[P+\left(a/V^2\right)\right]\left(V-b\right)=RT,\quad \text{recast to read as follows:}$$

$$V=b+\frac{RT}{P+\left(a/V^2\right)}. \tag{1.1}$$

Equation 1.1 delivers the desired functional dependence $V(P,T)$; as we recall the term $\left(a/V^2\right)$, introducing the gas non-ideality correction, reads in the first approximation as follows:

$$\left(\frac{a}{V^2}\right)=\left(\frac{a}{\left(\frac{RT}{P}\right)^2}\right).$$ Indeed, the latter could still obey the thermodynamic equation of state for an ideal gas and renders the van der Waals equation the "first non-ideality correction."

This leads to the following expression for $V(P,T)$ [18]:

$$V = b + \frac{RT}{P\left[1 + \left(aP/R^2T^2\right)\right]}. \tag{1.2}$$

Here we suggest continuing the above functional iteration up to its fourth step. Next, we place the third successive expression for $V(P,T)$ into Maxwell relationships to derive the functional dependencies of the internal energy, entropy, and enthalpy on P and T. This should bring us to the desired functional dependencies for the Helmholtz and Gibbs functions.

We employ an iterative solution to the van der Waals equation for a system's volume V as a function of P and T, with R standing for the universal gas constant:

$$V(i+1) = b + \frac{RT}{\left(P + \frac{a}{V(i)\cdot V(i)}\right)}, \quad \text{with } i = 0, 1, 2, \ldots n;$$

$$V(0) = \frac{RT}{P}. \tag{1.3}$$

By solving Eq. 1.3 up to i = 3, we get the following compact expression for the function $V(P,T)$ at i = 2:

$$V(P,T) = b + \frac{R^7T^7}{P\left(a^3P^3 + 2a^2P^2R^2T^2 + aPR^4T^4 + R^6T^6\right)}. \tag{1.4}$$

If we deal with systems where the number of particles, N, remains constant (e.g., in the absence of mass exchange and/or chemical reactions), we might consider the following Maxwell relationship:

$$\left(\frac{\partial S}{\partial P}\right)_{T,N} = -\left(\frac{\partial V}{\partial T}\right)_{P,N}. \tag{1.5}$$

Next, we need the first partial derivative of Eq. 1.4 by T, so the compact expression for the necessary derivative should read as follows:

$$\left(\frac{\partial V}{\partial T}\right)_{P,N} = \frac{R^7T^6\left(7a^3P^3 + R^6T^6 + 3aPR^4T^4 + 10a^2P^2R^2T^2\right)}{P\left(a^3P^3 + 2a^2P^2R^2T^2 + aPR^4T^4 + R^6T^6\right)^2}. \tag{1.6}$$

As per Eq. 1.5, to get the dependence of entropy S on P and T, we have to take Eq. 1.6 with a minus sign and integrate it over P. Before doing so, we consider that the van der Waals fitting parameter a, that is, the cohesion pressure, is of the order $(1 \div 600)\cdot 10^{-3}\left[\left(\mathrm{J}\cdot \mathrm{m}^3\right)/\mathrm{mol}^2\right]$ for realistic substances [19]. Then the terms containing $a^3$and a^2in Eq. 1.6 must be much smaller than its other terms, and we might safely neglect them before performing the integration. Thus, for the entropy $S(P,T)$, we get the following handy expression:

$$S(P,T) \approx R\left(2 - \frac{2}{1+\dfrac{R^2T^2}{aP}} + \ln\left[1+\frac{R^2T^2}{aP}\right]\right). \tag{1.7}$$

The functional form of Eq. 1.7 closely resembles the Boltzmann–Planck formula $S = k \cdot \ln W$ derived by Linhart when adopting the ideal gas model, namely $S = Const \cdot \ln(1 + x^{K})$; $x \equiv \dfrac{T}{T_{\mathrm{ref}}}$ [8, 12], with T_{ref} being some reference temperature. Remarkably, by taking the limit of Eq. 1.7 at $T \to 0$ or at $P \to \infty$, we find that S would tend to zero, as expected; otherwise entropy should tend to infinity with the logarithm's rate if $P \to 0$ and/or $T \to \infty$.

Hence, Eq. 1.7 guarantees entropy decrease at higher pressures and lower temperatures and its increase at lower pressures and/or higher temperatures. Such properties suggest choosing the pertinent scales for both T and P to deal with the entropy values, which must be finite if they are nonzero. This should then be in full accordance with the second basic law of thermodynamics in the Horstmann–Liveing representation: the entropy must be the summary of ubiquitous HOR with respect to driving forces for realistic processes [1]. Indeed, zero driving force is correspondent to zero entropy, while any increase in the former entails a corresponding increase of the latter up to its maximum.

Next, we try inferring the approximate functional dependence for the enthalpy, $H(P, T)$.

We note that the full differential of H and the relevant Maxwell relationship should be cast as follows:

$$dH = TdS + VdP = T\left(\frac{dS}{dT}\right)_P dT + \left(T\left(\frac{dS}{dP}\right)_T + V\right)dP; \left(\frac{dS}{dP}\right)_T = -\left(\frac{dV}{dT}\right)_P. \tag{1.8}$$

Then, after considering Eqs. 1.4, 1.6, and 1.7, while neglecting the terms containing $a^3$and a^2, for the partial derivatives of H, we finally get:

$$\frac{\partial H}{\partial T}=T\left(\frac{dS}{dT}\right)_P\approx\frac{2R^3T^2\left(3aP+R^2T^2\right)}{\left(aP+R^2T^2\right)^2}; \tag{1.9}$$

$$\frac{\partial H}{\partial P}=T\left(\frac{dS}{dP}\right)_T+V=V-T\left(\frac{dV}{dT}\right)_P\approx b+\frac{R^7T^7}{P\cdot\left(aPR^4T^4+R^6T^6\right)}$$
$$-\frac{2R^7T^6\left(3aPR^4T^4+R^6T^6\right)}{P\cdot\left(aPR^4T^4+R^6T^6\right)^2}; \tag{1.10}$$

To arrive at the final approximate expression for $H(P, T)$, we must integrate Eqs. 1.9 and 1.10 over T and P, respectively, and sum up the results. Before doing so, it is worth noting that the van der Waals fitting parameter b, that is, the covolume/excluded volume, is numerically lower than or equal to four times the volume of a separate microparticle, being the pertinent atom or molecule. This parameter is of the order $(17\div300)\cdot10^{-6}\left[\mathrm{m}^3/\mathrm{mol}\right]$ for realistic substances [19], and therefore the terms containing b as a coefficient might have significantly lower values than the other ones.

In Appendix 1, we discuss in detail the actual sense and consequences of the approximation method we employ.

Therefore, we get the following approximation for $H(P, T)$:

$$H(P,T)\approx RT\cdot\left(4+\frac{b}{a}RT-\frac{4}{1+\dfrac{R^2T^2}{aP}}\right). \tag{1.11}$$

Now we have practically everything to use Eqs. 1.7 and 1.11 to cast the approximation for the general Gibbs function $G(P, T)$ with a constant particles number, N:

$$G(P,T)=H(P,T)-T\cdot S(P,T)\approx RT\cdot\left(2+\frac{b}{a}RT-\frac{2}{1+\dfrac{R^2T^2}{aP}}-\ln\left[1+\frac{R^2T^2}{aP}\right]\right). \tag{1.12}$$

Of significant interest would also be to arrive at the approximate general function $A(P, T)$, that is, the Helmholtz function. First, we have to derive the relevant approximation for the internal energy $U(P, T)$.

The corresponding full differential might be cast as follows, employing the relevant Maxwell relationship $\left(\frac{dS}{dP}\right)_T = -\left(\frac{dV}{dT}\right)_P$:

$$dU = TdS - PdV = \left(T\left(\frac{dS}{dT}\right)_P - P\left(\frac{dV}{dT}\right)_P\right)dT - \left(T\left(\frac{dV}{dT}\right)_P + P\left(\frac{dV}{dP}\right)_T\right)dP. \quad (1.13)$$

Then, after considering Eqs. 1.4–1.7, while neglecting the terms containing a^3 and a^2, we finally get for the partial derivatives and their integrals:

$$\frac{\partial U}{\partial T} = T\left(\frac{dS}{dT}\right)_P - P\left(\frac{dV}{dT}\right)_P \Rightarrow \int \frac{\partial U}{\partial T} dT \approx \frac{R^3T^3}{aP + R^2T^2}; \quad (1.14)$$

$$-\frac{\partial U}{\partial P} = T\left(\frac{dV}{dT}\right)_P + P\left(\frac{dV}{dP}\right)_T \Rightarrow \int \frac{\partial U}{\partial P} dP \approx \frac{R^3T^3}{aP + R^2T^2}. \quad (1.15)$$

Now we must sum up the results in Eqs. 1.14 and 1.15 and get the approximate expression for the internal energy $U(P, T)$:

$$U(P,T) \approx 2\frac{R^3T^3}{aP + R^2T^2}. \quad (1.16)$$

Using Eqs. 1.7 and 1.16, we cast the approximation for the Helmholtz function $A(P, T)$ if the number of particles N is constant:

$$A(P,T) = U(P,T) - T \cdot S(P,T) \approx RT \cdot \ln\left[\frac{1}{1 + \frac{R^2T^2}{aP}}\right]. \quad (1.17)$$

A comparison of Eqs. 1.12 and 1.17 clearly shows that both Gibbs and Helmholtz free energies are possessed of largely similar functional dependencies on P and T, as might be expected, whereas some mathematical difference between them is concentrated in the algebraic addition to their logarithmic-algebraic parts, which is explicitly present in G, thus underlining its sheer absence in A.

Next, we must check whether Eqs. 1.7 and 1.11 might in principle describe the EEC. We introduce the appropriate reference pressure and temperature values—as discussed above—and thus the proper scales for both P and T so that $T \equiv \dfrac{t}{t_{\text{ref}}}$; $P \equiv \dfrac{p}{p_{\text{ref}}}$ in Eqs. 1.1–1.17, which are just relevant coefficients not affecting the physical dimensions of T and P, respectively. Then we choose some reasonable values for the parameters a and b, as well as for the variable P, to compare the entropy and enthalpy function trends using the graph in Fig. 1.1.

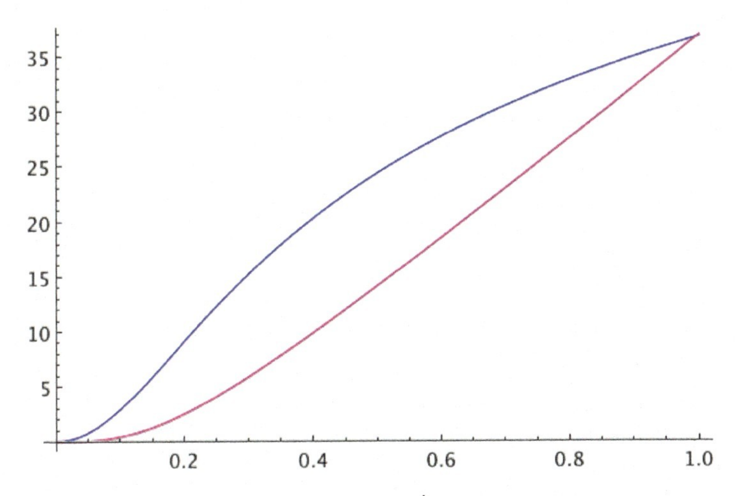

Figure 1.1 The relative temperature $T \equiv \dfrac{t}{t_{\text{ref}}}$ is abscissa, while the ordinate contains the values for entropy (Eq. 1.7, blue) and enthalpy (Eq. 1.11, magenta).

Figure 1.1 clearly shows that the enthalpy and entropy graphs are possessed of rather similar trends and cross each other at T = 1, that is, it should be throughout possible to treat the reference temperature t_{ref} as the EEC temperature. Now we would like to use the same parameter values to visualize the behavior of all the approximations we have here in trying to clarify the physical sense of the valid EEC. We get Fig. 1.2.

Figure 1.2 demonstrates that whereas enthalpy compensates entropy at the EEC temperature, Helmholtz energy reaches some value accordingly, Gibbs free energy comes to zero, and both free energy functions have zero values at zero temperature.

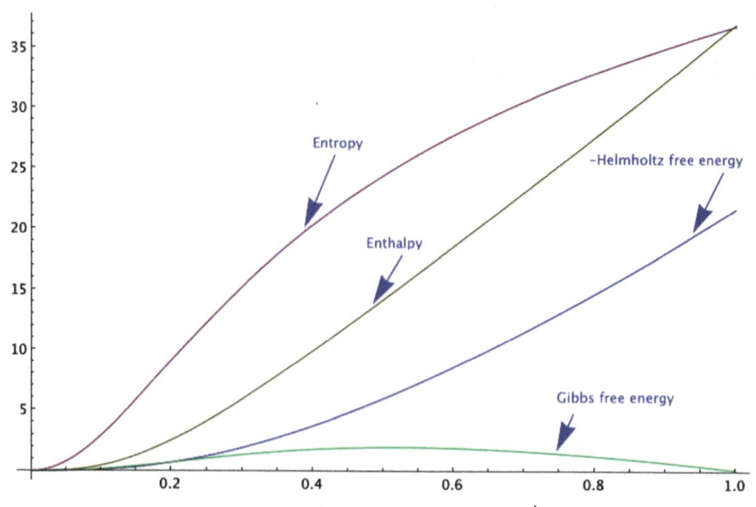

Figure 1.2 Again, here the relative temperature, $T \equiv \frac{t}{t_{\mathrm{ref}}}$ is an abscissa, while the ordinate contains the values for entropy (Eq. 1.7), enthalpy (Eq. 1.11), Gibbs free energy (Eq. 1.12), as well as Helmholtz free energy (Eq. 1.17). The values of the latter have been taken with a negative sign.

Therefore, as it is based on the behavior of the Gibbs function, it is okay to consider the EEC as a kind of phase transition, just as we have already suggested earlier [11, 14].

Remarkably, such a result is also in full accordance with our previous findings [15] on the physical sense of the parameters for a valid linear EEC. Indeed, the valid linear regression expression reads $H = T_c \cdot S + a$, where H stands for enthalpy and S for entropy. Then, the regression slope T_c is the EEC temperature and the regression intercept a is the relevant value for the Helmholtz free energy.

1.2.2 Correctness of Our Macroscopic-Thermodynamic Approach

The poser immediately arising after reading the above story would be:

If Eqs. 1.7, 1.11, 1.12, 1.16, and 1.17 are correct indeed, then how about the correctness of the relationships in Eq. 1.18 (introducing the thermodynamic functions by definition)?

$$H(P,T) \equiv U(P,T) + P \cdot V(P,T);$$
$$G(P,T) \equiv A(P,T) + P \cdot V(P,T). \tag{1.18}$$

To perform the check, we compare Eq. 1.11 with the proper combination of Eq. 1.4 and Eq. 1.16. Accordingly, Eq. 1.12 should be comparable with the proper combination of Eq. 1.4 and Eq. 1.17. Finally, after considering our above-mentioned approximation guidelines for the coefficients a and b, we arrive at the following conclusion:

$$H(P,T)_{\text{Eq.1.11}} - U(P,T)_{\text{Eq.1.16}} - P \cdot V(P,T)_{\text{Eq.1.4}} = -bP + \frac{bR^2T^2}{a} + \frac{R^2T^2}{aP + R^2T^2} = 0;$$
$$G(P,T)_{\text{Eq.1.12}} - A(P,T)_{\text{Eq.1.17}} - P \cdot V(P,T)_{\text{Eq.1.4}} = -bP + \frac{bR^2T^2}{a} + \frac{R^2T^2}{aP + R^2T^2} = 0. \tag{1.19}$$

Both equations in Eq. 1.19 are true if $aP\sqrt{b} \equiv R^{3/2}T^{3/2}\sqrt{a + bRT}$. Hence, our approach must introduce some functional relationship between the relative pressure and the relative energy variables. After substituting the expression for the $P(T)$ function just revealed in the logarithmic-algebraic term of Eq. 1.7, we get $\ln\left(1 + \sqrt{\frac{T}{\frac{a}{bR} + T}}\right)$, which is Linhart's logarithmic expression for the entropy function $\ln\left(1 + \left(\frac{T}{T_{\text{ref}}}\right)^K\right)$. The poser is then, what is the parameter K? We discuss this below.

Next, let us compare Eq. 1.11 to Eq. 1.12 and Eq. 1.16 to Eq. 1.17. We reveal the intrinsic difference between the Gibbs and Helmholtz functions, which is clearly visualized by Fig. 1.2.

1.2.3 What Is the Actual Difference between Gibbs and Helmholtz Functions?

While the former does contain both algebraic and algebraic-logarithmic terms, the latter contains only the logarithmic-algebraic term. The mathematical reason for such a result is based upon the functional forms of the enthalpy and internal energy, respectively. The functional form of enthalpy is quite different from that of the algebraic part of entropy, so their combination in the Gibbs function results in some residual algebraic term in addition to the logarithmic-algebraic one explicitly introduced by the entropy function.

Instead, the functional form of the internal energy is identical to that of the algebraic part of the entropy, so their combination in the Helmholtz function results in their cancelling each other. Thus, the Helmholtz function remains with the sole logarithmic-algebraic term introduced by the entropy function.

To try deciphering the physical sense behind this math, we would like to recall here the conceptual differences between the enthalpy and internal energy notions.

Indeed, in the well-known handbook on thermodynamics by Zemansky [20], we find a clear physical definition of what enthalpy is: "Enthalpy comprises a system's internal energy, which is the energy required to create the system, plus the amount of work required to make room for it by displacing its environment and establishing its volume and pressure." In other terms, enthalpy must mean nothing more and nothing less than the heat content of a system, that is, the total amount of energy within a substance as it is, being as a result both kinetic and potential energy of the ubiquitous thermal motion, if the system does not take part in some externally or internally driven process. Meanwhile, internal energy must be just the same, but *without the amount of work against the surrounding*, which is but inherent in enthalpy.

Nevertheless, the ubiquitous thermal motion as it is, like any realistic physical phenomenon, should anyway possess both the relevant driving forces and the HOR to overcome, which should be *intrinsically* present in the enthalpy function, but apparently *in the implicit form*. On the contrary, the Helmholtz function explicitly separates all the contributions due to the internal energy U from the actual "work required for making room for the system by displacing its environment and establishing its volume and pressure." Physically, the latter phenomenon must be the work against the influence of the system's surroundings.

Hence, we might assume that in performing such a work, the system is counteracting the HOR due to its surroundings. Consequently, it is this part of the work that should be inherently connected to the system's entropy, which is then duly introduced as the explicit entropic term $-T \cdot S(P,T)$ into the Helmholtz function but not into the explicit expression for enthalpy.

Thus, the Gibbs function does contain both implicit and explicit entropic contributions, whereas the Helmholtz function contains only some part of them.

To sum up this discussion, it is expedient to open the book by Karl Trincher (1910–1996), entitled *Die Gesetze der biologischenThermodynamik* (*The Laws of Biological Thermodynamics*; Urban und Schwarzenberg, Wien-München-Baltimore, 1981).

Among a lot of insightful information, the book contains Section 3.11, entitled "Die Umwandlungsformen der Energie: Arbeit und Wärme"("The Energy Conversion Forms: Work and Heat").

Trincher invites us to recast a glance into the seminal ideas of Robert Julius Mayer, who never separated the conservation of energy from its transformation. Instead, his unique approach was to try consequently answering three basic posers concerning how to treat the systems under study (Mayer was mostly dealing with biological systems, but his approach is definitely general):

- Is the system solely an objective working body?
- Is the system solely an objective consumer of heat?
- Is the system playing the both above-mentioned roles?

A positive answer to the first poser: It is throughout possible only when a finite volume of ideal gas obtains a heat portion from some infinitely large thermostat so that the temperature of the latter is equal to that of the former. In such an ideal case, any transfer of heat from the thermostat to the gas would not lower the temperature of the latter. Hence, the entire heat portion thus transferred will be 100% transformed into the isothermal work of the gas: $\partial Q_T = p \times dv$, with the heat portion on the left and the product of the pressure p and the volume change dv on the right.

A positive answer to the second poser: It is throughout possible only when the ideal gas having temperature T contacts a thermostat having temperature $(T + dT)$ so that obtaining a heat portion from the latter would solely increase the internal energy of the former without changing the gas volume. In such an ideal case, the increase in the gas's internal energy U_G will just be proportional to the temperature difference dT: $\partial Q_{T+dT} \equiv U_G \equiv C_V \cdot dT$, with C_V being constant isochoric heat capacity.

In comparing the first two trains of thoughts, we notice that in the unrealistic ideal gas case, temperature might solely be an adjustable parameter. In effect, the latter is an intensive thermodynamic variable. Therefore, any realistic consideration must deal with the extensive thermodynamic variables and all the relevant concepts as the proper functions of the intensive variables.

A positive answer to the third poser: It is throughout possible only when the ideal gas having temperature T contacts a thermostat having temperature $(T + dT)$ so that obtaining a heat portion from the latter would both increase the gas's internal energy and be working at a constant pressure. This ideal case may well describe the realistic situations if we design some device to ensure the control of pressure in the system. This way we may regulate the pressure on the gas and thus the internal pressure of the latter as well. As a result, the heat coming from the thermostat will be spent not only to increase the gas's internal energy U_G but also to work: $\partial Q_{p,T+dT} \equiv dU_G + p \cdot dv \equiv C_p \cdot dT$, with C_p being isobaric heat capacity.

The mathematical difference between the trains of the second and third thoughts consists in the properties of the heat capacities C_V *and* C_p. Whereas C_V is just a constant for an ideal gas, the C_p becomes a function of pressure, even in the unrealistic case of an ideal gas. To be fully rigorous, we notice that being solely parameters in the latter case, pressure and temperature are intensive thermodynamic variables in fact.

To analyze the physical sense of Helmholtz and Gibbs functions was not Trincher's goal, but thinking over this section of his monograph ought to be throughout helpful for our present discussion. We see that the Helmholtz function resulted from following the trains of the first and second thoughts by separating them from each other. Meanwhile, the Gibbs function was an attempt to follow the train of the third thought, *which does stress the ultimate significance of the* ***actual interplay between being both an objective working body and objective consumer of heat***.

Remarkably, this is exactly the very idea by Carnot, and this is just how we might consequently consider both energy conservation and energy transformation without cutting the former from the latter.

Moreover, it is just this way that we do follow the original train of thought by Robert Julius Mayer without falling into the trap by Rudolf Clausius, who did erroneously produce two separate "fundamental" laws from the actually unique one.

How could Clausius commit such a fault, and what are the actual consequences/implications of it? These are good posers, which we shall analyze elsewhere.

Coming back to the actual methodological difference between Helmholtz and Gibbs functions, we see that the Helmholtz free energy function would deliver a perfect description of the realistic situation when the system under study is in equilibrium by itself and with its environment so that nothing noteworthy might happen in it, except for the ubiquitous thermal motion. Instead, the Gibbs free energy function must be used to analyze specific processes beyond the latter.

1.2.4 The Actual Physical Sense of the EEC

Now, what should be the pertinent physical sense of the valid EEC?

The EEC must be an inherent thermodynamic feature, for the enthalpy function anyway contains implicit entropic contributions, which are just added to all the energy terms already inherent in the internal energy function.

Purely mathematically, Eq. 1.7 reveals that the entropy function should possess both algebraic and logarithmically algebraic parts. Though both must be physically important, it is the latter one that seems to bear an intrinsic relationship to the ubiquitous thermal motion in any realistic physical-chemical system. Indeed, in considering the ideal gas model, Boltzmann has ingeniously introduced the notion of some "magic probability W" as the very intrinsic physical basis for the entropy being its logarithmic function. Solely, Linhart could persuasively demonstrate [1, 8, 12] that the W in question should be just a handy algebraic function of the absolute temperature to be placed under the logarithmic sign. Hence, the W does stand for the proper general description of the ubiquitous chaotic thermal motion.

There is much more to the story, for in their ingenious works, neither Boltzmann nor Gibbs or Linhart could have exceeded the ideal gas model framework. This fact has caused us to try and drive their common train of thoughts to its next logical station, namely, to consider realistic non-ideal physical-chemical systems.

We can see that the approximate mathematical form of the logarithmic term in the entropy function to describe non-ideal systems does largely mimic Linhart's result. Meanwhile, we need to make an important amendment under the logarithmic sign. The algebraic function of both temperature and pressure should be situated here, but not the function of temperature only.

Further detailed comparison between Linhart's logarithmic term and ours reveals that the former expression being cast as $S = \frac{C_\infty}{K} \cdot \ln\left(1 + x^K\right); C_\infty \approx 3 \cdot R$, with R being the universal gas constant, does contain a further parameter K, which Linhart introduces as a scalar fitting parameter. For general non-ideal systems, K must be a function of both temperature and pressure. For realistic physical-chemical systems, there must be some nontrivial mathematical transformation of Linhart's expression that brings us to something like the approximation cast in Eq. 1.7.

We could mathematically reveal here a plausible logarithmically algebraic term of entropic relevance. This term shows up in different conventional thermodynamic functions. The approximations we introduce here can demonstrate the actual interplay among the driving forces (i.e., enthalpic factors) and the relevant HOR (i.e., entropic factors).

Indeed, the latter compensation should be inherent in any realistic physical-chemical process, be it even just the stationary thermal motion, whose "immensely chaotic stationarity" must anyway result from the mutual compensation of some basic enthalpic and entropic effects. The presence of some externally or internally driven "non-equilibrium/irreversible" process should anyway affect the stationary thermal motion experienced by the system(s) taking part in such a process.

Therefore, to properly investigate the mechanism of any realistic process, it is necessary to reveal how we might characterize the compensation just mentioned. A valid suggestion can be to follow the EEC using all the pertinent thermodynamic functions, as discussed here.

To sum up, it is then throughout clear as well why there must be some EEC and why the Helmholtz function and enthalpy do exhibit quite relevant functional trends, as seen on Fig. 1.2. Moreover, the latter figure also reveals why the Gibbs function is much more suited than the Helmholtz function for analyzing the actual conditions and mechanisms of realistic physical-chemical processes, just as practical physical chemistry could already persuasively demonstrate.

Indeed, it is the Gibbs function, not the Helmholtz function, that must indicate the actual equilibrium between the progress of some realistic process and the combination of all the external and internal hindrances/obstacles to the latter: It is the Gibbs function that reaches its zero values both at the zero temperature and at the EEC compensation temperature.

The former minimum of G corresponds to the initial equilibrium situation, when there are no driving forces, and consequently no entropic effects, whereas the latter minimum introduces the temperature at which the driving forces could have successfully compensated the maximum of all the entropic effects.

Finally, the temperature at which entropic effects do arrive at their maximum value must be the one at which the Gibbs function reaches its maximum value, whereas the maximum value of the Helmholtz function reveals the total energy amount the driving forces require *both* to compensate the maximum entropy *and* to achieve the actual aim of the process (i.e., the total energy of the combined enthalpic or entropic effects).

The Helmholtz function is hence right for studying the modalities of some physical-chemical system with respect to its surrounding, but in the absence of any external or internal realistic process unidirectional in time, that is, the one having its definite beginning and end states.

Instead, the Gibbs function must help study the modalities *in the presence* of the latter kind of situations.

At first glance, both of the above sound obvious. They are necessary, especially in view of the practically meaningless and even sheer misleading—though already conventional/apologetic—presentation of both functions as the "important state functions of the physical-chemical systems, with the Helmholtz function being valid under the isochoric-isothermal conditions, while the Gibbs function being valid under the isobaric-isothermal ones." By the way, the notorious handbook's reference to the enthalpy, Gibbs and Helmholtz functions, as the "functions of the system's state" is just a handy tautology to mean, "Look, people, here are some very useful functions, but we have not the faintest idea about their actual mathematical formulations."

And last but not the least, both equations presented by Eq. 1.19 clearly visualize the intrinsic interrelationship between the Gibbs and Helmholtz functions, as well as entropy and enthalpy. Both former ones should preferably be committed together, whereas both latter ones do compensate each other. It is by committing the whole functional communion just named that we might manage to extend the accessible informational horizon when studying the mechanisms of realistic (bio)physical-chemical processes.

Attentively analyzing the original works by Massieu (1832–1896), who had introduced a unique energy function, plus the immediate resonance his works caused in the French professional community [21–23], as well as the well-known contributions by Gibbs and Helmholtz [24–30], could provide us with truly detailed descriptions of the mathematical properties and the actual physical sense of this function. This way, we might find firm support for our present train of thoughts.

1.2.5 Statistical-Mechanical Standpoint

The above arouses a poser as to why the authors would not like to follow the good old statistical-mechanical way to calculate or simulate Helmholtz and Gibbs functions.

We have already started discussing this point in detail elsewhere [1, 31]; here we would like to briefly sum up our standpoint.

Nowadays, statistical mechanics in its conventional formulation remains at the conceptual border reached by eminent colleagues Boltzmann and Gibbs in representing a useful and multifaceted computational tool of unquestioned power and efficiency. However, employing it for a conceptual deepening into the modalities of realistic physical-chemical systems may be debatable.

Indeed, Boltzmann could just start his seminal work by publishing several important and seminal conjectures, but for purely objective reasons, he had no time to duly investigate the details. His numerous followers could clearly demonstrate the actual giant scale of Boltzmann's ideas and visions, but the necessary conceptual work in connection with them was and still is not clearly visible.

This same pertains to the seminal studies by Gibbs, who could start his work by considering in detail the simplest possible case of an ensemble consisting of a vast number of statistically independent microparticles. To our regret, for purely objective reasons, he had no time to go on with the pertinent elaborations and refinements. The fate of Gibbs' train of thoughts is by and large similar to that of Boltzmann.

Meanwhile, the actual productive follower of both eminent colleagues was Linhart, about whom we have already published several works [1, 8, 12].

In our most recent review paper devoted to the ideas by Carnot [32], we have posed the following question for the statistical interpretation of thermodynamics: Was it just plain success or sheer despair? To our mind, the correct answer would be, both. Why?

Indeed, the first basic thermodynamics' law cannot be considered a kind of statistical regularity, whereas it is well known that the second one does obey the statistical regularities. If so, how could it be possible to combine this fact with the dialectic interrelationship between both laws (see Refs. [1, 15] for a more detailed discussion of this important topic)?

The first pioneering suggestion [33] concerning a viable way out of the above-mentioned logical blind alley came as early as in 1911 from van der Waals, Jr., the son of the author of the equation of state and a Nobel Prize winner. In his brief note, van der Waals, Jr., had discussed the interrelationship between the notions of probability and entropy, as it appeared from the considerations by Boltzmann and Gibbs. What he had suggested was not just a speculation but a clearly formulated, substantiated, justified, and fully competent suggestion to employ the Bayesian approach in deriving the relationship between the notions of entropy and probability. The only serious and effectual obstacle to the practical embodiment of that ingenious suggestion was the fact that the Bayesian approach to the probability notion was not really "trendy" at the time of publication. Even after the Second World War, it had to succumb to the "frequentist" train of thoughts.

Fortunately, van der Waals, Jr.'s, suggestion has not experienced a traceless dissolution in the whirlpool of the "trendy" medium. It was employed by Linhart [34–37], and, moreover, it is being further developed in recent time (see Ref. [38] plus all the recent work from this very active and successful group [39–43]), as well as the references therein). Remarkably, neither Linhart nor Johal and his colleagues refer to the work [33].

Meanwhile, the "frequentist" statistical-physical approach is based upon the long-known and over-all accepted atomistic representation of the matter, which makes us operate with an "enormous number of atoms/molecules," because of which we have seemingly no other reasonable way than just to apply the conventional statistical treatment.

It is extremely important to note that the very possibility to avoid the explicit microscopic consideration could be rather helpful, for notions like "an enormous number of X," whatever X might be, are in fact fuzzy, in accordance with the well-known sorites paradox.

Indeed, it is a practically challenging task to achieve a universally strict definition of what exactly the "enormous number" is, and hence operating with such notions and their derivatives as, for example, the "thermodynamic limit," is not really a productive approach. Thus, the aim of our present communication should be to demonstrate how the Bayesian approach might be handy and useful in circumventing the sorites paradox, as well as to show the way of establishing the mathematical/logical interconnection between its results and the well-known, tried-and-true products of the "frequentist" train of thoughts.

Basically, we are using here Linhart's standpoint concerning macroscopic thermodynamics and statistical mechanics.

Our next move would be to cast a detailed look at how our macroscopic-thermodynamic inferences, introduced and discussed in the above paragraphs, might be analyzed statistically-mechanically by employing Linhart's Bayesian approach.

1.2.6 What Is the Actual Probability Distribution behind the Statistical Mechanics?

To attack this poser, we would like to use the approach suggested in Ref. [38] and do the following steps:

We consider a thermodynamic system having two subsystems, one at temperature T_1 and one at temperature T_2, so that the system's total temperature depends on two parameters T_1 and T_2, where each parameter lies somewhere within a specified temperature range $[T_-, T_+]$ so that $T_- < T_+$. The subsystems are somehow coupled to each other; hence we might assume a one-to-one relationship between their temperatures, $T_1 = F(T_2)$, which assigns a unique value of T_1 for a given value of T_2 and vice versa, but there is no a priori information about any of the two.

In following the Bayesian approach to the statistical analysis of our thermodynamic system, we assign the probabilities Pr for those values of T_1 and T_2, and such probabilities must generally be the functions of both P and T (but not only of T) possessed of the same functional form:

$$\Pr(P,T_1)\mathrm{d}T_1 = \Pr(P,T_2)\mathrm{d}T_2. \tag{1.20}$$

Specifically, Eq. 1.20 quantifies the natural assumption that in the face of incomplete knowledge, the degree of belief that a certain value of T_1 lies in a small range $[T_1, T_1 + \mathrm{d}T_1]$ should be the same as the degree of belief that the corresponding value of T_2 lies in a small range $[T_2, T_2 + \mathrm{d}T_2]$, whereas some given function $F(\cdot)$ must define an univocal relation between T_1 and T_2.

Then, our task should be to try finding the actual mathematical form of the probability function Pr on the basis of the available prior information contained in the function $F(\cdot)$.

To find the above prior information is not a trivial task, but having the proper prior probability distribution function opens the way to a systematic statistical-mechanical analysis of the thermodynamic modalities.

The entities having temperatures T_1 and T_2 might be either two subsystems of some larger system or the total system of our interest, as it is, plus its unmediated surroundings taken separately. Nevertheless, let both entities be interacting with each other during some process. Then it is important to know what the exact conditions of achieving the equilibrium state in the total system under study are, that is, when all the relevant driving forces would perfectly equilibrate all the pertinent HOR on the way to progress of some realistic process advancing in our system.

To this end, it would be reasonable to consider the equality of the entropies of the entities involved, explicitly considering the actual positions of their temperatures T_1 and T_2 within (or with respect to) their definitive temperature interval $[T_-, T_+]$. Let us assume that T_1 is approaching T_+, while T_2 is striving for T_-. We consider the situation when $S(P,T_1) - S(P,T_+) = S(P,T_2) - S(P,T_-)$.

This way, both T_1 and T_2 should approach T_+ and T_- from below, so $T_2 < T_- < T_+$, whereas there are two different possibilities for T_1: $T_2 < T_- < T_1 < T_+$ and $T_2 < T_1 < T_- < T_+$. Which possibility should be put into effect?

The correct answer is unclear a priori. Still, whatever the answer to the latter poser, T_+ and T_- are the initial temperatures, whereas T_1 and T_2 should be the final ones, regarding the process in the system of our study.

Such an entropy condition suggests employing Eq. 1.7 to arrive at the following relationship:

$$\frac{2}{1+\frac{R^2T_1^{\,2}}{aP}}+\ln\left[1+\frac{R^2T_1^{\,2}}{aP}\right]-\frac{2}{1+\frac{R^2T_+^{\,2}}{aP}}-\ln\left[1+\frac{R^2T_+^{\,2}}{aP}\right]$$
$$=\frac{2}{1+\frac{R^2T_2^{\,2}}{aP}}+\ln\left[1+\frac{R^2T_2^{\,2}}{aP}\right]-\frac{2}{1+\frac{R^2T_-^{\,2}}{aP}}-\ln\left[1+\frac{R^2T_-^{\,2}}{aP}\right]. \quad (1.21)$$

We divide Eq. 1.21 into two parts (the algebraic one and the logarithmic-algebraic one):

$$\frac{1}{1+\frac{R^2T_1^{\,2}}{aP}}-\frac{1}{1+\frac{R^2T_+^{\,2}}{aP}}-\frac{1}{1+\frac{R^2T_2^{\,2}}{aP}}+\frac{1}{1+\frac{R^2T_-^{\,2}}{aP}}=0 \quad (1.22)$$

and

$$\ln\left[1+\frac{R^2T_1^{\,2}}{aP}\right]-\ln\left[1+\frac{R^2T_+^{\,2}}{aP}\right]-\ln\left[1+\frac{R^2T_2^{\,2}}{aP}\right]+\ln\left[1+\frac{R^2T_-^{\,2}}{aP}\right]=0. \quad (1.23)$$

If we assume that the temperature T_1 lies somewhere near either T_+ or T_-, Eqs. 1.22 and 1.23 do enable us to express T_2 as the following function of T_1, or vice versa:

$$T_2 = \pm \frac{\sqrt{R^2 T_-{}^2 T_+{}^2 + aP\left(-T_1{}^2 + T_-{}^2 + T_+{}^2\right)}}{\sqrt{\left(aP + R^2 T_1{}^2\right)}};$$

$$T_1 = \pm \frac{\sqrt{R^2 T_-{}^2 T_+{}^2 + aP\left(-T_2{}^2 + T_-{}^2 + T_+{}^2\right)}}{\sqrt{\left(aP + R^2 T_2{}^2\right)}}. \quad (1.24)$$

Next, we may differentiate T_2 by T_1 or vice versa and then rearrange the resulting expression to arrive at the explicit form of Eq. 1.20 by obtaining the following relationship for casting the Bayesian prior probability in our case:

$$\frac{dT_2}{dT_1} = -\frac{aPT_1}{\sqrt{\left(aP + R^2 T_1{}^2\right)}\sqrt{R^2 T_-{}^2 T_+{}^2 + aP\left(-T_1{}^2 + T_-{}^2 + T_+{}^2\right)}}$$

$$-\frac{R^2 T_1 \sqrt{R^2 T_-{}^2 T_+{}^2 + aP\left(-T_1{}^2 + T_-{}^2 + T_+{}^2\right)}}{\left(aP + R^2 T_1{}^2\right)^{3/2}}$$

$$= -\frac{aPT_1}{\left(aP + R^2 T_1{}^2\right)T_2} - \frac{R^2 T_1 T_2}{\left(aP + R^2 T_1{}^2\right)} = -\frac{aP\left(\frac{T_1}{T_2}\right) + R^2 T_1^2 \left(\frac{T_2}{T_1}\right)}{\left(aP + R^2 T_1{}^2\right)}. \quad (1.25)$$

Considering Eq. 1.20 and Eq. 1.24, we recast Eq. 1.25 as follows, to get the ultimate expression for the Bayesian prior probability distribution Pr as a function of P and T:

$$\frac{dT_2}{dT_1} = -\left(\frac{T_1}{T_2}\right)\frac{\dfrac{\left(aP + R^2 T_-{}^2\right)\left(aP + R^2 T_+{}^2\right)}{\left(aP + R^2 T_1{}^2\right)}}{\dfrac{\left(aP + R^2 T_-{}^2\right)\left(aP + R^2 T_+{}^2\right)}{\left(aP + R^2 T_2{}^2\right)}} = -\frac{\dfrac{T_1}{\left(aP + R^2 T_1{}^2\right)}}{\dfrac{T_2}{\left(aP + R^2 T_2{}^2\right)}} \Rightarrow$$

$$\Pr\left(P, T_2\right) = \frac{T_2 dT_2}{\left(aP + R^2 T_2{}^2\right)}; \quad \Pr\left(P, T_1\right) = -\frac{T_1 dT_1}{\left(aP + R^2 T_1{}^2\right)}. \quad (1.26)$$

To this end, if we assume that $dT_1 = -dT_2$, then the desired prior as a function of pressure and temperature must be unique for all the subsystems under study, and we might therefore cast it in its general form as follows:

$$\Pr(P,T) = \text{Const} \cdot \frac{TdT}{\left(aP + R^2T^2\right)}. \tag{1.27}$$

The functional form of Pr immediately suggests considering Eq. 1.27 in Linhart's representation. Indeed, if we introduce a coefficient of efficiency K (we shall refer to it as Linhart's coefficient from now on), as well as a dimensionless variable x [8, 12, 34–37], with R being the universal gas constant, we arrive at the following expression:

$$\Pr(P,T) = \text{Const} \cdot \frac{TdT}{\left(aP + R^2T^2\right)} \equiv C_\infty \cdot \frac{x^{K-1}}{1+x^K} dx;\, x \equiv \frac{T}{T_{\text{ref}}};\, C_\infty \approx 3R. \tag{1.28}$$

Our next step should then be to clarify the actual meaning of the constants K and T_{ref}. From Eq. 1.28, we immediately deduce that Linhart's coefficient K must be some specific function of both temperature and pressure, $K[P,T]$, so that we should refer to it as Linhart's function instead.

Indeed, as we formally solve the equation $x^{K[P,T]} = \frac{R^2x^2}{aP}$ for $K[P,T]$ for the set of real numbers, we get the following result for $K[P,T]$:

$$\left(\ln[x] \neq 0 \wedge \left(R \neq 0 \wedge a > 0 \wedge P > 0 \wedge x > 0 \wedge K[P,T] = \frac{\ln\left[\frac{R^2}{aP}\right] + 2\ln[x]}{\ln[x]} \right) \vee \left(R = \sqrt{aP} \wedge x = 1\right)\right) \tag{1.29}$$

Physically, Eq. 1.29 means that as soon as we have not achieved the perfect EEC ($\ln[x] \neq 0 \Leftrightarrow x \neq 1 \Leftrightarrow T \neq T_{\text{ref}}$, if $T_{\text{ref}} \equiv T_c$, with T_c being the EEC temperature), then Linhart's function should have values around 2, depending on the actual value of the pressure. Nevertheless, on the verge of the perfect EEC (i.e., if $R = \sqrt{aP} \wedge x = 1$), Linhart's function tends to infinity.

Our result is, therefore, in full accordance with Linhart's findings, as he fitted experimental data to his theoretically derived formulae [1, 8, 12, 34–37], and with his idea, that physically, "the proportionality factor" K must be the efficiency constant of the process under study, because the smaller the value of K, the greater the hindrance and the slimmer the chance for the process to reach its final aim/state/destination [1, 8].

Indeed, achieving the perfect EEC means that "the hindrance," that is, the entropy, has reached its maximum value. Meanwhile, the driving force intensity has been enough to compensate the latter. Nonetheless, it is the very fact of entropy's arriving at its maximum value that should anyway correspond to the zero of the "efficiency function" introduced by Linhart.

Meanwhile, Eq. 1.28 suggests that, in Linhart's approximation, the desired prior as a function of temperature must read as a specific heat capacity at constant volume divided by temperature:

$$\Pr(P,T) = C_\infty \cdot \frac{x^{K-1}}{1+x^K} dx = \left(\frac{C_V(P,T)}{T} \right).$$

Hence, considering Eq. 1.29 allows us to think of Linhart's theory as the one dealing with isochoric cases (with volume being not ultimately constant, as is maintained conventionally, but being an externally adjustable/controllable parameter).

Interestingly, when the expectation (average) temperature is calculated, using the latter form of prior, we get the same form of the Gaussian hypergeometric function ${}_2F_1$ as we obtain when calculating the internal energy within Linhart's approach (see Eq. 1.8 in Ref. [44]). Thus, we readily see that the Aneja–Johal approach fits well into the general framework of Linhart's "Bayesian" statistical mechanics. Meanwhile, there is also a clear technical (a purely mathematical) complication—to work out some closed formulae for such thermodynamic notions as the work amount, the work efficiency, etc.—for we should approximate the available special transcendent functions. Nevertheless, this should not constitute any principal complication. In what follows, we will illustrate the applicability of the algorithm chosen here.

1.2.7 Bayesian Statistical Thermodynamics of Real Gases

To find a generalized thermodynamic expression for the work done during realistic processes, we start with the formulation by Aneja–Johal [38]. We may then readily evaluate the amount of work (W) done during some process involving two subsystems, one at temperature T_2 and the other at temperature T_1, as we just discussed. The work amount in question should be equal to the decrease in the internal energy (U) of the total system, namely $W = -\Delta U$, where $\Delta U = U_{\text{fin}} - U_{\text{ini}}$.

We shall base our further work upon the expression for the internal energy U in Linhart's approximation [44]:

$$U(P,T) = \int_0^T C_V(P,T)dT = T_{\text{ref}} C_\infty \int_0^x \frac{x^K}{1+x^K} dx$$

$$= T_{\text{ref}} C_\infty x \left[1 - {}_2F_1\left(1, \frac{1}{K}, 1 + \frac{1}{K}; -x^K\right)\right]. \quad (1.30)$$

Worth noting is the fact that Eq. 1.30 is valid for an ideal gas. An attentive reader might then immediately ask, "What does this equation have to do with the real gases dealt with in the present work?"

Linhart was the actual follower of Gibbs, who was trying to build up the statistical-atomistic base of Horstmann's energetics (see Ref. [1] for details). Since Gibbs, like Boltzmann, was working on the purely mathematical model of an ideal gas, it is of practical interest to try to identify the interconnections between the logics of statistical-mechanical inferences and energetics. Here we just go the initial step. The aim of the story must be the universal thermodynamic equation of the state by Engelbrektsson and Franzén, which is also based upon different thermodynamics (see Ref. [1] for details). With this in mind, we employ Linhart's approximation.

Above we have demonstrated that the functional forms derived by Linhart are of general validity. In going from the ideal gas to the real gas, we need to check mathematical details of the results thus inferred. For example, some constants in the ideal gas representation might be functions (or even functionals) of (at least) P and T—the actual forms of these functions should be determined.

The paper by Aneja–Johal [38] suggests the following algorithm to estimate the work amount involved: $W = -U_{\text{fin}} + U_{\text{ini}}$. As we have assumed here that T_+ and T_- are initial temperatures, with T_1 and T_2 being the final ones, we consider Eq. 1.30 to get the expression for W as follows:

$$W = C_{\infty} T_{\text{ref}}$$
$$\left[\begin{array}{l}\left(x_+\left(1 - {}_2F_1\left(1,\frac{1}{K},1+\frac{1}{K};-x_+{}^K\right)\right) + x_-\left(1 - {}_2F_1\left(1,\frac{1}{K},1+\frac{1}{K};-x_-{}^K\right)\right)\right) \\ -\left(x_1\left(1 - {}_2F_1\left(1,\frac{1}{K},1+\frac{1}{K};-x_1{}^K\right)\right) + x_2\left(1 - {}_2F_1\left(1,\frac{1}{K},1+\frac{1}{K};-x_2{}^K\right)\right)\right)\end{array}\right],$$

$$\text{where } x_+ = \frac{T_+}{T_{\text{ref}}};\, x_- = \frac{T_-}{T_{\text{ref}}};\, x_1 = \frac{T_1}{T_{\text{ref}}};\, x_2 = \frac{T_2}{T_{\text{ref}}}. \tag{1.31}$$

We see that Eq. 1.31 is formulated in terms of the special function, namely the Gauss hypergeometric function ${}_2F_1$, which complicates a lot its practical usage. Hence, we would surely need to simplify Eq. 1.31. First, we use one of the Pfaff transformations:

$${}_2F_1\left(1,\frac{1}{K},1+\frac{1}{K};-z^K\right) = \left(1+z^K\right)^{-\frac{1}{K}} {}_2F_1\left(\frac{1}{K},\frac{1}{K},1+\frac{1}{K};\frac{z^K}{1+z^K}\right). \tag{1.32}$$

Second, we employ the well-known interconnection between the Gauss hypergeometric and incomplete beta functions:

$$px^{-p} B_x(p,q) = {}_2F_1(p,1-q,1+p;x) \Rightarrow \frac{1}{K}\left(\frac{z^K}{1+z^K}\right)^{-\frac{1}{K}} B_{\frac{z^K}{1+z^K}}\left(\frac{1}{K},1-\frac{1}{K}\right)$$
$$\equiv {}_2F_1\left(\frac{1}{K},\frac{1}{K},1+\frac{1}{K};\frac{z^K}{1+z^K}\right). \tag{1.33}$$

At this point, we recall the definition of the incomplete beta function:

$$B_x(a,b) = \int_0^x t^{a-1}(1-t)^{b-1}\,dt; a>0; b>0. \tag{1.34}$$

Hence, it is possible to recast the right-hand side of Eq. 1.32 using the function defined by Eq. 1.34, as follows:

$${}_2F_1\left(1,\frac{1}{K},1+\frac{1}{K};-z^K\right) = \left(1+z^K\right)^{-\frac{1}{K}} \frac{1}{K}\left(\frac{z^K}{1+z^K}\right)^{-\frac{1}{K}} B_{\frac{z^K}{1+z^K}}\left(\frac{1}{K},1-\frac{1}{K}\right)$$
$$\equiv \frac{z^{-1}}{K} B_{\frac{z^K}{1+z^K}}\left(\frac{1}{K},1-\frac{1}{K}\right). \tag{1.35}$$

As a result, we may express the Gaussian hypergeometric functions in Eq. 1.31 via regularized incomplete beta functions I_x as soon as we notice that:

$$I_x(a,b) = \frac{B_x(a,b)}{B_1(a,b)}. \tag{1.36}$$

Here B_x is the incomplete beta function and B_1 the conventional complete beta function. Then, taking Eq. 1.36 into account, we may rewrite the right-hand side of Eq. 1.35 as follows:

$${}_2F_1\left(1,\frac{1}{K},1+\frac{1}{K};-z^K\right) \equiv \frac{z^{-1}}{K} B_{\frac{z^K}{1+z^K}}\left(\frac{1}{K},1-\frac{1}{K}\right) \equiv$$

$$\frac{z^{-1}}{K} \frac{B_{\frac{z^K}{1+z^K}}\left(\frac{1}{K},1-\frac{1}{K}\right)}{B_1\left(\frac{1}{K},1-\frac{1}{K}\right)} B_1\left(\frac{1}{K},1-\frac{1}{K}\right) \equiv \frac{z^{-1}}{K}\left(\pi\csc\left(\frac{\pi}{K}\right)\right) I_{\frac{z^K}{1+z^K}}\left(\frac{1}{K},1-\frac{1}{K}\right), \tag{1.37}$$

where $\csc \equiv 1/\sin$ stands for the conventional cosecant function.

To sum up, Eq. 1.31 might be essentially clarified in the following way:

$$W = C_\infty T_{ref}$$

$$\left[\begin{array}{l} \left(x_+\left(1-\frac{x_+^{-1}}{K}\left(\pi\csc\left(\frac{\pi}{K}\right)\right) I_{\frac{x_+^K}{1+x_+^K}}\left(\frac{1}{K},1-\frac{1}{K}\right)\right) + x_-\left(1-\frac{x_-^{-1}}{K}\left(\pi\csc\left(\frac{\pi}{K}\right)\right) I_{\frac{x_-^K}{1+x_-^K}}\left(\frac{1}{K},1-\frac{1}{K}\right)\right)\right) \\ -\left(x_1\left(1-\frac{x_1^{-1}}{K}\left(\pi\csc\left(\frac{\pi}{K}\right)\right) I_{\frac{x_1^K}{1+x_1^K}}\left(\frac{1}{K},1-\frac{1}{K}\right)\right) + x_2\left(1-\frac{x_2^{-1}}{K}\left(\pi\csc\left(\frac{\pi}{K}\right)\right) I_{\frac{x_2^K}{1+x_2^K}}\left(\frac{1}{K},1-\frac{1}{K}\right)\right)\right) \end{array}\right]. \tag{1.38}$$

The important mathematical point here is that the function $I_Y(a, b)$, the regularized incomplete beta function, represents just the cumulative probability function for some random number Y obeying the beta distribution. Thus, we have formally proven the parallel between Linhart's theory and the Bayesian approach to statistics, just as originally guessed, for it was Reverend Thomas Bayes himself who started employing the continuous beta distribution function as a prior for the discrete binomial distributions.

Physically, this means that we must consider the heat capacity at a constant volume to be a random variable. Would the latter conclusion be a kind of heresy—and should the competent researchers' community immediately condemn us? Not at all, God bless, for such a seemingly unexpected standpoint could still be throughout plausible (see, e.g., the work in Ref. [45] and the references therein, as well as the corresponding discussion in Ref. [44]).

Indeed, the statistical-mechanical sense of heat capacity consists in showing how the "sum" of all the possible elementary excitations in the system would describe its macroscopic state. Linhart's formula (see Eq. 1.5 in Ref. [44]) is just the mathematical expression of this property. Moreover, it shows how the experimentally measurable heat capacity, that is, in our notations $Y \equiv C$, might well be expressed in terms of the standard probability theory: $\left(\dfrac{C}{C_\infty}\right)$.

To sum up, Eq. 1.38 here expresses the work to be done via the probabilities to achieve some definite value of the heat capacity at some temperature values.

And last but not the least, it is of interest to check how the Bayesian statistical-mechanical approach outlined above, which is based upon the ideal gas model, might correspond to the macroscopic train of thoughts relevant to real gases we are discussing here.

1.2.8 Applicability of Linhart's Approach to Real Gases

Instead of using only Eq. 1.30 to estimate the work amount $W = -U_{\text{fin}} + U_{\text{ini}}$ in the system of our interest, we now consider the entropic waste during the working process explicitly. We shall then look for $W = -U_{\text{fin}} + (TS)_{\text{fin}} + U_{\text{ini}} - (TS)_{\text{ini}}$ because $(TS)_{\text{fin}} - (TS)_{\text{ini}}$ should be equal to zero, as we have considered above. Physically, the latter condition means the equilibration of the process under study. Equilibration here means the "entropy-enthalpy equilibration" (the unique result of the EEC).

Therefore, it could be possible to make use of Eq. 1.17 and recast Eq. 1.31 as follows:

$$W = RT_{\text{ref}}\left[\left(T_1 \cdot \ln\left(1+\frac{R^2T_1^2}{\alpha P}\right)+T_2 \cdot \ln\left(1+\frac{R^2T_2^2}{\alpha P}\right)\right)\right.$$
$$\left.-\left(T_+ \cdot \ln\left(1+\frac{R^2T_+^2}{\alpha P}\right)+T_- \cdot \ln\left(1+\frac{R^2T_-^2}{\alpha P}\right)\right)\right]. \tag{1.39}$$

Remarkably, Eq. 1.39 is the difference between the initial and final values of the Helmholtz function, but not solely those of internal energies, which should be equal to the work amount necessary to achieve the entropy-enthalpy equilibration, just what we are looking for herewith.

Meanwhile, Eq. 1.39 might be duly transformed if we recall that

$$\ln\left(1+\frac{R^2T^2}{\alpha P}\right)=\frac{\left(\frac{R^2T^2}{\alpha P}\right)}{1+\frac{R^2T^2}{\alpha P}}\cdot {}_2F_1\left(1,1;2;\frac{\left(\frac{R^2T^2}{\alpha P}\right)}{1+\frac{R^2T^2}{\alpha P}}\right)$$

$$=a\cdot B(a,b)\cdot\frac{\left(1-\frac{\left(\frac{R^2T^2}{\alpha P}\right)}{1+\frac{R^2T^2}{\alpha P}}\right)^a}{\left(\frac{\left(\frac{R^2T^2}{\alpha P}\right)}{1+\frac{R^2T^2}{\alpha P}}\right)^b}\cdot I_{\frac{\left(\frac{R^2T^2}{\alpha P}\right)}{1+\frac{R^2T^2}{\alpha P}}}(a,b). \quad (1.40)$$

Nota bene: The parameters a and b of the resulting regularized incomplete beta function should not be confused with the van der Waals fitting parameters a and b, used in the previous paragraphs. Therefore, van der Waals a and b in Eqs. 1.1–1.17 stand for α and β in Eqs. 1.39 and 1.40, respectively. Mathematically, the parameters a and b of the $I_x\,(a,\,b)$ (Eq. 1.36) should be chosen in such a way that $a + b \equiv 1$ and $a + 1 \equiv 2$. We may then choose a and b to ensure a valid comparison of Eq. 1.38 and Eq. 1.39.

To sum up, we do arrive at the following condition for the parameters a, b, and K:

$a=\frac{1}{K}; b=1-\frac{1}{K}; a+b\equiv 1$ and $a+1\approx 2$, to render $K\approx 1$, that is, $K\in[1,2]$, when $K\to 1$.

Provided the latter condition, Eq. 1.40 might be recast to read:

$$\ln\left(1+\frac{R^2T^2}{\alpha P}\right)$$

$$\approx\left(\frac{1}{K}\right)\cdot B\left(\frac{1}{K},1-\frac{1}{K}\right)\cdot\left(\frac{\alpha PR^2T^2}{\left(\alpha P+R^2T^2\right)^2}\right)^{\frac{1}{K}}\cdot\left(1+\frac{\alpha P}{R^2T^2}\right)\cdot I_{\frac{\left(\frac{R^2T^2}{\alpha P}\right)}{1+\frac{R^2T^2}{\alpha P}}}\left(\frac{1}{K},1-\frac{1}{K}\right). \tag{1.41}$$

Equation 1.41 builds up an approximate conceptual bridge between Linhart's ideal gas theory in Eq. 1.38 and the macroscopic thermodynamic result for real gases in Eq. 1.39. Of interest and importance would be looking for the actual conditions, under which $\ln\left(1+\frac{R^2T^2}{\alpha P}\right)$ may read as $\ln\left(1+\left(\frac{T}{T_{\text{ref}}}\right)^K\right)$.

We substitute the solution to Eq. 1.19 into the logarithmic-algebraic term of the $S(P, T)$, cf Eq. 1.7. We get:

$$\ln\left[1+\frac{R^2T^2}{\frac{R^{3/2}T^{3/2}\sqrt{\alpha+\beta RT}}{\sqrt{\beta}}}\right]\equiv\ln\left[1+\sqrt{\frac{\beta RT}{\alpha+\beta RT}}\right]. \tag{1.42}$$

Therefore, we should investigate in detail the solution to the following equation regarding K:

$$\sqrt{\frac{\beta RT}{\alpha+\beta RT}}=\left(\frac{T}{T_{\text{ref}}}\right)^K. \tag{1.43}$$

Nota bene: Above and below, α and β stand for van der Waals fitting parameters a and b, respectively, as compared to those in Eqs. 1.1–1.17.

We get the following formal solution for Eq. 1.43:

$$\left\{T_{\text{ref}}\neq 0\wedge T=0\wedge \text{Re}[K]>0\right\}\vee$$
$$\left\{\text{Const}\in\text{Integers}\wedge T_{\text{ref}}\neq 0\wedge\sqrt{\beta RT}\sqrt{\alpha+\beta RT}\neq 0\wedge K\right.$$

$$= \frac{2 \cdot i \cdot \pi \cdot \text{Const} + \ln\left[\sqrt{\frac{\beta RT}{\alpha + \beta RT}}\right]}{\ln\left[\frac{T}{T_{\text{ref}}}\right]} \wedge \ln\left[\frac{T}{T_{\text{ref}}}\right] \neq 0 \Bigg\}. \qquad (1.44)$$

In letting the integer Const be equal to zero in Eq. 1.44 and comparing the latter with Eq. 1.29, we notice that it is throughout possible to use Linhart's theory to study the modalities of the EEC in detail. Indeed, Eq. 1.44 suggests that at $T \in \left[0, T_{\text{ref}}\right]$, we might consider K to be some positive real number, without taking care of its actual functional dependence on P and T; Linhart was managing it in a similar manner.

Moreover, Eq. 1.44 suggests that, in effect, just like we have seen above, when discussing the physical sense of Eq. 1.29, $K \to \infty$ if $T \to T_{\text{ref}}$. Hence, it is in full accordance with Linhart's suggestion that K must be the process's efficiency so that the lower the K values, the less probable would be the dominance over the hindrances/obstacles through process driving forces. The fact that K tends to infinity at $T = T_{\text{ref}} = T_{\text{c}}$ should then be in full accordance with the actual physical sense of the very notion of the EEC, which denotes the situation when the process driving forces (the enthalpic effects) fully compensate and equilibrate the summary of the relevant HOR (entropic effects).

Nonetheless, at any combination of the temperature and pressure values that is not correspondent to the EEC, we must consider Linhart's process efficiency measure K a univocal function of both P and T.

Finally, Eq. 1.43 and Eq. 1.44 show that it is throughout possible to mathematically operate with the set of real numbers when considering the function $K(P, T)$.

To sum up our present story, we might conclude that Linhart's variant of statistical mechanics is in full accordance with the energetics initially introduced by Horstmann in Germany and independently by Liveing in the United Kingdom, whereas Gibbs, in the United States, could pick it up from Horstmann when he was in Heidelberg. Linhart was, therefore, the only immediately consecutive successor of Gibbs.

It is worth noting that energetics has been introduced, thought over, and experimentally verified in Sweden, but quite independently of the colleagues just named, by Engelbrektsson and Franzén. In parallel to and independently of the Swedish colleagues, M. B. Weinstein, in Germany, was trying to bridge the conceptual gaps among thermodynamics, statistical physics, and physical kinetics. The details of the very roots of this whole story can be seen in Ref. [1] and the references therein.

1.2.9 Is There Some Physical Connection between Boltzmann's and Gibbs' Entropy Formulae?

An attentive reader might ask at this point, well, if Linhart was a direct successor of Gibbs, then why could he derive Boltzmann's formula

$$S(X) = k \cdot \ln(W(X)), \tag{1.45}$$

instead of Gibbs expression

$$S(X) = \sum_{x \in \aleph} w_X(x) \cdot \ln(w_X(x))? \tag{1.46}$$

Indeed, in the flood of literature, we might find a remarkable paper (see Ref. [46]) published by a team of theoretical physicists and mathematicians from Mexico and tackling the problem of the actual relationship between these two well-known tried-and-true formulae. The authors in Ref. [46] pose the problem as follows:

> In the traditional statistical mechanics textbooks, the entropy concept is first introduced for the microcanonical ensemble and then extended to the canonical and grand-canonical cases. However, in the authors' experience, this procedure makes it difficult for the student to see the bigger picture and, although quite ingenuous, the subtleness of the demonstrations to pass from the microcanonical to the canonical and grand-canonical ensembles is hard to grasp.

In deriving the classical-mechanical entropy definition, the authors adapt Schrödinger's approach to introduce the entropy definition for quantum mechanics. The resulting entropy formula is valid for all ensembles, being in complete agreement with the Gibbs entropy. Remarkably, such an agreement could be achieved in the continuum limit, for discretized equations like Eq. 1.46 are not delivering consistent definitions of entropy, because then its value depends on the "cell system" [46]. The cell system is just the way of taking the system's total phase volume into cells of arbitrary dimensions. Thus, if we deal with a system of N particles, then the cell number i should physically correspond to the state of the i-th particle (meaning mathematically the communion of the particles' three Cartesian coordinates and three Cartesian impulses). The authors present a careful mathematical analysis of such a picture and then draw their above-cited conclusion. To arrive at some general conclusion, we must anyway go over to the continuum representation of Eq. 1.46.

The cell system the authors employ to arrive at the continuous form of Eq. 1.46 is that suggested by Schrödinger, which he used to build up the well-known, tried-and-true quantum mechanics. Mathematically, the authors represent the sum (Eq. 1.46) as a Riemannian sum.

To sum up, there are no questions to the authors of the work [46], for they could conclusively arrive at the expected result.

The main question we pose is to ourselves. Indeed, Eq. 1.45 does contain the probability W, which is just some magic notion, the probability of "I do not know what *scilicet*." Interestingly, Schrödinger substitutes W by ψ and refers to this latter as "the wave function" (having some fancy "probabilistic meaning," that is, no immediately clear physical sense).

Although there is no immediately clear physical sense for either W or ψ, this is just the point where it is important to recognize that both statistical mechanics and quantum mechanics are correct, useful, and productive theories.

Meanwhile, if we are performing scientific research and not working as engineers in some industry, the poser that should excite us is, "So, what is the actual physical sense of both W and ψ?" In trying to answer this poser, we would not only satisfy our own curiosity but also create prerequisites for further refinement of good old tried-and-true tools for their even more successful employment by industrial engineers and, *summa summarum*, for our common well-being.

Most recently, it has finally become clear that Eq. 1.45 is not just an ingenious guess but possessed of a clear-cut physical sense.

Linhart could demonstrate that the magic probability W is just a handy algebraic function of the absolute temperature (for detailed discussions on this theme, see Ref. [1]).

What remains for us here would be to find out the exact mathematical representation for the function $w_x(x)$ if we consider Linhart's result for the function $W(X)$.

To address this problem, we have to first recast Eq. 1.46 as a Riemannian sum. Linhart's form of $W(X)$ is $W(X) = \ln(1 + X^k)$, where X is a dimensionless absolute temperature, like in Eq. 1.28.

Therefore, in Eq. 1.46, we are looking for $w_X(x)$, which should be some function of the absolute temperature as well. In mathematically preparing the pertinent Riemannian sum, we are now working not with the "phase volumes," which are just model representations and might thus be of restricted validity, but with the experimentally verifiable axis of the absolute temperature, which is properly scaled. Even in such a case, we are encountering the nontrivial problem of "quantizing" the axis of consideration [47, 48] to arrive at the proper widths and heights of the Riemannian rectangles.

Hence, in our case, Eq. 1.46 after the variable change $x \Rightarrow \frac{X \cdot i}{n}$ might be recast as follows:

$$S(X) = \sum_{x \in \aleph} w_X(x) \cdot \ln\left(w_X(x)\right) \Rightarrow \lim_{n \to \infty} \sum_{i=1}^{n} w\left(\frac{X \cdot i}{n}\right) \cdot \frac{X}{n} \cdot \ln\left(w\left(\frac{X \cdot i}{n}\right)\right). \quad (1.47)$$

Mathematically, we introduce the continuous axes of entropy S and absolute temperature t. Then, we quantize this coordinate system by introducing an infinite number of Riemannian rectangles. After performing the limit operation, we arrive at the following definite integral, which might be taken:

$S(X) = \int_0^X w(t) \cdot \ln(w(t)) \cdot dt$; after the variable change $y = w(t)$, we get $S(X) = X \cdot w(X) \cdot \ln(w(X))$. (1.48)

Now we compare Eqs. 1.46 and 1.48 to arrive at the equation

$$a \cdot \ln\left(1 + X^K\right) = X \cdot w(X) \cdot \ln\left(w(X)\right), \quad (1.49)$$

which has the following solution:

$$w(X)=e^{\text{ProductLog}\left(\frac{a\cdot\ln\left(1+X^K\right)}{X}\right)}\wedge X\neq 0\wedge\left[a\cdot\ln\left(1+X^K\right)\right]\neq 0. \quad (1.50)$$

Remarkably, from the theory of the so-called product logarithm (Lambert's W-function), it is well known that the following relationship is true:

$$Y^Y=Z\Rightarrow Y=\frac{\ln(Z)}{\text{ProductLog}(\ln(Z))}=e^{\text{ProductLog}(\ln(Z))}. \quad (1.51)$$

Hence, in our case, $Z\equiv\left(1+X^K\right)^{\frac{a}{X}}\Rightarrow Y\equiv Y^Y\Rightarrow w(X)\equiv w(X)^{w(X)}$. The latter relationship dictates the following condition for Linhart's parameter K:

$$\left(1+X^K\right)=\frac{a}{X}.\text{ This is correct if } K=\frac{2\cdot\pi\cdot i\cdot C[1]+\ln\left(\frac{a-X}{X}\right)}{\ln(X)} \quad (1.52)$$

$$\wedge\, \text{C}[1]\in\text{Integers}\wedge \text{X}\neq 0\wedge(a-X)\neq 0\wedge\ln(X)\neq 0.$$

Hence, the K is not constant; it is but a function of temperature, like entropy itself. This is in accordance with Eq. 1.44, which represents K as a function of pressure and temperature.

It is worth noting that Eqs. 1.48–1.52 mathematically represent the expression for the so-called differential entropy, well known in information theory. The latter one, while representing a successful and extremely useful theory, also demonstrates a way to easily downgrade the very important and fundamental entropy notion to a sheer misnomer.

Thus, truly surprising is the stubborn and still prevalent emotional excitement around the entropy notion. Especially surprising are the statements of such prominent specialists in their respective fields as Léon Nicolas Brillouin (1889–1969) [49–51] and Norbert Wiener (1894–1964) [52], to give just two brightest examples. Still more surprising is never taking seriously the voices of other, no less prominent specialists who were dealing thoroughly with the topic, like the Ehrenfest couple (Paul Ehrenfest [1880–1933] and Tatiana Afanassjewa-Ehrenfest [1876–1964]) [53, 54] and Edwin Thompson Jaynes (1922–1998) [55], though they were vocal.

To our sincere regret, due to reasons independent of our will, the seminal work by Boltzmann could not reach some logically consistent point. Hence, trying to uncritically follow his legacy might lead to such interesting "physical discoveries" as the "entropy conservation law" [56]. This same pertains to the seminal work by Gibbs, whose time could not allow him to bring his statistical-mechanical studies to any point beyond treating statistically independent systems.

Hence, we arrive at the serious methodological problem in connection with properly using the powerful tools of the probability theory and mathematical statistics. This deserves a separate detailed discussion, which we will take up in Appendix 3 in this book. Moreover, Appendix 3 tells about the colleague who was trying to extend the statistical-mechanical approach by Gibbs, independently of Linhart's work. He was very close to proving the general validity of the EEC. Strikingly, he could show that when following Gibbs' train of thoughts, it is possible to avoid introducing quantum mechanics. While demonstrating the undoubted relevance of the latter, this assures us that the latter is by far not the only valid way of thinking.

To sum up, the probability theory and mathematical statistics are good old tried-and-true tools of mathematical analysis. The point is, how to properly employ them. Our suggestion is to productively treat the microscopic modalities of the EEC/equilibration using the exploratory factor analysis of correlations, which is a method of multivariate statistics. Correlations (coupling) among atoms/molecules is the source of potential energy. The latter may give rise to kinetic energy, which is the source of the driving force for realistic processes. Entropic agents counteract the driving force. It is the dialectic interplay between the former and the latter that drives the process to finally come to its desired end (the entropy-enthalpy equilibrium).

Hence, proper analysis of the pertinent correlations must help analyze the microscopic mechanisms of the process under study.

1.2.10 Can Our Approach Be Really Productive?

Our main result here is the formal inference of several handy approximations for entropy and enthalpy, for Helmholtz free energy, and for Gibbs free energy as functions of intensive variables temperature and pressure.

The resulting expressions describe the basic EEC and reveal the difference in the physical sense of the Helmholtz and Gibbs functions. Indeed, the former delivers the value for work spent to achieve equilibration among the relevant driving forces (enthalpic factors) and HOR (entropic factors) during a realistic physical-chemical process, whereas the zeros disposition of the latter introduces a microscopic "phase transition" correspondent to the enthalpic-entropic compensation/equilibration under study.

Furthermore, our considerations here show that we may handle all the statistical-mechanical inferences within the Bayesian approach. This is possible, for we are following the approach by Linhart [34–37].

The immediate profit here consists in our not restricting our considerations to the fuzzy notion of some "vast number" of atoms/molecules, as is normally the case for the conventional treatment (see, for example, Ref. [57] and the references therein). This profit is achievable within the framework of the Bayesian approach, which helps get rid of the fuzziness arising from straightforward and blindly adopted atomistic hypothesis. So, here we might get an opportunity to significantly widen the actual horizons of the conventional statistical thermodynamics to include not only strictly macroscopic but also meso- and even nanoscopic levels of studies.

Meanwhile, there is only one serious and crucial poser that remains unanswered:

What must be the formal logical links between the Bayesian representation outlined above and the conventional Gibbs ensembles described by the normal (Gaussian) distribution function, the tried-and-true Boltzmann exponential distribution, etc., which have been known for very long to be faithful and useful mathematical instruments of theoretical physics?

The answer is, we do not have to overthrow or completely refurbish all the conventional statistical mechanics! We just have to widen the horizons of the latter and carefully look after the detailed conceptual links between its various aspects and applications.

The above reads like a truly long story, but to succinctly sum it up here, we would like to point out the following well-established facts:

The beta probability distribution is of extreme importance in the formal mathematical derivation of the conventional canonical distribution [58]. Moreover, beta and normal probability distributions are tightly interrelated mathematically, though sometimes, the former turns out to be even more suitable for practical applications in comparison to the truly ubiquitous good old latter one [59–61].

Finally, the beta probability distribution does enable constructing the tried-and-true method of successful statistical evaluation in the cases where fuzzy sets are at work [62, 63]. Physical-chemical problems involving myriads of interacting atom/molecules do represent such a case.

1.2.11 A Methodological Perspective

Finally, we must work out the proper methodology of further research efforts. It is necessary to reconcile the wide conceptual gap between the macroscopic and microscopic representations of the story.

Indeed, a thorough analysis [1] shows that macroscopic thermodynamics is by far not a unique branch of knowledge.

The conventional discipline, the so-called equilibrium statistical thermodynamics, which results from a genuinely frequentist probabilistic/statistical approach to the problem of the atomic/molecular structure of the matter, is well known. The ultimate conceptual basis of the equilibrium thermodynamics is statistical physics. The latter is based upon an ingenious conjecture of Boltzmann, $S = k \cdot \ln(W)$, which he had no time to investigate in more detail. His immediate followers did not bother with such details and could successfully arrive at quantum physics after clarifying that Boltzmann was right in this point. Remarkably, either the exact physical sense of W in $S = k \cdot \ln(W)$ or, consequently, that of Ψ in $S = k \cdot \ln(\Psi)$ remained largely unclear for a long time. Meanwhile, using Gibbs' statistical-mechanical approach, it is possible to prove that the quantum-physical way of thinking is not unique (refer to Appendix 3).

Nevertheless, the proponents of both statistical mechanics and quantum physics could consequently work them out and clearly proved their validity.

They have opposed all the different standpoints, especially those related to energetics, aiming at completely eradicating any dissident idea from coming into focus [1].

Fortunately, the latter kind of result is impossible, but the trend just revealed still tremendously complicates any further research work. The proper trend would be to combine statistical mechanics and quantum physics with energetics, and we should look for the proper ways to arrive at a successful solution.

A basic difficulty consists in recalling the actual methodology of energetics in detail, which seems practically lost for us [1]. Therefore, we have decided to present two remarkable essays by Marius Ameline, a French psychiatrist, published in 1898 and 1908. These works suggest using energetics to produce a proper classification of the hereditary mental disorders. Please refer to Appendix 2 here for their English translation and our comments.

On the other hand, we should not forget the good old tried-and-true tools of probability theory and mathematical statistics in solving (bio)physical-chemical problems. This is achievable by systematic employment of the EFAC. In the atomistic representation of the matter, the correlations do immediately come to surface as soon as we try to properly consider all the wealth of interatomic coupling.

Aside from definite help provided by EFAC in consistently and fruitfully interpreting diverse experimental data, this approach might be very useful in properly interpreting the results of computer-aided simulations.

1.2.12 What Is the Actual Zest of Our Approach?

Our present story is about the thermodynamic equation of the state for real gases.

An attentive reader might immediately ask about the sense in writing so many pages on the theme, which might in principle be described in a truly compact way (see, for example, the recent marvelous monograph in Ref. [64] and the references therein).

Our answer is straightforward: It is true that engineering thermodynamics has achieved a high level of development and can help successfully solve many practical problems and be compactly described. But a number of fundamental questions still remain unanswered.

Indeed, macroscopic thermodynamics and statistical mechanics were and are still moving in largely parallel ways because the actual physical sense of the entropy notion is not properly clarified.

Engineering thermodynamics successfully used and still uses energetics, which is the actual form of macroscopic thermodynamics. Energetics clarifies the entropy notion. Still, the usage of energetics in engineering thermodynamics was and still is purely operational, that is, without diving into the conceptual depths. This is clear because the aims of engineering thermodynamics are purely practical.

Chemical thermodynamics could also use energetics quite productively if it followed the trains of thought triggered by Horstmann (1842–1929) and Liveing (1827–1924).

Meanwhile, the revolutionary development of physics in the twentieth century, with the introduction of the powerful tool of quantum physics, has produced a strange combination of equilibrium thermodynamics and statistical mechanics.

Being mathematically impeccable, the latter combination produces no clear insight into the physical sense of the crucial entropy notion. Employing the conceptual work of energetics might, therefore, immensely help in promoting research activities in the whole natural sciences field.

To our mind, the optimum way of proceeding in such a direction would be to try taking advantage of both Her Majesty Energetics and the holy trinity of quantum physics–equilibrium thermodynamics–statistical mechanics.

Our present essay should represent the very first step in this direction.

Our ultimate theoretical aim is to build up a statistical-mechanical basement for the universal thermodynamic equation of state by Engelbrektsson and Franzén [1].

It is impossible to start with a clean slate, so we consider the results available on the theme. We suggest starting with the Frost–Kalkwarf approach to integrate the Clapeyron–Clausius equation of state. Physically, this relies upon van der Waals approximation.

Meanwhile, there is another approach to the integration of the Clapeyron–Clausius equation [65]. This work reports a skillful mathematical trick: it rigorously transforms the original Clapeyron–Clausius formula into an exact differential equation by employing the proper integration factor. The result abides by Euler's criterion for exact differentials and allows us to formally obtain analytical solutions producing a number of useful results for engineering thermodynamics (see Ref. [66] and the references therein).

The approach in Ref. [65] demonstrates that we may achieve thermodynamically grounded inferences of a number of well-known empirical equations of state, including that by Frost–Kalkwarf.

The approach in Ref. [65] is thus of significance for engineering/ applied thermodynamics and clearly demonstrates that the Frost-Kalkwarf approach might in principle be used in trying to solve fundamental problems as well.

The Frost–Kalkwarf formalism is truly well known. Most recently, several reports on this topic have been published [67, 68].

To productively solve the implicit and transcendent Frost-Kalkwarf vapor pressure equation, the authors [67] succeeded in skillfully approximating the latter in terms of the Lambert function. This enabled them to derive a handy iterative procedure of great importance in interpreting the wide variety of relevant experimental data.

Instead, the most recent work [68] employs two skillful mathematical tricks to arrive at the explicit analytical solution of the Frost–Kalkwarf vapor pressure equation. The authors [68] helped avoid using iterative procedures, which might in principle diverge.

To sum up, the current work in the field is mainly concentrated on applied-mathematical issues of undoubted importance—being unrelated to the fundamental research of our interest here. From the above minireview, we borrow the applicability of the Frost–Kalkwarf approach to our aims.

1.3 Conclusions

To shortlist our main results, we might conclude that:

- Using a combination of Frost–Kalkwarf and Linhart approaches, we could derive the statistical-mechanical expressions for the Helmholtz and Gibbs thermodynamic potentials applicable to real gases.
- The result of the above is the very first step in building-up the atomistic basement of the universal thermodynamic equation of state by Engelbrektsson and Franzén.
- It is the expression for the Gibbs thermodynamic potential derived here that enables us to treat the EEC as a microscopic phase transition.

- It is possible to productively employ the meanwhile notorious statistical standpoint bringing quantum physics and statistical mechanics into severe conceptual difficulties: The recipe is to use exploratory factor analysis when interpreting the results of experiments and computer simulations. This suggestion is based upon the universal applicability of energetics.

1.4 Outlook

We have already applied the approach outlined here to tackle the detailed molecular mechanisms of enzymatic reactions, by properly analyzing the trajectories of the relevant molecular dynamical simulations. The work is still in progress, and we report our preliminary results in Chapter 2. Our approach allows us to pick up the regular physically meaningful dynamics from the swamp of chaotic thermal motion. Moreover, we might get the full dynamical picture of allosteric effects in proteins.

We plan to apply this same approach in trying to reveal the microscopic mechanisms of hydrophobicity.

We plan to continue building up the statistical-mechanical basement of the universal thermodynamical equation of state by Engelbrektsson and Franzén.

References

1. E. B. Starikov (2019). *A Different Thermodynamics and Its True Heroes*, Pan Stanford Publishing, Singapore.
2. E. B. Starikov, B. Nordén (2007). Enthalpy-entropy compensation: a phantom or something useful?, *J. Phys. Chem. B*, **111**, pp. 14431–14435.
3. E. B. Starikov, I. Panas, B. Nordén (2008). Chemical-to-mechanical energy conversion in biomacromolecular machines: a plasmon and optimum control theory for directional work. 1. General considerations, *J. Phys. Chem. B*, **112**, pp. 8319–8329.
4. E. B. Starikov, D. Hennig, B. Nordén (2008). Protein folding as a result of 'self-regulated stochastic resonance': a new paradigm?, *Biophys. Rev. Lett.*, **3**, pp. 343–363.
5. E. B. Starikov, B. Nordén (2009). Physical rationale behind the nonlinear enthalpy-entropy compensation in DNA duplex stability, *J. Phys. Chem. B*, **113**, pp. 4698–4707.

6. E. B. Starikov, D. Hennig, H. Yamada, R. Gutierrez, B. Nordén, G. Cuniberti (2009). Screw motion of DNA duplex during translocation through pore. I. Introduction of the model, *Biophys. Rev. Lett.*, **4**, pp. 209–230.
7. E. B. Starikov, B. Nordén (2009). DNA duplex length and salt concentration dependence of enthalpy-entropy compensation parameters for DNA melting, *J. Phys. Chem. B*, **113**, pp. 11375–11377.
8. E. B. Starikov (2012). George Augustus Linhart as a 'widely unknown' thermodynamicist, *World J. Condens. Matter Phys.*, **2**, pp. 101–116.
9. B. Nordén, E. B. Starikov (2012). Entropy-enthalpy compensation may be a useful interpretation tool for complex systems like protein-DNA complexes: an appeal to experimentalists, *Appl. Phys. Lett.*, **100**, p. 193701.
10. B. Nordén, E. B. Starikov (2012). Entropy–enthalpy compensation as a fundamental concept and analysis tool for systematical experimental data, *Chem. Phys. Lett.*, **538**, pp. 118–120.
11. B. Nordén, E. B. Starikov (2012). Entropy-enthalpy compensation: is there an underlying microscopic mechanism?, in: *Current Microscopy Advances in Science and Technology*, A. Mendez-Vilas (ed.), Formatex Research Center, Spain, Vol. 2, pp. 1492–1503.
12. E. B. Starikov (2013). 'Entropy is anthropomorphic': does this lead to interpretational devalorisation of entropy-enthalpy compensation?, *Monatsh. Chem.*, **144**, pp. 97–102.
13. E. B. Starikov (2013). Valid entropy–enthalpy compensation: fine mechanisms at microscopic level, *Chem. Phys. Lett.*, **564**, pp. 88–92.
14. E. B. Starikov (2013). Entropy-enthalpy compensation and its significance in particular for nanoscale events, *J. Appl. Sol. Chem. Model.*, **2**, pp. 126–135.
15. E. B. Starikov (2013). Valid entropy-enthalpy compensation: its true physical-chemical meaning, *J. Appl. Sol. Chem. Model.*, **2**, pp. 240–245.
16. E. B. Starikov (2014). 'Meyer-Neldel rule': true history of its development and its intimate connection to classical thermodynamics, *J. Appl. Sol. Chem. Model.*, **3**, pp. 15–31.
17. E. B. Starikov (2015). The interrelationship between thermodynamics and energetics: the true sense of equilibrium thermodynamics, *J. Appl. Sol. Chem. Model.*, **4**, pp. 19–47.
18. A. A. Frost, D. R. Kalkwarf (1953). A semi-empirical equation for the vapor pressure of liquids as a function of temperature, *J. Chem. Phys.*, **21**, pp. 264–267.

19. R. C. Weast (1972). *Handbook of Chemistry and Physics: A Ready-Reference Book of Chemical and Physical Data*, Chemical Rubber Company, CRC Press.
20. M. W. Zemansky, R. H. Dittman (1997). *Heat and Thermodynamics: An Intermediate Textbook*, The McGraw-Hill Companies, Inc., New York, USA.
21. F. J. D. Massieu (1869). Sur les fonctions caractéristiques des divers fluides et sur la théorie des vapeurs, *Comptes rendus*, **69**, pp. 858–862, 1057–1061.
22. F. J. D. Massieu (1876). Thermodynamique, mémoire sur les fonctions caractéristiques des divers fluides et sur la théorie des vapeurs, *Mémoires présentés par divers savants à l'Académie royale des sciences de l'Institute national de France*, Tome XXII, Numéro 2.
23. E. Bouty (1877). 'Sur le travail de F. Massieu,' Journal de physique théorique et appliquée, par Joseph Charles Almeida, Société française de physique, t. VI, pp. 216–222.
24. J. W. Gibbs (1876). On the equilibrium of heterogeneous substances, *Trans. Conn. Acad. Arts Sci.*, **3**, pp. 108–248, 343–524.
25. J. W. Gibbs (1878). On the equilibrium of heterogeneous substances, *Am. J. Sci. Arts*, **XVI**, pp. 441–458.
26. H. von Helmholtz (1877). Über galvanische Ströme, verursacht durch Concentrations-Unterschiede; Folgerungen aus der mechanischen Wärmetheorie, *Monatsberichte der Kgl. Preuß. Akad. des Wissensch. zu Berlin*, Seiten 713–726.
27. H. von Helmholtz (1882). Die Thermodynamik chemischer Vorgänge. Erster Beitrag, *Sitzungsberichte der Kgl. Preuß. Akad. der Wissensch. zu Berlin*, 1. Halbband, Seiten 22–39.
28. H. von Helmholtz (1882). Zur Thermodynamik chemischer Vorgänge. Zweiter Beitrag, *Sitzungsberichte der Kgl. Preuß. Akad. der Wissensch. zu Berlin*, 1. Halbband, Seiten 825–836.
29. H. von Helmholtz (1883). Zur Thermodynamik chemischer Vorgänge. Dritter Beitrag, *Folgerungen die galvanische Polarisation betreffend*, Wiedemanns Annalen Bd. 5, S. 182; Bd. 16 S. 561; Bd. 17, S. 592.
30. H. von Helmholtz (1884). Studien zur Statik monocyklischer Systeme, *Sitzungsberichte d. Preuß. Akad. d. Wissensch. zu Berlin*, Phys.-Math. Kl., Seiten 159–177.
31. E. B. Starikov (2018). Bayesian statistical mechanics: entropy-enthalpy compensation and universal equation of state at the tip of pen, *Front. Phys.*, **6**, p. 2.

32. E. B. Starikov (2014). What Nicolas Léonard Sadi Carnot wanted to tell us in fact?, *Pensée J.*, **76**, pp. 171–214.
33. J. D. van der Waals (1911). Über die Erklärung der Naturgesetze auf statistisch-mechanischer Grundlage, *Phys. Z.*, **XII**, pp. 547–549.
34. G. A. Linhart (1922). Note: The relation between entropy and probability: the integration of the entropy equation, *J. Am. Chem. Soc.*, **44**, pp. 140–142.
35. G. A. Linhart (1922). Correlation of entropy and probability, *J. Am. Chem. Soc.*, **44**, pp. 1881–1886.
36. G. A. Linhart (1922). Additions and corrections – correlation of entropy and probability, *J. Am. Chem. Soc.*, **44**, pp. 2968–2969.
37. G. A. Linhart (1933). Correlation of heat capacity, absolute temperature and entropy, *J. Chem. Phys.*, **1**, pp. 795–797.
38. P. Aneja, R. S. Johal (2013). Prior information and inference of optimality in thermodynamic processes, *J. Phys. A*, **46**, p. 365002.
39. R. S. Johal (2006). Models of finite bath and generalized thermodynamics, in: *Studies in Fuzziness and Soft Computing*, A. Sengupta (ed.), Vol. 206, pp. 207–217.
40. P. Aneja, R. S. Johal (2012). Prior probabilities and thermal characteristics of heat engines, *Cent. Eur. J. Phys.*, **10**, pp. 708–714.
41. R. S. Johal (2015). Efficiency at optimal work from finite reservoirs: a probabilistic perspective, *J. Non Equilibr. Thermodyn.*, **40**, pp. 1–12.
42. P. Aneja, R. S. Johal (2015). Form of prior for constrained thermodynamic processes, *Eur. Phys. J. B*, **88**, pp. 129–138.
43. R. S. Johal, R. Rai, G. Mahler (2015). Reversible heat engines: bounds on estimated efficiency from inference, *Found. Phys.*, **45**, pp. 158–170.
44. E. B. Starikov (2010). Many faces of entropy or Bayesian statistical mechanics, *ChemPhysChem*, **11**, pp. 3387–3394.
45. R. L. Scott (2006). The heat capacity of ideal gases, *J. Chem. Educ.*, **83**, pp. 1071–1081.
46. M. Santillán, E. S. Zeron, J. L. Del Rio-Correa (2008). Derivation of the Gibbs entropy with the Schrödinger approach, *Eur. J. Phys.*, **29**, pp. 629–638.
47. D. K. C. MacDonald (1952). Information theory and its application to taxonomy, *J. Appl. Phys.*, **23**, pp. 529–531.
48. F. P. Tarasenko (1968). On the evaluation of an unknown probability density function, the direct estimation of the entropy from independent observations of a continuous random variable, and the distribution-free entropy test of goodness-of-fit, *Proc. IEEE*, **56**, pp. 2052–2053.

49. L. Brillouin (1951). Maxwell's Demon cannot operate: information and entropy. I, *J. Appl. Phys.*, **22**, pp. 334–337.

50. L. Brillouin (1951). Physical entropy and information. II, *J. Appl. Phys.*, **22**, pp. 338–343.

51. L. Brillouin (1951). Information theory and most efficient codings for communication or memory devices, *J. Appl. Phys.*, **22**, pp. 1108–1111.

52. N. Wiener (1989). *The Human Use of Human Beings: Cybernetics and Society*, Free Association Books, London, UK.

53. T. Ehrenfest-Afanassjewa (1958). *On the Use of the Notion "Probability" in Physics, Am. J. Phys.*, **26**, pp. 388–392.

54. P. Ehrenfest, T. Ehrenfest-Afanassjewa (1990). *The Conceptual Foundations of the Statistical Approach in Mechanics*, Dover Publications, Mineola, USA.

55. E. T. Jaynes (1965). Gibbs vs. Boltzmann entropies, *Am. J. Phys.*, **33**, pp. 391–398.

56. S. K. Godunov, U. M. Sultangazin (1971). On discrete models of the kinetic boltzmann equation, *Russ. Math. Surv.*, **26**, pp. 1–56.

57. A. R. Khokhlov, A. Yu. Grosberg, V. S. Pande (1997). *Statistical Physics of Macromolecules (Polymers and Complex Materials)*, AIP Press, New York, USA.

58. B. H. Lavenda (1991). *Statistical Physics: A Probabilistic Approach*, Wiley-Interscience Publication, New York, Chichester, Brisbane, Toronto, Singapore, pp. 224–230.

59. F. Tajima (1989). Statistical method for testing the neutral mutation hypothesis by DNA polymorphism, *Genetics*, **123**, pp. 585–595.

60. J. B. McDonald, Y. J. Xu (1995). A generalization of the beta distribution with applications, *J. Econom.*, **66**, pp. 133–152.

61. C. Forbes, M. Evans, N. Hastings, B. Peacock (2011). *Statistical Distributions*, 4th ed., Wiley & Sons, Chichester, Hoboken.

62. T. J. Ross, J. M. Booker, W. J. Parkinson (2002). *Fuzzy Logic and Probability Applications: Bridging the Gaps*, Society for Industrial and Applied Mathematics, Philadelphia, Pennsylvania, USA, American Statistical Association, Alexandria, Virginia, USA.

63. M. Smithson, J. Verkuilen (2006). *Fuzzy Set Theory: Applications in the Social Sciences*, Sage Publishing, Thousand Oaks, CA, USA, pp. 41–43.

64. K. H. Lüdtke (2004). *Process Centrifugal Compressors: Basics, Function, Operation, Design, Application*, Springer Verlag, Berlin, Heidelberg, New York, pp. 47–70.

65. C. Mosselman, W. H. van Vugt, H. Vos (1982). Exactly integrated Clapeyron equation: its use to calculate quantities of phase change and to design vapor pressure-temperature relations, *J. Chem. Eng. Data*, **27**, pp. 246–251.

66. L. Q. Lobo, A. G. M. Ferreira (2001). Phase equilibria from the exactly integrated Clapeyron equation, *J. Chem. Thermodyn.*, **33**, pp. 1597–1617.

67. C. F. Leibovici, D. V. Nichita (2013). A quasi-analytical solution of Frost-Kalkwarf vapor pressure equation, *Comput. Chem. Eng.*, **58**, pp. 378–380.

68. H. Fatoorehchi, R. Rach, H. Sakhaeinia (2017). Explicit Frost-Kalkwarf type equations for calculation of vapor pressure of liquids from triple to critical point by the Adomian decomposition method, *Can. J. Chem. Eng.*, **95**, pp. 2199–2208.

Appendix 1 to Chapter 1

Using the pertinent value ranges of van der Waals fitting parameters a and b enables us to construct physically relevant analytical approximations.

Let us now consider in more detail the functional iterative process given by Eq. 1.3.

Indeed, we have the following results for the five consecutive steps of the process in question (in the main text, we have started this process at $i = 0$ and stopped it at $i = 2$):

$$V(0)=\frac{RT}{P};V(1)=b+\frac{RTV(0)}{PV(0)^2+a}=b+\frac{RT}{P\left[1+\left(aP/R^2T^2\right)\right]}$$

$$=b+\frac{R^3T^3}{P\left[R^2T^2+aP\right]}; \tag{A1.1}$$

$$V(2)=b+\frac{RTV(1)}{PV(1)^2+a}=b+\frac{R^7T^7}{P\left(a^3P^3+2a^2P^2R^2T^2+aPR^4T^4+R^6T^6\right)}$$

$$\approx b+\frac{R^3T^3}{P\left[R^2T^2+aP\right]}; \tag{A1.2}$$

$$V(3)=b+\frac{RTV(2)}{PV(2)^2+a}$$

$$=b+\frac{R^{15}T^{15}}{\left(P\left[R^2T^2+aP\right]\right)\cdot\left(2a^2P^2R^8T^8+3a^3P^3R^6T^6+3a^4P^4R^4T^4+R^{12}T^{12}\right)}$$

$$\approx b+\frac{R^3T^3}{P\left[R^2T^2+aP\right]}; \tag{A1.3}$$

Entropy-Enthalpy Compensation: Finding a Methodological Common Denominator through Probability, Statistics, and Physics
Edited by Evgeni B. Starikov, Bengt Nordén, and Shigenori Tanaka

ISBN 978-981-4877-30-5 (Hardcover), 978-1-003-05625-6 (eBook)
www.jennystanford.com

$$V(4) = b + \frac{RTV(3)}{PV(3)^2 + a} = b + \frac{R^{31}T^{31}}{P\left(R^{30}T^{30} + aPR^{28}T^{28} + \sum_{i=2}^{i=15} a^i \cdot F_i[P,T,R]\right)}$$

$$\approx b + \frac{R^3T^3}{P\left[R^2T^2 + aP\right]}. \tag{A1.4}$$

From Eqs. A1.1–A1.4, we see that if we skip the terms containing a^2 and a^3, as well as higher powers of a, we conclude that our functional iteration might be considered approximately convergent at $i = 1$ and up to the step $i = 5$. The resulting functional dependence of the system's volume on its pressure and temperature $V(P,T)$ should then be consistent with the approximations inherent in the van der Waals equation of state. Would this be enough for our purposes? Of course, but we still must remember that the van der Waals equation of state is long and well known to be not an entirely satisfactory approximation as it is and below we would like to cite the wise and insightful words by Sydney Young (1857–1937) on this theme, published at the very beginning of the last century [1]:

> The researches of Andrews, Amagat, and Ramsay and Young, however, led to the conclusion that the equation of van der Waals, although reproducing the general form of the isothermals, did not give results in sufficiently close agreement with the experimentally determined data; and physicists have proposed numerous modifications of the formula, but there is no completely satisfactory equation as yet.
>
> On the other hand, van der Waals pointed out that by expressing the pressures as fractions of the critical pressure, the absolute temperatures as fractions of the absolute critical temperature, and the volumes of liquid and of vapor as fractions of the critical volume, a "reduced" equation could be derived from the original formula which should be applicable to all substances.
>
> Applying the reduced equation to the special case of liquids at their boiling points and saturated vapors, it would follow that if the pressures of any two substances are proportional to their critical pressures, their boiling points (expressed as absolute temperatures) should be proportional to their critical temperatures, and their volumes, both as liquid and as saturated vapor, should be proportional to their critical volumes.

To our sincere regret, the above conclusions still hold, although other attempts to successfully derive and verify the universal thermodynamic equation of state have been made but remain widely unknown. We mean here the relevant work by Engelbrektsson and Franzén (see Ref. [1] of Chapter 1 for details).

This fact urges us to recognize that our present approximation might be improved if we would follow the train of thoughts triggered by Engelbrektsson and Franzén.

And last but not the least ought to be the problem of how to analytically integrate the Clapeyron–Clausius equation in its complete generality, mentioned in Ref. [18] of Chapter 1, and still attracting the attention of colleagues [2].

Meanwhile, Clapeyron had derived an equation in 1834 when the second basic law of thermodynamics, together with the notion of entropy inherent in it, had not yet been stipulated [3], while Maxwell was only three years old.

In turn, Clausius was trying to integrate this equation using the approximation of the ideal gas (see the expression for $V(0)$ in Eq. A1.1), whereas Maxwell could skillfully generalize the original Clapeyron–Clausius equation by putting it in differential form and thus triggering in such a way the move to analytically integrate it, which remains attractive [2].

Our approximation does allow us to check the possibility of integrating the Clapeyron–Clausius equation analytically, by employing the pertinent Maxwell relation.

Indeed, we write this equation in its different equivalent forms, which all describe the simplest case of two coexisting phases/aggregate states in the simplest example of a one-component system:

$$\frac{dP}{dT}=\frac{s^{(2)}-s^{(1)}}{v^{(2)}-v^{(1)}}=\frac{L}{T\cdot\Delta v};L=\frac{s^{(2)}-s^{(1)}}{T}\Leftrightarrow\left(\frac{\partial P}{\partial T}\right)_{V,N}=\left(\frac{\partial S}{\partial V}\right)_{T,N}. \quad \text{(A1.5)}$$

In Eq. A1.5, s stands for the molar entropy, v for the molar volume, S for the entropy, P for the pressure, T for the temperature, and L for the molar latent heat of the phase transition in question (i.e., the enthalpy of the phase transition under study) and it will adopt positive values if we choose the "first" aggregate state to be more condensed (liquid or solid) than the "second" one (gas or liquid, respectively). The number of particles in our system N remains constant.

Whereas the recent work, in Ref. [2], considering the integration of Eq. A1.5 remains within the "holy realm" of the ideal gas, here we would like to employ our "van der Waals" approximation for reanalyzing the Maxwell relation in connection with Eq. A1.5. To do so, we first solve Eq. A1.1 for P and get the following functional dependence $P(V, T)$:

$$P = \frac{-bR^2T^2 + R^2T^2V \pm \sqrt{-4RT(ab - aV) + \left(bR^2T^2 - R^2T^2V\right)^2}}{2(ab - aV)};$$

or, neglecting the product $a \cdot b$:

$$P = \frac{bR^2T^2 - R^2T^2V \mp \sqrt{4aRTV + \left(bR^2T^2 - R^2T^2V\right)^2}}{2aV}. \tag{A1.6}$$

Physically, there can be no negative pressure, so we must choose only one of the two solutions mentioned above, namely the following one:

$$P = \frac{bR^2T^2 - R^2T^2V + \sqrt{4aRTV + \left(bR^2T^2 - R^2T^2V\right)^2}}{2aV}. \tag{A1.7}$$

Our next step is to clarify the functional dependence $S(V, T)$. We substitute Eq. A1.7 into Eq. 1.7 to get the following expression:

$$S(V,T) \approx R^2T\frac{(b+V) - \sqrt{(-b+V)^2 + 4\frac{aV}{RT}}}{-a + bRT} - R\ln\left[\frac{(b+V) + \sqrt{\left((-b+V)^2 + 4\frac{aV}{RT}\right)}}{(b-V) + \sqrt{\left((-b+V)^2 + 4\frac{aV}{RT}\right)}}\right]. \tag{A1.8}$$

Finally, we must guarantee that our system obeys the pertinent Maxwell relation. That is, we must take the corresponding partial derivatives of P and S, using Eqs. A1.7 and A1.8, and ensure that the difference between both is equal to zero. As a result, we arrive at the following expression:

$$\begin{pmatrix} bR\left(4a^2RTV\right)-RT\left(b-V\right)^2\left(R^2T^2\left(b-V\right)+\sqrt{R^3T^3\left(RT\left(b-V\right)^2+4aV\right)}\right)+ \\ a\left(R^2T^2\left(b-5V\right)\left(b-V\right)+\left(b-3V\right)\sqrt{R^3T^3\left(RT\left(b-V\right)^2+4aV\right)}\right) \end{pmatrix}$$

$$/\left(a\left(a-bRT\right)V\left(RT\left(b-V\right)^2+4aV\right)\right). \tag{A1.9}$$

Equation A1.9 is equal to zero at $T = 0$ and/or at $a \to 0$ and/or $b \to 0$. Apart from this, there are no other solutions to Eq. A1.9, nor even any handy relationships between the variables of states V and T in the space of real numbers, whereas only branched solutions in the space of complex numbers might still be possible.

The above consideration suggests that any prompt attempt to rigorously integrate the Clapeyron–Clausius equation using the conventional equilibrium thermodynamics would hardly lead to some handy mathematical formulae describing phase transitions. To achieve this goal, skillful mathematical tricks are definitely necessary. This in no way deprives this equation of its empirical validity and practical usefulness.

Remarkable in this respect is the recent paradoxical introduction of the "negative absolute temperature" [4, 5], which has no definite physical meaning except for a sheer conceptual affront to the address of the equilibrium thermodynamics. This example demonstrates the logical inconsistency of the latter and supports our doubts that it might deliver any reliable conceptual foundation.

References

1. S. Young (1909). The vapour-pressures, specific volumes, heats of vaporisation, and critical constants of thirty pure substances, *Sci. Proc. R. Soc. Dublin*, **12**, pp. 374–443.
2. D. Shilo, R. Ghez (2008). New wine in old flasks: a new solution of the Clapeyron equation, *Eur. J. Phys.*, **29**, pp. 25–32.
3. J. Wisniak (2000). Negative absolute temperatures, a novelty?, *J. Chem. Educ.*, **77**, pp. 518–522.

4. J. Wisniak (2001). Historical development of the vapor pressure equation from Dalton to Antoine, *J. Phase Equilib.*, **22**, pp. 622–630.
5. J. Wisniak (2002). The thermodynamics of systems at negative absolute temperatures, *Indian J. Chem. Technol.*, **9**, pp. 402–406.

Appendix 2 to Chapter 1

Methodological Roots and Significance of Energetics

A2.1 Introduction

Are we introducing some "breaking methodological news"?

Of course, not!

The story is long and well known—especially among the engineers. The basic references are presented in Table A2.1 in a chronological/evolutionary order.

Table A2.1

Author/Editor	Article/Book	Publication details
Josiah Willard Gibbs	A Method of Geometrical Representation of the Thermodynamic Properties of Substances by Means of Surfaces	*Transactions of the Connecticut Academy of Arts and Sciences II*, pp. 382–404, December 1873
Louis Georges Gouy	Sur l'Énergie Utilisable	*Journal de Physique, Deuxième Série*, **8**(1), pp. 501–518, 1889
Aurel Stodola	Die Kreisprozesse der Gasmaschine	*Zeitschrift des VDI*, **32**(38), pp. 1086–1091, 1898
Jacques Charles Émile Jouguet	Remarques sur la Thermodynamique des Machines Motrices	*Revue Mécanique*, **19**, p. 41, 1906

Entropy-Enthalpy Compensation: Finding a Methodological Common Denominator through Probability, Statistics, and Physics
Edited by Evgeni B. Starikov, Bengt Nordén, and Shigenori Tanaka

ISBN 978-981-4877-30-5 (Hardcover), 978-1-003-05625-6 (eBook)
www.jennystanford.com

George Alfred Goodenough	*Principles of Thermodynamics*	Henry Holt and Company, 1911
Max Born	Kritische Betrachtungen zur Traditionellen Darstellung der Thermodynamik	*Physikalische Zeitschrift*, **22**, pp. 218–224, 249–254, 282–286, 1921
William Lane DeBaufre	Analysis of Power-Plant Performance Based on the Second Law of Thermodynamics	*Mechanical Engineering*, ASME, **47**(5), pp. 426–428, 1925
Joseph Henry Keenan	A Steam Chart for Second Law Analysis	*Mechanical Engineering*, ASME, **54**(3), pp. 195–204, 1932
Fran Bošnjaković	Technische Thermodynamik. Band I.	Dresden, Germany: Verlag Theodor Steinkopff, 1935
Zoran Rant	Exergie, Ein Neues Wort für "Technische Arbeitsfähigkeit"	*Forschung auf dem Gebiete des Ingenieurwesens*, **22**, p. 36, p. 37, 1956
M. J. Moran and E. Sciubba, Eds.	Second Law Analysis of Thermal Systems	Fourth International Symposium on Second Law Analysis of Thermal Systems. Rome: American Society of Mechanical Engineers, May 25–29, 1987
A. Bejan	*Entropy Generation Minimization: The Method of Thermodynamic Optimization of Finite-Size Systems and Finite-Time Processes*	Boca Raton: CRC Press, 1996

Remarkably, Gibbs and Gouy (1854–1926) were physicists.

In his work we cite in Table A2.1, Gibbs could develop the formalism of the idea he could borrow from the lectures by Horstmann he attended in Heidelberg.

This idea consists in representing entropy as ubiquitous obstacles/hindrances on the way of progress guaranteed by any realistic driving force. In fact, the latter is spent to overcome (to equilibrate/compensate) the former. This compensation is the physical prerequisite for the successful termination of any realistic process under study.

Gibbs' compatriot engineers Goodenough, DeBaufre, and Keenan, among the truly international worldwide engineers' team, could mathematically embody this idea to work out a handy concept of the useful energy/exergy. In fact, this is the very result by Carnot he had no time to work out in detail himself.

At the same time, the worldwide physicists' community had apparent difficulties with grasping this idea, which is 100% apparent from the work by Born we have cited in Table A2.1. Born is the actual author of the probabilistic interpretation of quantum mechanics, which results from the probabilistic interpretation of the famous Boltzmann–Planck formula $S = k \cdot \log(W)$.

To sum up, the actual difficulty consists in the methodological dilemma, should we treat entropy S implicitly, just by playing mathematical games with the Boltzmann–Planck formula, or do we explicitly take S into account, by considering its actual physical sense?

Nowadays, we know that both approaches are valid and fruitful, but they still exist conceptually/methodologically separated from each other.

Is the latter approach possible in physics/chemistry/biology? Yes, of course (see Ref. [1] of Chapter 1). Is there any contradiction between both approaches? Of course, no: the work by Linhart demonstrates full accordance between both (see Chapter 4 in Ref. [1] of Chapter 1).

And last but not the least, whereas the usefulness of the former approach is nowadays undoubtful, could any perspective be recognizable when using the latter approach in physics/chemistry/ biology? Our answer is positive: this is just what we would like to discuss later in this chapter.

Meanwhile, the difficulties caused by the apparent/alleged incompatibility of both approaches just named are still persistent. Moreover, they look like being rather stubborn.

A typical illustration of this fact could be, for example, the work by Tolman [8], who was one of the outstanding theoretical physicists of his time and whom we also know (see Ref. [1] of Chapter 1) from Linhart's diaries as one of central actors who made heavy weather of Linhart's approach:

We cite here the abstract of the work [8]:

> The importance of irreversible entropy production is emphasized. The first and second laws of thermodynamics are both expressed by equalities by giving due recognition to the irreversible production of entropy. By specialization and combination of these equations, an "efficiency equation" is derived, applicable to cyclical and steady state processes, and useful in studying the efficiency of practical processes. The thermodynamic theory involved in determining amounts of entropy produced by irreversible processes is investigated, and six cases of irreversible entropy production are treated in detail. The "efficiency equation" is applied to four practical processes to illustrate its usefulness in analyzing the causes of inefficiency in practical situations. The first and second law equations are directly applied to two theoretical problems so selected as to illustrate differences in the theoretical importance of being able to express the second law in the form of equality. The concept of temperature is discussed, and the extent of its validity under non-equilibrium situations is analyzed.

Remarkably, Prof. Tolman was actively collaborating with Prof. DeBaufre and citing his above-mentioned 1925 paper. In the final part of his writing, Prof. DeBaufre does clearly acknowledge his "indebtedness for the general idea of preparing a second-law balance to Dr. R. C. Tolman, of the Californian Institute of Technology."

It is noteworthy that the mid-twenties of the last century was just the time of Linhart's activity in the field. On attentively reading Prof. Tolman's paper, we find no sign of clarity as to the dilemma we are just discussing. The posers are then:

- Were more than 20 years of careful thinking over the problem not enough for an outstanding theoretical physicist to recognize the solution?
- What was it exactly—an ultimate hatred with respect to Linhart or an overall imprudence/conservatism?

Nevertheless, the "different thermodynamics" (i.e., the energetics) was/is/will be a valid approach the revolutionary active successful physicists were stubbornly trying to eradicate (see Ref. [1] of Chapter 1).

Regretfully, this is just one of the points where they could have undoubted success.

This is why we shall present here our English translations of two French works demonstrating the apparent deficiency of forwardly eradicating something if we deal with the scientific research work. Moreover, the works to follow clearly show how energetics could be successfully used for solving stubbornly intricate biophysical problems.[a]

A2.2 Energetics Is a Generally Applicable Concept

The following translation (Sections A2.2.1–A2.2.8) of the original French essay *Énergie, Entropie, Pensée: Essai de Psychophysique Générale Basée sur la Thermodynamique, Avec un Aperçu sur les Variations de l'Entropie dans Quelques Situations Mentales*, by Marius Ameline,[b] might help establish the future directions of the research work on the foundations and applications of thermodynamics.

Energy, Entropy, Thought: A General Psychophysical Essay Based on Thermodynamics, with a Glimpse into Variations of Entropy in Some Mental Situations

A2.2.1 Foreword

Since the year 1895, I had the good fortune of being able to attend the well-known clinical lessons of Prof. Joffroy and thus to initiate the fascinating study of mental diseases. It would delight me if Prof. Joffroy would allow me to express my most profound gratitude to him and accept the dedication of this modest work.

First, it was just attending the work place of Prof. Joffroy and then attending those of Mr. Magnan and Mr. Bouchereau, who could teach me to recognize the importance of various madness types in the hereditary degenerates. It was this opportunity that suggested to me the idea of applying the fundamental principles of energetics to study the latter.

From the works of Mayow, Lavoisier, Liebig, and other colleagues, we already know that it is possible to apply the physical principles

[a]We have reformatted our translations as compared to the original works' structure in order to increase their readability.

[b]Bachelor of Physical Sciences, former external trainee at the hospitals of Paris, now internal trainee at the asylums of the Seine; April 25, 1914, to October 27, 1931, Ameline was the director of medicine at the Specialized Hospital Center Ainay-le-Chateau, the Colony for the Therapeutic Family Reception in Matters of Psychiatry.

and approaches in the field of biology, for example, in the field of nutrition physiology.

Still, the current situation is not the same for other areas of physiology. Indeed, the physiology of muscles has made a lot of progress in this direction due to the work by Mayer and Béclard, unlike the nervous system physiology. Remarkably, to cite just one example, which defeats an absolute authority, it is well known how Prof. Gautier rises from time to time against any attempt to assimilate the notion of thought to some form of energy.

Nevertheless, I believe that not only should energetics have applications in physiology or in laboratory research but also it ought to guide clinicians in the study of nosography itself. To this end, Prof. Leduc, of Nantes, published some indications on the possible results when studying typhoid fever, cerebral palsies, diabetes, intoxications, etc.

Indeed, it is not only the principle of equivalence that might have fruitful consequences but also the principle of Carnot, though it is still full of unknowns, despite its apparent significance. Nonetheless, the proper combination of these principles would surely change the face of biological sciences after the proper bouleversement of the physical sciences.

The reason that has led me to write this work is straightforward: I have started my medical studies, while still being strongly impressed by the vigorous teaching our masters at Sorbonne were delivering. To be mentioned here, in particular, are the contributions by Profs. Bouty and Lippmann.

Sure, when I was facing the radical change of methods, passing without transition from the realm of experimentation into that of pure observation, there were only very general principles that could help me to overcome the confusion that had momentarily persisted.

One of the serious faults of the observational sciences is their compelling too often the observer to classify facts per fortuitous analogies or just coincidences.

Indeed, every time the reasoning is attractive for us, which follows the principle *post hoc, ergo propter hoc* (after this, therefore because of this). Such a standpoint is inherent to the very method and does not depend in any way on the observer.

Moreover, in the choice of the multitude of results offered to us, we had to use our previously acquired knowledge.

As for me personally, I must add to the above that in acquiring this knowledge, it is just thermodynamics that I have used as the base. Hence, it is not surprising for me that I have preserved such a strong impression, that I have sought later to find a relation between the principles of energetics and the normal versus pathological processes occurring in the nervous system.

Before beginning the presentation of my reports, I cannot forget that I was an apprentice of Lancereaux.

I shall always remember having received his instruction during the year before, while at his work place. Everything was so clear and so fruitful and even so penetrated by truly scientific methods. I want to express here my truly lasting, sincere, and devoted gratitude to him.

Moreover, I must sincerely thank Mr. Reynier for having accepted me as a working student at his surgery department in Lariboisière Hospital in Paris: I am deeply grateful for the advice he provided me and for the kindness he showed to me during the time I spent with him, which was but too short, to my sincere regret.

I also hope that Mr. Descroizilles and Mr. Jalaguier would accept my sincere thanks for having guided me with as much benevolence as they did during my study of infantile pathology.

Concerning the asylums of the Seine, I would like to express my deepest gratitude to Bouchereau for the affectionate interest in my activities over last year at his work place at Saint-Anne.

I shall always have excellent memories of my internship at the work places of Drs. Blen and Boudrie, the chief doctors of the asylum of Vaucluse; I hope they would accept my sincere thanks for the kindness they have shown me on several occasions.

Finally, I shall not forget the extreme kindness Dagonet bestowed on me by giving me his advice; I owe him my heartfelt thanks.

I hope Prof. Joffroy, once again, accepts the homage of my sincere gratitude for the honor he has shown me in accepting the presidency of my thesis.

A2.2.2 The First Definition of Entropy

The word "entropy" means a tendency to stability. Entropy indicates a movement; it stands for evolution. Entropy is not stability, for a stable state should be immobile. It is not instability either because of an unstable state, as it might be immobile as well.

Entropy means passing from a state of some stability to a more stable state.

To increase the entropy of a system is to increase the stability already available to it.

We know that in the final analysis, it is possible to treat various known phenomena as some energy transformations. Thus, one form of energy becomes another form of energy.

Now, all kinds of energy—electricity, chemical affinity, heat, mechanical work, light, Newtonian gravity—are not equally capable of being modified, so some ought to be more stable than the others. Hence, from this point of view, there is a particular hierarchy in the varieties of energy.

The most stable variety, the one that is the most difficult to transform, the one that hence occupies a special place among all those we know, ought to be heat energy.

So, transforming forms of energy stepwise into heat means to gradually increase their entropy. When a system is at its maximum entropy, all its energies are in the state of heat.

Heat and entropy finally appear as synonyms, and when a system has released heat, we will be entitled to say that it has acquired entropy.

As to the meaning of the entropy notion, it should just specify both synonymous aspects accurately so we may conceive entropy as the thermal factor of evolution.

This way, the entropy notion indeed indicates both the tendency itself and the purpose of this tendency to transform energy.

Here, it is also necessary to make a provocative remark. Would not monitoring a tendency in the physical and mostly inorganic phenomena underlie the exclusive character of living beings? Indeed, the latter ones tend to self-development, the promise of a future in an individual state being the announcement of an improvement to come. To introduce, as we do, energetics into the study of psychical phenomena should no longer be equal to materializing the vital phenomena. On the contrary, it would correspond to finding life in the manifestations of the so-called inanimate matter.

In these two different cases, both perfecting and a kind of progress are possible, so we might think of these improvements and progress as the processes of acquiring the maximum of entropy. Achieving the latter aim should be necessary both to possess the maximum of stability and to release the maximum of calorific energy.[c]

Consequently, the entropy notion does deserve a thorough and detailed study, so the next part of this work will be devoted to an exposition of the novel ideas that constitute the science of energy, the energetics.

A2.2.3 Introduction and Preliminary Concepts

Summary

- Evolution of general ideas by thesis, antithesis, and synthesis (Hegel)
- Examples: bacterial poisons, Landry's paralysis, and alcohol toxicity
- Time and space, the local sign, and the temporal sign
- Homogeneity of space
- Evolution of time: Science finds invariants excluding possibilities and proceeds by successive approximations
- Mass and force invariants and their combinations with speed
- The livening force and the work: A fusion of these two notions into the one energy notion (Rankine)
- Balance, stability, and the principle of Maupertuis

---

[c]**Our immediate comment**: Here it is important to notice the unexpected slant Ameline is following to introduce the entropy notion. His train of thought looks like being in apparent contradiction to other authors of "Different Thermodynamics" (see Ref. [1] of Chapter 1). Indeed, Horstmann, Liveing, and all their colleagues and followers do consider the physical sense of entropy as hindrances/obstacles/resistances (HOR) to be overcome by livening/driving forces.

We may ask here, How should we introduce "stability/sustainability" like Ameline is insisting?

Our suggestion would be to consider the full story in detail:

1. Livening/driving forces result from kinetic energy.
2. Kinetic energy results from coupling among the subsystems of a system under study, that is, from the system's potential energy.
3. Livening/driving forces compensate/equilibrate HOR.
4. The latter processes exhaust the necessary kinetic energy, thus driving potential energy to some of its minima.
5. Minima of potential energy denote stability, whereas always reaching minima denotes sustainability. This is how we may consider the enthalpy–entropy compensation/equilibration.

To sum up, there might indeed be two equivalent slants in considering the entropy notion.

Of clear pedagogical interest is the way Ameline introduces his very interesting slant.

We will precede this presentation with a few words on the evolution of the general notions that served as the basis of the grand theories of physics.

A general feature of this evolution is that it comes in the form of successive approximations. Hence, it is by gradually modifying the previously available general notions acquired that we may obtain the present notions. However, the mechanism used to perform such modifications is not as simple as it seems at first sight.

Indeed, at any given moment, a sum of knowledge that we will represent, for example, by the stretch OA (see Fig. A2.1), can serve us as a sign, a summary, or as a general notion resulting from this knowledge.

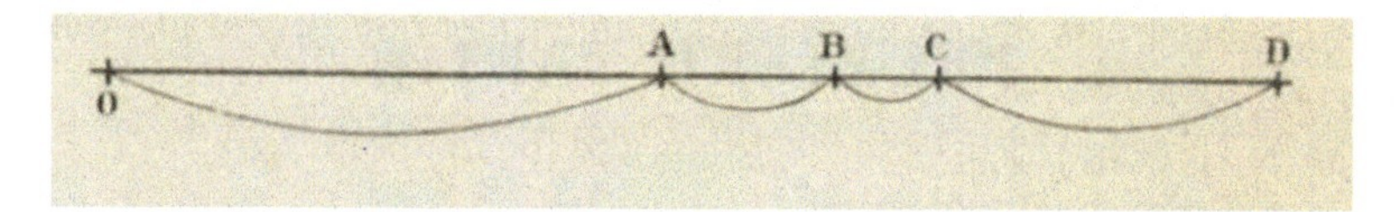

Figure A2.1

The modifications to be made to OA will no longer consist in just adding the summaries AB, BC, and CD to OA, with the successively acquired knowledge portions, AB, BC, and CD being arbitrary and without any relation among themselves.

The actual result should be that the modifications are going to produce fewer and fewer changes to the sum acquired previously by the overall modification; therefore, we should have BC < AB, CD < BC, etc. Thus, instead of Scheme 1, we will obtain Scheme 2 (Fig. A2.2).

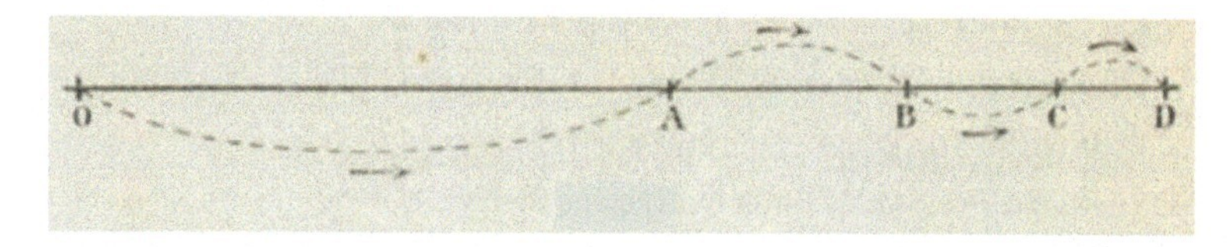

Figure A2.2

We may describe this process by saying that the final sum tends to a limit, as we put it in mathematical terms. We note, however, that the sole condition of diminishing the importance of the modifications is not sufficient to assure us of the tendency toward a limit.

Indeed, let us observe, as remarkably expressed by Hegel, that knowledge proceeds by thesis, antithesis, and synthesis, which means that the modifications do not always have the same meaning, that is to say, that they do not successively add to each other. On the contrary, they add and subtract themselves alternately. As soon as we arrive at the first sum of knowledge, we have to modify it as a result of a new acquisition that might contradict the first result. Further, the change that follows will be contrary to the immediately preceding amendment, being but still in agreement with the latter, and so on. Moreover, it will be intermediate among all the steps already over. In other words, the actual progress ought to be alternately positive and negative. To sum up, there should be a continual oscillation around a limit to be approached as much as we want, without sacrificing the results already achieved, without any bankruptcy or possible defeat of a scientific conception. We illustrate this process in Fig. A2.3 or Fig. A2.4, where the distances at the limit L, Oo, Aa, Bb, Cc, and Dd are shown to decrease gradually.

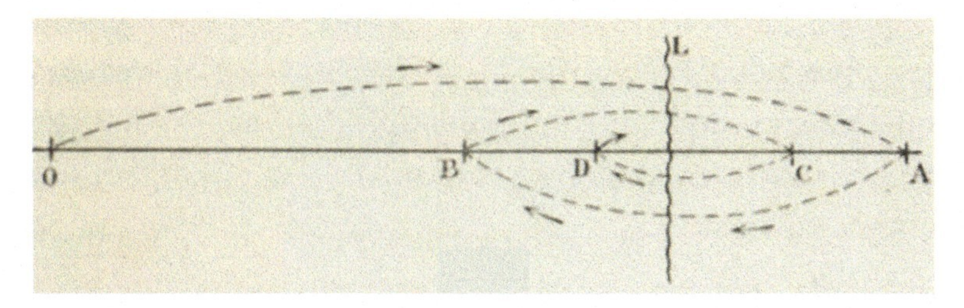

Figure A2.3

To properly use the term borrowed from mathematics, we would say that a convergent series (having an alternation of positive and negative terms and finally a definite limit) would adequately describe the evolution of theories.

We reserve the right to show that this limit does exist and that it does agree with the notions acquired most recently.

However, from now on, not forgetting that here we are dealing with medicine, we would like giving some illustrations of the scientific development always going through thesis, antithesis, and synthesis, while borrowing some examples from the actual medical themes.

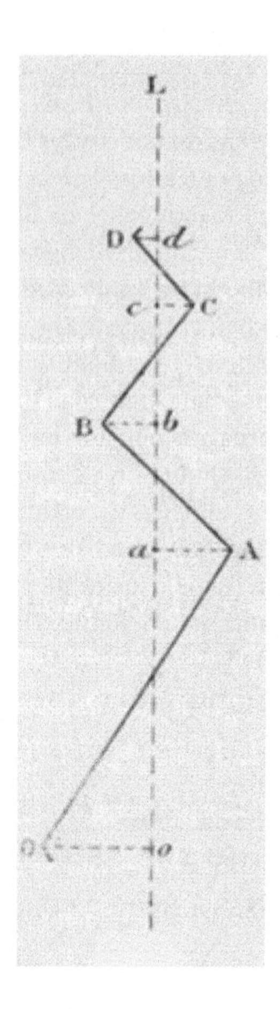

Figure A2.4

***

The history of research on the origin of bacterial poisons clearly shows us this oscillation in the opinions of different authors. At the very beginning, the research works of Seybert, Gaspard, and Stich could establish the symptoms and lesions of experimental sepsis. The results by Panum, Bergmann, and Schmiedeberg, who could determine that these lesions are due to a chemical poison, taken together with the alkaloid treatments performed by Hiller and his followers, first led to the intoxication model of the infection, with the poisons formed in the decaying tissues being everything and the microbes themselves being nothing during the disease.

Later, the specialists reversed the importance of these two factors: intoxication was a commonplace phenomenon that deserved no special attention, while specialists thought of the entire disease as being produced by the progressive invasion of the living animal by the microbes. This victory of the vitalism doctrine over the chemical doctrine (that of the contagionists Davaine, Pasteur, Cohn, etc.) was only momentary, because, as Pasum stated, "Depending on the particular case, the possible pathogens playing a considerable role in diseases might be either some externally supplied chemical poison or the production of the poison by the microorganisms."

Indeed, at present, intoxication has finally reclaimed its rights.

An infection is considered a form of intoxication, but it is a special kind of intoxication by a poison specific for a relevant pathogenic bacterium. Especially studies of the three diseases diphtheria, cholera, and tetanus helped support this opinion. The final convergence of the concept was achieved in 1872, as the results of Profs.Gautier and Selmi concerning ptomaines (bacterial alkaloids) and leucomaines appeared, with the latter being the decomposition products of the former if the decomposition could be carried out by adding oxygen (i.e., using physiological alkaloids).

We might also find a second simple example still capable of demonstrating the oscillatory mode of conceptual evolution, which becomes apparent if we consider the so-called Landry disease.

Just as Prof. Raymond could show, we tended initially to relate the latter to myelitic disorders only. However, as the polyneuritis finally attracted a more significant deal of attention, it was acute polyneuritis that specialists considered as the cause of this disease.

Finally, Mr. Raymond showed that we should consider two forms in this affection and that we should unite them as a peripheral neuronal affection emanated from the marrow. In summary, specialists presume that the Landry disease boils down to an anterior polyneuritis of some toxic or infectious origin.

Hence, in this case, there was the first central stage, then some peripheral stage, and finally, reconciliation, with the help of the neurons theory.

Moreover, considering the general evolution of the research methods in clinical medicine reveals that pure observations arose, at first, as the only method used. On the contrast, later, laboratory science came, which tried to monopolize the truth. For example, the study of alcoholism, first of all of the alcohol toxicity, was mostly clinical at the beginning. It was based only on the approximation of the symptoms observed and on the results of the patients' anamnesis.

Then, during the analysis of ingested liquids, toxic substances were revealed, and specialists revealed the most toxic impurities—aldehydes, furfural, etc.—to be the fundamental reason of intoxication.

It was only by comparing precisely the effects of each of the incriminated substances on animals and by combining all the methods to achieve the most stable conditions that Prof. Joffroy could show that we should incriminate just the alcohol.

Remarkably, the theory of hallucinations has undergone similar fluctuations, being first a purely sensory event as per Esquirol, it became a purely psychic disorder as per Falbet, to appear finally as a psychosensory disorder, thanks to Baillarger and the following authors.

Here we would like to confine ourselves to these examples, with daily observations playing a significant role, with the plausible mechanistic doctrines appearing only because of some historical development of the research on the realistic phenomena. This story ought to demonstrate to us the idea that we have just arrived at—that scientific research proceeds by successive approximations.

***

We shall now examine in more detail, though in a broader outline, several general notions, together with the principles of evolution and transformation of their cognizance.

However, first, we should observe that it is challenging to explain rigorously the modalities of such fundamental notions as space, time, mass, and force, which have long served as a basis for scientific research.

These notions are becoming so imperfect that they are being replaced by others, which seem to be genuinely adequate regarding the present ideas of modern science, and it is just because their lack of precision has become so intolerable for some time that now we have admitted new ones. Nevertheless, we will give some indications on the genesis of the notions formerly used to represent only the phenomena known at the respective periods.

Science consists in putting our sensations in order, whereas the first and foremost characteristic of a sensation is what specialists have denoted as its local sign (Lotze), that is, a specific result of our perceiving them somehow, which is the very starting point: perception by eye, ear, skin, muscle, etc.

In our opinion, we must understand this notion of the local sign as follows: Every sensation contributes to forming within us the notion of space. We should not regard this notion as having its exclusive origin in the exercise of vision or touch or of one of these two senses combined with the muscular sense, as various psychologists have argued. On the contrary, every sensation ought to contain a definite spatial element. Conversely, we can use space as a means of classifying sensations.

There is but another element contained in all sensations, which is the notion of duration: this notion seems rather intimately related to the preceding one. One reason may be that it is related to the muscular sense. But then, a muscular contraction should always accompany the sensations (Wundt, Féré etc.). Or, more philosophically, we ought to base ourselves on the fact that the notion of duration has its origin in the variation of the local signs. It is, therefore, just because the local signs do not remain of the same nature at all (there are definite changes in them) that we possess the notion of time, which we might express as follows: time is the conciliation of being and of not being. Thus, we see that the time notion should still be necessary as a means of ordering sensations.

Nevertheless, despite all the attempts to rationalize them, these concepts still lack clarity. Because what should be the local sign, strictly speaking? This topic seems to be obscure, and this is just what we would like to try to clarify in the rest of the chapter.

Two scientific fields are mainly concerned with the notions mentioned above: geometry, which primarily studies the properties of space, and the more complex kinematics, because it combines time and space by studying the properties of movement, which was and is still regarded as a primordial concept by some physicists.

Let us now say a couple of words about the properties of space and movement.

Our space enjoys a dominant property: it is possible to draw figures of varied sizes while their shapes resemble each other. Furthermore, it is possible to transport a figure from one position to another without significant distortion.

We can obtain paintings and drawings that do not change their shape while changing their position. Hence, there is a conservation of form: in a word, space is homogeneous (Delboeuf). The facts that the resulting geometry expresses are as follows: the sum of the angles of a triangle is equal to 180°, one can always lead a unique parallel from a point taken out of a straight line, and it is possible to trace similar triangles (or other figures, according to Euclid's postulate).

As for the movement, we will just drop a few hints necessary to understand what will follow.

Of all the movements, only two kinds will be of interest to us here.

First, these are uniform motions by which equal spaces are traveled in equal periods by a mobile (point, sign). The amount of space covered during the unit of time is the speed/velocity of the mobile. If we denote the latter by v, then the amount of space e covered by the mobile at the end of time t will be t times v and the expression we will obtain for this space is $e = vt$.

There might also be uniformly varied motion, in which the speed/velocity is not always the same but varies consistently. Then, v will increase by the same quantity, called acceleration, for equal periods, so if g denotes this acceleration, at the end of time t, we will have

$$v = gt.$$

As for the space traveled during each instant (or speed/velocity) in such a case, it is equal to g during the first second, $2g$ during the second one, and so on, being equal to gt during the last second. The average speed will then be such as if the mobile body has just left its resting point,

$$V = \frac{0 + gt}{2} = \frac{gt}{2},$$

and the total space covered by the motion will then read

$$E = V \cdot t = \frac{gt}{2} \cdot t = \frac{gt^2}{2}.$$

Recall that this formula just describes the movement taking a body down to the surface of the earth because of free fall.

***

A significant remark is that in practice, we never observe rigorously uniform movements. Now, as Ostwald has judiciously said, all phenomena of the real world, despite their variety, are only particular cases of all the possibilities we can conceive. To distinguish among the possible cases and the real ones ought to be the actual signification of natural laws, and all of them usually reduce to the same form: to find an invariant, which makes it possible to exclude particular possibilities. Science progresses by successive exclusions of possible phenomena. The mind goes from the indefinite to the definite (Ribot).

An example borrowed from very elementary mathematics will explain the term "invariant" and its role.

Let us consider the well-known equation

$$p^2 + p \cdot x + q = 0.$$

We know that the above can, in principle, admit two values of x for which the equation is satisfied, but sometimes it might admit only one value or even none.

To explicitly and compactly write this down, we know that usually, it is enough to calculate the quantity

$$\left[\frac{p^2}{4} - q\right]$$

and depending on whether it is greater than, equal, to or less than zero, the number of solutions would be 2, 1, or 0, respectively.

Well, the quantity

$$\left[\frac{p^2}{4} - q\right]$$

is the invariant of equations of the second degree with respect to an unknown; this makes it possible to exclude the possibility, for example, of one of the two roots satisfying the proposed equation.

Mass and force are the invariants introduced by mechanics concurrently with space and time to account for the practical impossibility of uniform motion—an exclusion made necessary by experience.

Although we share the opinion that it is impossible to provide entirely satisfactory notions of mass and force, while accepting both these as nothing more than some convenient coefficients (Hertz, H. Poincaré, Ostwald, Stallo, etc.), we might still state that these two notions are not independent of each other. Hence, sometimes, we first define the force to deduce the notion of mass, but sometimes we do the opposite. Of these two ways of conceiving mechanics, the former was particularly popular in France, and one prefers the latter in Great Britain. This situation could even boil down to the well-known impression that the French colleagues always had a spiritualistic way, while the British ones had a materialistic way of working out the same mechanical theory (Bouty).

Indeed, the notion of force originates from the subjective sensations of effort that we experience in many circumstances. However, to make it a physical quantity, it is necessary to measure it, that is, to have some means of comparing various forces. We may accomplish this, for example, by measuring the space traveled by the same body at the same time but subject to one or another force—by comparing the accelerations the same body achieves because of each force under study.

Thus, it is possible to define mass as the quotient resulting in force after multiplying it by acceleration; then we arrive at the following expression:

$$f = m \cdot g,$$

with *m* designating mass.

The notions of mass and force are, therefore, inseparable, and they can be used to define each other.

Let us now look at the meaning of work and *vis viva* (the livening force).

The work of a force for a given time is the product of the force's intensity when acting on the body and the path traveled by this body in the direction of that force's action. Then, the path traveled by the motile body, or its trajectory, is never arbitrary. It is only that traversed in the direction of the applied force. In other words, the work performed by the acting force does not depend on the actual shape of the trajectory (i.e., on any arbitrary position of the motile body on this trajectory) but only on the initial and final positions of the motile body on this trajectory.

From yet another standpoint, mechanically, we should never exclude all the possible trajectory forms by introducing the notion of work in addition to the real actions of forces. Hence, mechanically, we should consider the imperfection degrees of the latter.

Let us take as an example the force of the action to which we attribute the fall of bodies toward the surface of the earth, and we denote the weight of the falling body to be p; so, we have

$$p = m \cdot g,$$

and the work of gravity on a body falling from a height E will read as

$$\vartheta = p \cdot E.$$

On the other hand, we have seen that for a uniformly varied movement the following should be correct:

$$E = \frac{g \cdot t^2}{2}.$$

And then we must cast the velocity at the time point t as

$$V = g \cdot t,$$

wherefrom

$$t = \frac{V}{g} \quad \text{and} \quad t^2 = \frac{V^2}{g^2}.$$

So, we can recast the value of E as:

$$E = \frac{g}{2} \cdot \frac{V^2}{g^2} = \frac{V^2}{2 \cdot g}.$$

To sum up, the above suggests that

$$V^2 = 2 \cdot g \cdot E.$$

After multiplying both sides of the above by m and dividing the results by *2*, we arrive at the following expression:

$$\frac{m \cdot V^2}{2} = m \cdot g \cdot E = p \cdot E\,,$$

Where $p \cdot E$ should stand for the work performed by gravity. As for $\frac{m \cdot V^2}{2}$, it is just *vis viva* (Leibnitz) that is acquired by some body having a velocity V and a mass m.

Hence, we arrive at the following relation:

$$\vartheta = \frac{m \cdot V^2}{2}$$

Alternatively, we have the following fundamental law: 1/2 of the increase of the livening/driving force of a body is equal to the work of the actual force, or the resultant of all the forces acting on this body.

Then, at time point 1 and in position/state 1, the motile body has a speed V_1, and then we have for the work done at this moment:

$$\vartheta_1 = \frac{1}{2} \cdot m \cdot V_1^2 .$$

At time point 2, with the motile body being in position/state 2, we should then arrive at

$$\vartheta_2 = \frac{1}{2} \cdot m \cdot V_2^2 .$$

We get

$$\vartheta_2 - \vartheta_1 = \frac{1}{2} \cdot m \cdot V_1^2 - \frac{1}{2} \cdot m \cdot V_2^2 .$$

Or, in a different formulation,

$$\vartheta_1 + \frac{1}{2} \cdot m \cdot V_1^2 = \vartheta_2 + \frac{1}{2} \cdot m \cdot V_2^2 .$$

Meanwhile, ϑ_1 and $m \cdot V_1^2$ should be always the same. Indeed, as V might become equal to V_2, V_3, etc., ϑ might in turn be equal to ϑ_2, ϑ_3, etc., and the following should always hold:

$$\vartheta + \frac{1}{2} \cdot m \cdot V^2 = \vartheta_1 + \frac{1}{2} \cdot m \cdot V_1^2$$

and hence,

$$\vartheta_1 + \frac{1}{2} \cdot m \cdot V_1^2 = \text{Constant}.$$

Rankine has called the sum of these quantities "energy" and has thus introduced a new scientific branch, namely energetics, which has now dethroned conventional mechanics and which results from the fusion of the two derived notions, the notion of mass and the notion of force.[d]

[d]A Scottish mechanical engineer William John Macquorn Rankine (1820–1872) occupies a very special position in the history of thermodynamics (see, for example, Ref. [1] of Chapter 1 for more details). Noteworthy, being the true pioneer of energetics, Rankine earned furious criticism from Max Planck for that part of his legacy (see Section 2.3.1., pages 678 ff. in Ref. [1] of Chapter 1). Of course, not everything in Rankine's legacy deserves our immediate laudation (see Section 5.2., especially pages 612 ff. in Ref. [1] of Chapter 1), but Ameline's essay at hand does help separate "the seeds" from "the chaff," so to speak.

Energy should, therefore, have forms of movement and mechanical work. Note that the movement notion here is synonymous with *vis viva* and not just with some arbitrary displacement in space.[e]

Therefore, here we have a new invariant that has resulted from a row of successive generalizations to encompass all the other invariants acquired by physics and chemistry. Now we must indicate how this evolution took place, and then we shall start with the notion of energy to derive a new notion that excludes certain phenomena still compatible with mechanical theories but not interesting to us here.

***

Since we shall often use the words "equilibrium" and "stability," before presenting the details of the aforementioned derivation, it is essential to give rigorous definitions of these notions that appear to be well known to us from our experience.

A body is said to be in equilibrium when the resulting sum of the forces applied to the body does not accomplish any work.

The equilibrium is said to be stable or unstable, depending on the possibility of the body's displacement, which would lead to a negative or a positive work performed by the resulting sum of all the acting forces, respectively.

For example, let us consider a substantial body. We apply gravitational force to its center of gravity. If this center of gravity can move only by rising above its position, gravity, which directs a body toward the ground, does a negative work. So, the body will be in stable equilibrium.

If, on the other hand, the center of gravity can move only by lowering the body, gravity will be able to work in the direction of the ground and the work done will thus be positive and the body will not be in stable equilibrium.

Here, we have just formulated the principle of virtual work or, in other words, the principle of possible work.

Moreover, when equilibrium is stable, the force that tends to carry out actual work and that keeps a body in a stable position will react against any other force that would disturb the body from its equilibrium position (e.g., a spring and a pendulum).

We have just formulated the principle of Maupertuis (1744).

[e]Ameline could here clearly show what the sense of the Vis Viva notion is, in fact: Vis Viva stands for the energy of the movement, whereas only movements may drive something. This is why Vis Viva ought to be just the realistic source of any kind of the driving/livening force.

A2.2.4 Succinct Presentation of Thermodynamic Principles

Summary

- Preliminary notions on heat and caloric (principle of Black), temperature, and the quantity of heat: The caloric varying only with the initial and final states (conservation of the caloric)
- Transformation of work into heat
- Relationship between work and the amount of heat
- Joule–Mayer principle
- Conservation of internal energy: The internal energy varying only with the initial and final states
- Compatibility with the impossibility of perpetual motion: Example borrowed from chemistry
- Transformation of heat into work: Relationship between work and temperature
- Maximum yield
- Principle of Carnot
- Reversibility
- Absolute temperature
- Entropy: Null only for reversible transformations but in practice, can only increase (Clausius)

In the study of the heat, we encounter the notions of temperature degree and the quantity of heat.

The notion of temperature is due to vague sensations that deliver us the sense of touch on hot and cold items. This notion acquires a little higher degree of precision if, instead of only considering the sensations produced on our organism, we would in experience dwell on the modifications produced on a given body and take its temperature as a term of comparison with those on the other bodies.

The most straightforward phenomenon, and almost always in agreement with our sensations, is the change in volume: when a body A seems warmer to us than a body B, this means that body A will be superior to body B in increasing the volume of a body X taken for comparison. It will, therefore, be by the volume that we will designate the temperature. So for a volume V_1, the temperature will be 1°, for a volume $2V_1$, it will be 2°, etc., without conveying the meaning that, in effect, the temperature of 2° is twice that of 1°. It is only the volume that has doubled. We, therefore, have a scale of temperature given only by the variation in the volume of the body *X*, which is said to be a thermometer, a seemingly improper name, although the bearer of this name does seem to be measuring heat at first glance.

However, even if nowadays this term is recognized as improper, it was not so at the beginning of our acquaintance with it. Colleagues firmly supposed that the variation in volume ought to be proportional to the variation in heat, or instead, as one then said, of caloric. Moreover, we can measure the caloric by volume, and the unit of heat quantity (calorie) was the amount of heat required to vary the unit of volume, the primitive volume of the body chosen as a thermometer.

In other words, when a body, now in state 1, is transferred to some other state 2, it has acquired a quantity of heat Q. When it is brought from state 2 back to state 1, it yields a quantity of heat that is precisely equal to Q. Then, the body, which had left state 1 to finally return to the same state 1, having thus performed a closed cycle, as we say in physics, should retain the same quantity of caloric. Therefore, the amount of caloric might vary only with the initial state and/or the final state of the bodies under study (Black).[f] However, such a standpoint is incorrect, and it was necessary to amend it by coming closer to the truth, and finally, we could arrive at the proper results.

A2.2.4.1 Joule–Mayer principle

Suppose we spend a quantity of work Θ to produce a certain amount of heat Q. The experience shows that under certain conditions, we should cast the following relationship:

$$\Theta = E \cdot Q,$$

with E being here a quantity called the mechanical equivalent of heat or, better to say, calorie.

This way, the vague notion of a caloric substance disappears and heat is considered as a form of energy, just as defined above.

[f]Here Ameline comes back to the legacy of the Scottish physicist and chemist Joseph Black (1728–1799), who was the mentor of James Watt (1736–1819) and might thus be viewed as one the true pioneers of thermodynamics in the English-speaking sphere.

What are the conditions under which the previous equality is true?

Experience shows that, for the same rise in temperature, when we heat some body (a gas, for example) at some constant volume (we are leaving it at the same volume while heating it), the number of calories the body absorbs is less than what it would absorb when heated at some constant pressure.

Since the specific heat ought to be the number of calories necessary to raise the temperature of a mass unit of the body by 1°, we will say that the specific heat at a constant volume is smaller than that at constant pressure and we call their difference the latent heat of dilatation of the body considered.

Now, what amount of work does a body produce? It is the work used to overcome external forces (the pressure upon a gas, for example). It is, therefore, just to this amount of work that the quantity of heat produced must be equivalent. However, within this amount of heat, one part remains latent and cannot be measured directly whereas the remainder ought to be sensitive to the thermometer. Only the last one would be equal to the quantity Q.

How do we define latent heat? It is straightforward; since it does not intervene in the formula, its variation must be zero in the experiment considered, but it arises at a given moment and disappears at another time, that is, if the body passes from state 1 to state 2, the latent heat varies from U_1 to U_2, that is, the variation is $U_1 - U_2$. When the body goes from state 2 to state 1, if the variation is $U_2 - U_1$, the total variation $U_1 - U_2 + U_2 - U_1$ will be zero.

Therefore, the pertinent condition for us should remain in its original form

$$\Theta = E \cdot Q,$$

which means that after the process is over, the body must return to its initial state, or that it must go through a closed cycle.

The amount of latent heat U has an energy equivalent equal to $E \cdot U$, just like the sensible heat Q has one, that is, $E \cdot Q$, so the energy amount we will have for performing the work Θ, when the body will go from state 1 to state 2 but without returning to state 1, reads

$$\Theta = E \cdot Q - E \cdot U.$$

Therefore, the quantity $U = U_1 - U_2$, or internal energy, is dependent only on the initial state and the final state of the body considered. It is no longer the sensible caloric that enjoys this property, as was formerly supposed.

In what follows, we shall see that the latter statement is not yet entirely correct and that what depends solely on the initial and final states of a body is not only the internal energy, which we have just discussed.

To legitimize the introduction of our new notion, we must show that it excludes certain phenomena compatible with the notion of caloric conservation (Black, 1840).

Indeed, in a machine, the work expended is always more significant than the work used, and a quantity of heat occurs at specific points of the machine (due to friction, for example), without cooling being apparent in other points. This fact shows that the caloric does not retain or destroy itself. If it were otherwise, the so-called perpetual movement would be possible, since then the work expended would be equal to the work used added to the work equivalent of the quantity of heat produced, that is, a work that we could use in its turn. Hence, the bottom line is that the work used would be higher than the work expended—a phenomenon absolutely incompatible with the impossibility of perpetual motion. The principle of equivalence imposes a new notion that excludes the possibility of a phenomenon that was not excluded by the principle of materiality or the conservation of caloric.

The heat then becomes a form of energy and the principle of the conservation of energy, instead of applying to the central forces only (gravity, for example), could be applied to other phenomena as well, that is, it starts to be even more general (Joule, Mayer).

In another chapter, we will treat all the extensions that we may apply to the latter.

Before we finish, please note that it is, in principle, possible to choose another unit of heat quantity than the conventional calorie. The novel unit ought to be even more related to the units of diverse energy forms.

The calorie, which means the amount of heat required to raise the temperature of the unit of water mass from 0° to 1°, will be replaced by the amount of heat equivalent to the unit of work.

The new unit is called the "thermie," and if we express the amount of heat in thermies, then we have numerically

$$\Theta = Q \quad \text{or} \quad \Theta - Q = 0,$$

owing to the principle of equivalence, when the body goes through a closed cycle, but when it undergoes another transformation, we will have

$$\Theta - Q = U.$$

***

To accurately illustrate this notion of energy and internal work, we might recall one of the well-known principles of thermochemistry. It is the principle of the initial state and the final state (Hess, 1840). If a system of composite or unique bodies taken under specified conditions undergoes physical or chemical changes capable of bringing it to a new state (without giving rise to any mechanical effect external to the system), the quantity of heat released or absorbed by the effect of these changes depends solely on the initial state and the final state of the system. This principle holds whatever the nature and/or sequence of the intermediate states.

If there is no external work, the heat released corresponds only to the internal energy of the system and consequently this internal energy does not depend on the changes that would have succeeded one another between state 1 and state 2 of the system in question (Thomsen, Berthelot).

As we shall see later, since chemical affinity should also be a form of energy, it is not the body's internal energy that enjoys the property of being determined solely by the initial and final states of the body under study, or, in mathematical language, it is not the internal energy that admits a potential.

A2.2.4.2 Principle of Carnot–Clausius

The transformation of heat into work gives rise to fascinating remarks, unfortunately still little known, at least in biology.

We wish to apply this principle to the general modalities of the nervous system functioning, both from the normal and the pathological points of view.

We are going to study the relation between work and the quantity of heat expended to produce it. It is not a novelty that the principle of equivalence applies in this case. It is well known to be equally applicable to the opposite case—the transformation of work into heat. The second principle of thermodynamics consists of a relation between work and temperature, a relation obtained by the experimental study of heat engines used in the industry (Hirn, Régnault, Clausius).

Let us consider a thermal machine, a steam engine; for example, we see that we cannot produce work without employing some body capable of lowering the temperature. This conclusion is just what Carnot expresses in a general way as follows: In any thermal machine, there should be a temperature drop; there might be no work production without employing two sources at different temperatures.

However, this statement does not lend itself easily to calculations.

To achieve some analytical result, we must define the yield: it is the ratio between the work produced and the quantity of heat expended, supplied by the hot spring.

A machine has no minimum yield, but it has a maximum one determined by the principle of equivalence. Indeed, if all the heat is used, which is the maximum possible, we have, by evaluating the heat in thermies (using the heat units we have just introduced above):

$$\Theta = Q.$$

Hence, the maximum yield, in this case, would be equal to $\frac{\Theta}{Q}$, which is the yield that Carnot showed to be never attainable in practice.

What are then the conditions for the maximum performance of a realistic machine, which is never capable of producing any work without pertinent temperature drop?

Carnot, in his famous book *Reflections on the Motive Power of Fire* (1824), dealt thoroughly with this vital question.

First, to apply the principle of equivalence, as we have already done, we must ensure that the body would be brought back to its original state, that is, our system must go through some closed cycle.

Then, so no part of the work is lost, the work produced must be as near as possible to the work necessary, to what is called the work of external forces, in a word, that the mechanical conditions of the maximum efficiency are those of the full and perfect balance.

As for the thermal conditions, it is necessary to look for the temperature difference that needs to be achieved between the two relevant sources inherent to the body under study.

Let state 2 be the cold source and state 1 the hot source, indicating the two states that the body must have. It starts from state 1, reaches state 2, and continues its movement until it returns to state 1. This step closes the process cycle, just as we have noticed, to allow the application of the principle of equivalence. However, to go from the cold source to the hot spring, it will be necessary to transport the heat of a cold body to a hot body (Clausius). However, this is not achievable without performing some work, because according to the principle of Carnot in the opposite operation (to produce work), it is necessary to transport the heat from a hot body to a cold body.

To have maximum efficiency, it will be necessary to reduce this work expense as much as possible. For that, it will be necessary that states 1 and 2 are as close as possible to each other. From the thermal point of view, this means that the temperature difference between the two sources must be as low as possible (if there were no temperature difference, no work could be produced). Finally, the transformations should be without loss of heat because otherwise, there would be even fewer resources to produce work.

To indicate transformations occurring at the same temperature, physicists use the well-known word "isothermal" and for transformations at a constant heat, the word "adiabatic" (impermeable to heat).

To summarize, we might conclude that the conditions to achieve a maximum-yielding thermal machine are those very close to equilibrium, and that, consequently, it would be necessary to have minimal external and internal influences, for this would help to reverse the sequence of phenomena. Now, a series of phenomena taking place in a reversed order under weak influences are not only identical but also reversible.

Reversibility is, therefore, the condition of a perfect machine; it was this notion of Carnot that enabled him and would enable us to arrive at the mathematical expression of his principle.

Note also that the conditions mentioned above are, in effect, impractical, in that the perfect reversibility and hence the maximum yield ought never to be obtainable.

Finally, the notion of absolute temperature is still necessary to find the equation we are trying to find. So, let us consider a perfect machine: the bodies that compose it do not influence it because the performance does not involve this composition in its definition. So, either a machine is absorbing some heat amount Q_1 from the hot spring while returning Q_2 to the cold source or it is absorbing Q_1 while restoring Q_3 or Q_4 and so on till Q_n; concerning the cold source, the resulting ratios

$$\frac{Q_2}{Q_1}, \frac{Q_3}{Q_1}, \dots \frac{Q_n}{Q_1}$$

are characteristic of the sources that can absorb heat amounts of Q_2, $Q_3, \dots Q_n$ from the heat source while giving back the heat amount Q_1.

Thus, we might use the resulting ratios to define the temperature of the heat sources in question.

To start with, we have only to write $(\theta_1, \theta_2, \theta_3, \dots \theta_n)$ by designating the temperatures corresponding to the heat amounts $(Q_1, Q_2, Q_3, \dots Q_n)$, respectively, while the following holds:

$$\frac{Q_n}{Q_1} = \frac{\theta_n}{\theta_1}.$$

Then, let us consider a machine running between any two temperatures θ and θ^{I}. So we shall get

$$\frac{Q}{Q^{\mathrm{I}}} = \frac{\theta}{\theta^{\mathrm{I}}},$$

from where

$$\frac{Q}{\theta} = \frac{Q^{\mathrm{I}}}{\theta^{\mathrm{I}}}$$

and finally

$$\frac{Q}{\theta} - \frac{Q^{\mathrm{I}}}{\theta^{\mathrm{I}}} = 0.$$

If Q^{I} stands for a quantity of absorbed heat, it must then have the minus sign, $-Q^{\mathrm{I}}$, and after this correction, the above-mentioned ratio becomes $+\frac{Q^{\mathrm{I}}}{\theta^{\mathrm{I}}}$.

Thus, we get

$$\frac{Q}{\theta} + \frac{Q^{\mathrm{I}}}{\theta^{\mathrm{I}}} = 0.$$

To sum up, when a machine goes through a closed and reversible cycle, the sum of the quantities $\frac{Q}{T}$ should always be equal to zero if by T we denote here the absolute temperature.

Such is the mathematical statement of the principle of Carnot.

As we have seen, it is possible to put the principle of equivalence in the following form: the variation in internal energy is zero for a closed cycle, that is, the internal energy depends only on the initial and final states of the system. If we now compare both, we will say that for a closed and reversible cycle, the sum of the quantities analogous to $\frac{Q}{T}$ does not depend on the path traveled and that it depends only on the initial state and the final state of the system.

To correctly abbreviate, let us call $\frac{Q}{T}$ the entropy S of the system under study: in a state 1 the entropy ought to be S_1; then, in a state 2, the entropy becomes S_2.

Consequently, the entropy variation will be cast as

$$S = S_1 - S_2.$$

Hence, if the body returns to its initial state, $S_2 = S_1$ so $S = 0$.

It is just this quantity to which we were referring that must now replace the internal energy, just as internal energy had to replace caloric as something to preserve during the transformations of a body and, consequently, that can admit a potential in a reversible transformation.

It remains for us to classify the unrealistic phenomena that we might now exclude in considering the entropy, and for that purpose, we must look for the mode of variation that entropy experiences in the realistic phenomena that the universe reveals as irreversible transformations.

Let us now look at the sum of quantities $\frac{Q^{\mathrm{I}}}{\theta^{\mathrm{I}}}$ for the latter kind of transformations.

If $Q > 0$, there is heat absorption by the environment; therefore, if we designate by θ the temperature, it would be the same for the medium and the system. If there were reversibility, we should then have $\theta^{\mathrm{I}} > \theta$. Still, if $\frac{Q}{\theta}$ is the value of entropy for a reversible transformation,[g] we must have $\frac{Q}{\theta^{\mathrm{I}}} < \frac{Q}{\theta}$.

Meanwhile, if $Q < 0$, there would be heat release, and $\theta^{\mathrm{I}} < \theta$ should hold. In this case, we would arrive at the following relationship:

$$\frac{Q}{\theta^{\mathrm{I}}} > \frac{Q}{\theta},$$

for only Q is negative, so we have $-\frac{Q}{\theta^{\mathrm{I}}}$ instead of $\frac{Q}{\theta^{\mathrm{I}}}$. Then, like in the first case, we still must make this correction:

$$-\frac{Q}{\theta^{\mathrm{I}}} < \frac{Q}{\theta}.$$

Thus, in irreversible transformations, the sum of the quantities $\frac{Q}{\theta^{\mathrm{I}}}$ is less than the sum of the analogous amounts for a reversible transformation. Now, while for the latter transformation, it is equal to zero, for an irreversible transformation, it is always smaller than zero.

Then, if S_1 represents this sum at the beginning of the experiment it will be S_2 at the end of the experiment.

With the variation $S_1 - S_2$ being negative, that is to say

$$S_1 - S_2 < 0,$$

we have

$$S_2 > S_1.$$

In other words, the entropy of the system is increasing.

We can write

[g]In fact, Ameline does clearly demonstrate here that the (by and large, notorious) notion of *reversibility* is nothing more than just the purely mathematical consequence of the cyclic property of Carnot's process. Indeed, as we always end up at our starting point, it is all the same, whether we consider heat absorption or heat release, for there will be none of the former and none of the latter as a result. It is also no matter how exactly we travel along the cycle: *clockwise* or *counterclockwise*. Consequently, in Carnot's reversible cyclic process, it is but mathematically expedient to consider the *absolute value* of entropy, that is, Abs(Q/θ). Thus, we render apparent *both* the sense of Ameline's inference and the risks of *physicalizing* the ingenious model by Carnot. The *physicalization* starts when we try to endow the *reversibility* with some magic physical sense, which the latter does not have, in fact. Whatever the reasons of doing so, we end up with finding the basic perpetuum mobile in the form of Carnot's cycle. With this in mind, the alogism and hypocrisy of the conventional equilibrium thermodynamics become apparent.

$$S_2 - S_1 > 0,$$

or after designating by P an essentially positive number

$$S_2 - S_1 = P.$$

Clausius calls P the uncompensated transformation relative to a transition from a state 1 to a state 2.

To sum up, $S_2 - S_1$ is equal to zero for a reversible transformation, and Duhem suggests by symmetry that $S_2 - S_1$ should correspond to the *compensated* [emphasis added] transformation for the transition from a state 2 back to a state 1. Hence, we see that while S_2 and S_1 themselves can be 100% determined if we know only the initial state 1 and the final state 2 of the system, to determine P with pertinent rigor, it is necessary to learn about all the intermediate states through which the system is passing when it goes from state 1 to state 2 or vice versa.

Herewith, we finalize this summary of the thermodynamics foundations. We will now deal with the phases these foundations pass through when they are generalized and applied to some observable facts. The first principle dictates the conservation of energy, the second one the dissipation of energy. In particular, we are interested in the consequences of the latter principle (due to Thomson and secondly due to Helmholtz), especially regarding how to get rid of really extraordinary and untenable hypotheses that are difficult to verify and that we had forcefully introduced to ensure accordance between mechanical theories and the observable fact: that natural phenomena are not reversible. Meanwhile, reversibility does constitute the implicit basis of the so-called mechanical theories of whatever nowadays, or did so at least until now; we could not eliminate this unsustainable circumstance.

It would be a serious achievement, as Ostwald remarks, if one could introduce into mechanics a quantity endowed with the polarity property, that is, the one having a definite sign, such as entropy. Furthermore, there should also be a quantity being not always positive but capable of having the + sign and sometimes the – sign, depending on the situation.

Theoretical energetics seems to fulfill this condition, which means that we must prefer it, at least until we know more.

A2.2.5 Energy and the Forms of Sensitivity

Summary

- The principle of energy conservation
- The principle of conservation of mass
- Inutility of the dualism "force and matter"
- Monism
- The principle of conservation of space
- Energy and mass, where the extension can be reduced to a single notion
- The principle of energy dissipation
- Entropy and time, which are polar magnitudes and can be reduced to a single notion
- The thermal factor and evolution (Mouret)
- Atomo-mechanical theories and general energetics: Descartes, Newton, Boscowitch, and Faraday
- The irreversibility of phenomena
- The temporal sign (Ward), which is nothing more than the elementary variation of the entropy of the nervous system

The notion of energy may account for the phenomena previously attributed to caloric, and one could quickly extend it to explain luminous, chemical, and electrical phenomena; light, affinity, and electricity, just like heat, work, and Vis Viva are ultimately forms of the same thing, energy.

The principle of equivalence undergoes a transformation in parallel to the evolution of the energy notion, and as we generalize the latter, the former gets generalized too, ultimately leading to the principle of energy conservation.

What is the meaning of this principle? We shall try to tell you about this in what follows.

If all the natural agents hitherto inventoried are forms of energy, it becomes impossible to grant the energy notion a general definition, although one recognizes well in each case what energy ought to be. Consequently, if we wish to state this principle in all its generality by applying it to the universe, we should consider it, so to speak, to vanish or, to put it a better way, being reduced to this: there is always something that remains constant (Poincaré). Hence, concludes

Mr. Poincaré, this law is possessed of a framework flexible enough to be able to bring in everything you want. This flexibility is in no way a disadvantage. Instead, it ensures sustainability.

We will indeed show that this generality, which seems excessive, might nevertheless be fruitful in arriving at novel conclusions.

The idea of a constant element in the universe, an element that would take many forms, is susceptible to a higher generalization than the ideas of which we have spoken so far.

If energy is constant and if we find in the universe another constant element, then the sum of both these elements will be constant too.

Now, we have seen that Mother Nature does preserve its forms as well. At the very least, this results from experience. Experiments have shown space to be homogeneous.

However, is there any relationship between space and energy?

We know that we can use all the senses without exception to locate in space the starting point of a sensation, be it auditory, visual, tactile, or muscular. Appreciation of distances is always possible with a single organ, without intervention by the others.

Indeed, light, sound, heat, etc., vary in intensity with distance, while a surface can be perceived directly by touch or the muscular sense.

In short, the entire sensation brings a local sign with it, and this local sign varies depending on the form of energy for which the organ is adaptable, and, consequently, space as a form of energy ought to be an element common to all sensations, and it is, therefore, legitimate to regard spatial extension as a form of energy.

Everywhere, says Ostwald, it's all about energy, and if we separate the different forms of energy from matter, the latter vanishes. It does not even have the space it occupied because this space is known to us only by the expenditure of the energy necessary to penetrate it.

Energy has thus become a more general invariant than all the partial invariants hitherto considered (motion, work, space, etc.), and it should consequently replace them.

There is but still one of those partial invariants that we talked about at the beginning and that we did not represent as being, in fact, a form of energy as well. We want to talk here about the mass notion.

We know that Lavoisier has shown that mass does not vary in chemical reactions, whatever they are, so we stated two principles—

one dealing with mass conservation (Lavoisier) and one dealing with force or energy conservation (Helmholtz)—in a way putting them in opposition to each other.

As we pursue our unitary doctrine regarding the conception of the universe, here we would like to summarize in a few words the reasons that seem to us to militate in favor of the opinion bringing the principle of energy conservation and that of mass conservation under one conceptual roof and thus rejecting the notion of mass as useless and contradictory.

The opposition of the two principles originates in a distinction introduced long ago between the concept of mass and that of movement. In fact, within the Cartesian conception of the universe, matter would have no property; it would be inert, with all its diverse manifestations, if we consider that only the motional modes variety animate matter.

Matter would then be composed of atoms, which are both hard and elastic (which is already contradictory) while being inert. That is, their properties ought to be directed either by the nature of the movement they would have or again (as per Newtonian theories, in which the atoms are motionless) by the nature of the forces according to which they would interact.

In short, especially within these latter theories, it seems evident that the atom notion is useless and can be replaced by a center of force. First Boscovitch and then Faraday suggested this. Therefore, matter, to which Descartes had ascribed no other properties than extension, ought to be useless, to the extent that Leibnitz had suggested replacing the atoms of inert matter by the "monads," the "atoms of force," which ought to remain active and therefore useful, on the contrary.

The belief in the absolute solidity of matter is, moreover, entirely unscientific. Indeed, it is usual to explain the complicated stuff by simple notions, and it is the essence of the scientific method, as opposed to the mystical tendencies, that makes most people think first of the complex and the supernatural. Even nowadays, it is evident that the elasticity laws to describe solid bodies are not entirely clear, so shock is not a simple notion, whereas the laws of both physical and chemical properties of gases are much simpler. The posers are then: Why do we try to explain such simple, consistent patterns by the properties of colliding solid balls, which are still not correctly understood? Why do we strive to explain the simple by the complex?

As Cournot remarks: "Nothing obliges to conceive these atoms as small hard or solid bodies, instead of viewing them as small soft, flexible or liquid masses."

The reasons for such an unscientific standpoint, as per Cournot, are the prejudices of our education and the conditions of our animal life. In fact, perhaps, we ought to ascribe much more importance to sensations due to the muscular sense, due to touch, instead of following some metaphysical tendencies that remain constitutional to the extent that modern scientists have not been able to get rid of them. The reason lies not only in the "structural errors of the mind" (Stallo) but probably more so in the general structure of a human personality as well. Furthermore, the concept of hard and elastic atoms remains so contradictory that there have even been attempts to conceive atoms as a swirling motion of a continuous fluid filling the universe (Helmholtz, Thomson).

The latter attempts would mean taking steps toward the gaseous design of the universe, which is opposite to the concept of solid atomic balls, if I may call it so. Meanwhile, the supposition of partial movements of a continuous fluid is contradictory. To prove the latter, it will suffice to recall the famous arguments by Zeno of Elea against the possibility of a movement in the continuum.

To sum up, the concept of matter must disappear in face of the energy notion, and the principle of the conservation of matter must disappear in face of energy conservation. We cannot put them in opposition to each other. The dualism due to the kinetic theories has no reason to exist, as discussed in detail by Hirn. "The idea of dualism, also says Hæckel, represents the Spirit and the Force of the Matter as two essentially different substances, but that one of these both can exist apart of the other one and let itself be observable could not yet be proven experimentally."

Moreover, the theory of light called the emission theory supposes that light sources emit particles hitting bodies and thus transferring the energy (due to the force) necessary to make them visible. This theory had to retreat in face of the wave theory, where we must no longer assume any transport of matter (note that wave theory is also molecular theory, but it is Newtonian).

Noticeably, in this connection, both Newtonian and Cartesian theories are incapable of explaining why our universe is not reversible. In effect, they can only explain this fact by assuming either motions hidden for us (Helmholtz) or the presence of a demon (Maxwell) that would be capable of compensating for the energy dissipation (irreversibility). One might also simply start from the premise that our universe ought to be reversible in time (Breton). Thus, we assume that once the row of the current transformations is complete, this cycle will be reversed in the opposite direction, the current power becoming the resistance in the second series of transformations, and so on. The latter idea is just a counterpart to the Helmholtz hypothesis, which assumes that the transformation precisely opposite to what we see in our universe is simultaneously reproducing itself in another universe identical to ours. Still, taken together, all these hypotheses only show the original shortcoming of any mechanical theory. We might safely change the sign of the time variable without changing anything to the consequences of the formulae in question.

The mechanism we have just outlined is absolutely incompatible with the theorem of Clausius (principle of Carnot). This is Poincaré's conclusion in his handbook on thermodynamics and various articles. Authors like Stallo, Duhem, Ostwald, Hertz, Léchalas, and Robin, whom we quote at random, do arrive at the same conclusion.

We also see that a radical opposition exists between the notion of energy and that of entropy. The former one is conserved, whereas the latter one increases. Indeed, for entropy to remain constant, it would be necessary that absolutely no irreversible transformations exist, both in the past and in the future. Still, this circumstance does not seem to bother even the most convinced proponents of the universal atomo-mechanical theory.[h] Nevertheless, it seems logically impossible to formulate the same conservation principle for the *motive power* (if we borrow the words from Carnot's work, just as Le Chatelier and Mouret did), for the mass (the law of Lavoisier), for the livening force (the law of Descartes), for the law of Newton, for the quantity of electricity (the law of Faraday), and for entropy. There is no entropy conservation principle (the law of Clausius) since no realistic process can preserve entropy.

[h]Ameline's suggestion was absolutely vindicated at the time of writing his thesis. There was even more to the story: In Appendix 3 to this chapter, we shall meet Simon Ratnowsky, who could rigorously reveal the unphysical basement of the then atomo-physical theories. Nevertheless, today we are aware of the unconditional success and ultimate usefulness of the latter. Remarkably, the true reason for such a paradoxical development has been just the ingenious work to find out the ways of how to treat the inconvenient posers implicitly. It is this trend that was the very basement of Great Physical Revolution at the beginning of the past century. Still, fetishizing this successful revolutionary trend does hinder the scientific research work. This is why Ameline's deliberations are of immense importance for us.

And thus we arrive at the final stage of the successive conceptual generalizations that the principle of equivalence could undergo. The convergence point of this process is the energy conservation law, whose origin ought to be the transformation of livening force into work, and it was a decisive step taken after a careful analysis of the notion of heat. Meanwhile, such physical/chemical notions as electricity, affinity, and light could also be embraceable by the conservation principle, which is also true for the notion of mass. Finally, we have attempted to arrive at some conception similar in its properties to this latter notion. Nevertheless, for the same reason that we reject the atomo-mechanical theories, which put mass and energy in opposition to each other, we put entropy and energy in opposition to each other, and we will now seek to generalize the second principle of thermodynamics and to indicate its significance by applying it to the entire universe.[i]

***

Since the total energy is conserved, in every transformation one energy variety is converted into another one, and these two varieties can be distinguished as follows: the first type bears the generic name of potential energy, whereas the second type ought to be some currently manifestable, observable energy. Then, we must state the energy conservation principle as follows: the sum of the potential energy and the currently manifestable energy is constant in the universe.

Meanwhile, the above transformation is such that energy becomes less and less transformable. This critical observation is due to Thomson, who expressed it as the principle of the dissipation of energy.

---

[i]Remarkably, Ameline underlines here the fundamental vice of the atomic-mechanical theories. Meanwhile, in 1905, Einstein suggested his special relativity theory to introduce the famous relationship $E = mc^2$, while in 1926, Ameline's famous compatriot L. de Broglie suggested the microscopical principle of particle wave dualism. Along with the skillful mathematical deliberations by Heisenberg, Schrödinger, and many other colleagues, it has become possible to conceptually bypass the *apparent opposition of the energy and mass notions* and build up valid, meaningful, and useful physical theories.

We must but stress here one important point: Along with duly bypassing the latter opposition and being totally fascinated with the successes thus achievable, we do lose the way to analyzing the *opposition of energy and entropy notions*, which is but a fundamental physical principle. This is just the main lesson Ameline was trying to teach us... *Was it in vain*?

This dissipation of energy, or rather this tendency of energy to dissipate, is general for all energy forms, but here we shall give an example only for the energy form called chemical affinity.

> At the end of the XVIII-th century, in seeking the laws of double decomposition of dissolved salts, Berthollet could reveal in his detailed study a deficiency of the then concept of the so-called "elective affinities." Indeed, to Berthollet's mind, there had been an observable anomaly, which was more complicated than the others to deal with by any consistent principle. The tendency of one substance to combine with another one had turned out to decrease in proportion to the degree the latter could already saturate the former. Berthollet published several examples to show that a substance (A) can often displace from a second substance (B) a portion of another substance (C), which has a stronger genuine affinity for B because the power of combination of C for B gets weaker by the amount, with which the combination has already taken place. Conversely, the powerful attraction of a substance for a relatively small proportion of another explains why much more of a substance is required to combine thoroughly with a second substance by decomposing a compound containing it than to combine directly with it when the latter is in an isolated state. He could discover, for example, that during the extraction of saltpeter from crude nitric rock by dissolution in water, the increasing concentration of saltpeter in a solution made the remaining portions more difficult to dissolve, even though the salt never saturated the water. Further quantities could be dissolved with ease by substituting freshwater, but each successive washing yielded a smaller amount.

[**Our comment:** Earlier, we have just put Ameline's original description of Berthelot's seminal ideas in more detail by citing the work by Holmes (1932–2003) [1]. Ameline gives a truly concise account of Berthelot's results. Thus, to clarify Ameline's trains of thoughts, a more detailed statement concerning Berthelot's findings ought to be crucial.]

In general, we know nowadays that the solubility of bodies increases with increasing heat and that, consequently, the total or partial disappearance of the solubility corresponds to a release of heat. This is the germ of the more general theory of chemical reactions based on thermochemistry that we owe to the works of

Thomsen and Berthelot. The latter, under the name of the principle of maximum work, has enunciated the following law: any chemical change effected without the intervention of foreign energy tends to produce a system of bodies that release a considerable quantity of heat.

[**Our comment:** Now we know that in some realistic cases, even increasing the temperature could decrease the solubility of bodies; for more details on this exciting and vital topic, please see our work in Ref. [2] and the references therein. However, now we continue with Ameline's narration].

Nevertheless, such an expression of this fundamental principle is imperfect.

First of all, this imperfection of the maximum work principle consists in the lack of sharpness of the meaning of these words' combination: "intervention of foreign energy."

Indeed, Duhem points out that there cannot be "foreign energy," all energies being transformable into one another. However, here we would like to observe that this objection no longer applies if we consider this principle regarding the total energy quantity that the entire isolated and limited universe should contain.

The real objection rests on the nature of the "intervention" just mentioned in connection with this principle, all the more since we also contrast the principle of maximum work and the facts in flagrant contradiction of them.

There are also endothermic chemical reactions, that is, those entailing heat absorption without the possibility of invoking the coexistence of exothermic reactions rendering the final system exothermic. Here come examples of such reactions: the combination of carbon and hydrogen to produce acetylene (Berthelot) and the reduction syntheses that happen under the influence of light in the green parts of plants so that the carbonic acid excreted by animals as a result of the combination of the oxygen the animal inhales and the organic compounds the animal eats is reconvertible into the organic stuff by plants.

Stephenson had already remarked that the plant, of itself, cannot decompose the carbonic acid into its elements. If it were otherwise, we would on the most significant scale afford the embodiment of an equivalent perpetual motion dream.

The production of carbonic acid releases energy in the form of heat, and this energy could be usable. However, this energy is not even usable by the plant to redecompose carbonic acid products. Solar energy must intervene, and now we know that chlorophyll does accomplish its function under the influence of light rays that plants absorb most of (green rays).

Hence, there is transformation of luminous energy into chemical energy, whereas the latter is then convertible into heat energy.

Similarly, the production of acetylene can only occur if we introduce a voltaic arc into the mixture.

Likewise, a particular temperature increase in the immediate environment might trigger the processes relevant to metallurgy and based upon endothermic reactions.

The system (carbonic acid + heat) cannot go back to its original form. It is more stable than the coal and oxygen system that gave it birth, but not according to the following equation, in which we consider only the weight of the materials present:

$$\text{Coal} + \text{Oxygen} = \text{Carbonic acid.}$$

but following the thermochemical equation

$$\text{Coal} + \text{Oxygen} = \text{Carbonic acid} + \text{Heat.}$$

There must be degradation of light energy into heat energy to compensate for the loss of heat energy dispersed or dissipated by respiratory combustions.

The inferiority of the heat form of energy is also obvious in the following example: Joule took a solid vessel containing compressed air and placed it in communication with another vessel, in which he had created a vacuum. The two vessels were each immersed in a tank filled with water: the communication valve of the two receptacles was suddenly opened. The compressed air rushed into the empty vessel, and the pressure became equal in both vessels. There was some lowering of temperature in the vessel from which air was extracted. Thus, the air in the filled vessel spent some of its energy. It forced a definite amount of it to pass into the empty vessel. The empty vessel increased its temperature precisely by the same amount by which it got lower in the filled vessel. This result was just what placing the two vessels into the same water tank could help reveal.

This experiment shows that temperature does not vary; what one has lost is precisely what the other has won.

Meanwhile, it is possible to use the compressed air in the first vessel to produce some mechanical work (by making it act as a compressed air machine), but once it gets shared between the two vessels, it is no longer capable of doing so. However, there was no heat loss; the system performed no work, whereas the air had simply expanded its volume. Hence, the energy portion it had dissipated to achieve the latter result might no longer be usable.

Note that here is the crucial point of the experimental result just discussed.

Let us now try restoring this lost utility to the air portion in question by transferring it using a pump back into the vessel where it initially resided. When the second container gets emptied, this air portion is heated, and the quantity of heat it possesses is precisely equivalent to the work expended by the pump when restoring the utility of the same amount of air that was dissipated by expanding the air in the first part of the experiment.

We may thus sum up like this: It was necessary to transform mechanical work by a pump into heat to compensate for the dissipation of energy. Hence, we did not expend energy but we transformed one energy form into another.

Here comes the conclusion [emphasis added]: Since utility was finally restored to the air mass, it is a form of energy—mechanical work—that has now been lost by a pump during its work while being transformed into heat energy, and this heat turns out to be less usable than the mechanical work.

In other words, the decrease in the system's utility should ultimately correspond to the increase in its stability.

Therefore, the system, which gives off more heat than another one, is the most stable, and therefore the principle of maximum work must be reformulated by saying that the system, which tends to participate in some process, is the one that will have the greatest stability.

Finally, what is most improper is the use of the word "heat," which means here the number of calories.

Indeed, this implicitly supposes the indestructibility of the caloric, and likewise, in the statement of the principle of the initial and the final state, we were forced to change the term "heat" to that of "internal energy," so here we will have to say "entropy" instead of "heat."

This is just what Mouret and Duhem point out: What corresponds to the old word for "heat" or "caloric" is the new word for "entropy." We have accordingly insisted on the necessity of such a substitution.

To sum up, the principle of maximum work must then be restated as follows:

Any change that tends to occur is the one that produces the maximum increase in entropy [emphasis added].

Hence, energy tends to move from one form, which is relatively unstable, to another one, which ought to be more stable. This statement is just the principle of energy dissipation.

Note that the form formerly known as heat (or its synonym, entropy) is the ultimate last form of any energy transformations, being the least usable form of energy.

Clausius did mean the same when he said that we cannot pass the heat (entropy) of a cold body (which has a low entropy) to a hot body (which has more entropy).

This tendency of entropy to increase always legitimizes well the qualification of the "thermal factor of the evolution," which Mouret ascribed to the entropy notion.

Mouret says:

> In fact, whenever a change takes place, the total Entropy of the world always increases, for there are no strictly reversible phenomena. Indeed, everywhere in nature, there is an absence of equilibrium between the systems, which should be the right universal motor, without which neither life nor inorganic-chemical changes would be possible, for everywhere there are also internal frictions, which hinder the restoration of balance. Reversibility, like the uniform motion, is only a theoretical conception, whereas all the realistic phenomena are irreversible, they are all accompanied by an increase in total Entropy.
>
> Precisely this was Clausius idea when he noted that the Entropy of the World is continually tending to the maximum. Consequently, one must still add that the usable energies of motor forces do wear out incessantly as well, that they transform themselves into Heat and thus tend towards zero.

> Indeed, the entropy increase, or the dissipation of the usable Energy: These are just two sides of a significant fact discovered by the genius of William Thomson, the fact that regulates the evolution of substances and beings.

The above overview enables us to bring some precision into our hypotheses concerning the origin and end of the world.

If, as all the cosmogony models dictate, the initial state of the world was chaos or, that is to say, a general and universal absence of equilibrium, let us then also say that there was also a complete absence of heat, a zero entropy. Thus, considering everything that we already know, the final state ought to be the restoration of a general and universal equilibrium marked by the transformation of potential chemical energy and other energy types into the uniformly distributed heat.

The world will still exist as a result, but it will be motionless and lifeless.

***

Just as we have sought to generalize the principle of equivalence, we must seek to generalize the principle of energy dissipation, or, equivalently, the principle of entropy increase.

On the one hand, we have said that:

- Entropy is a polar variable endowed with a certain sense, as opposed to other quantities, that can indifferently be positive or negative and is said to be scalar (Hamilton).
- The Carnot–Clausius principle was an obstacle to the admission of mechanical (Hamiltonian) explanations as a hypothesis giving the key to all the realistic phenomena of our universe.

On the other hand, we have seen that atomo-mechanical theories allow us to indifferently endow the time variable with the + sign or the – sign in the appropriate formulae, that is to say, that these theories do not consider time to be a polar quantity. Léchalas states, "Immediately appears the consequence of suppressing 'before' and 'after' notions. Indeed, any state of the Universe may be the effect of another one, but it may also be its cause. Hence, it suffices to change [+ t] to [– t] in the formulae, in order to reverse the order of phenomena, and it seems that in the absence of a distinction alien to mechanics there will be no means for considering one of the orders of succession to be more credible than the other ones."

[**Our immediate comment:** This is just the point where the outrageous voluntarism might be the actual mastermind of the research activity, instead of the careful analysis of all the opportunities available. Refer to the monograph in Ref. [1] of Chapter 1 for some brightest examples of the voluntarism demonstrated by the adepts of the Hamiltonian atomo-mechanical theories. Remarkably, we might assume that it was voluntarism that could enable the physical revolution to occur at the border of the nineteenth and the twentieth centuries. The actual result is that stubbornly driving a voluntarily chosen train of thoughts might indeed lead to successful theories. Still, it is not okay to voluntarily discard all other trains of thought that might be plausible and finally lead to successful theories as well. This statement ought to be the core principle of scientific research work, and therefore, we have decided to publish the widely unknown and practically forgotten essay by Ameline.]

That is why Breton concluded that the universe ought to be reversible in terms of time, as we have already mentioned, and Mouret shares his conclusion. "Eternity," he says, "would then be the infinite series of grandiose oscillations between chaos and equilibrium, between movement and heat, the infinity of a long-period rhythm, punctuated by the lowering and raising of heat, by the ebb and flow of the immense thermal tide, whose entropy would then measure its insensitive progress."

Meanwhile, theoretical energetics dispenses with these difficulties, and there is no need for all these ingenious suppositions to agree with theoretical reversibility and practical irreversibility. It adopts categorically and in a straightforward manner that some mixture of chaos and balance ought to be present in our universe.

Moreover, we believe that we must achieve the generalization of the Carnot–Clausius principle by assimilating the time to entropy that might be accomplishable in a straightforward way. Remarkably, just as the assimilation of mass and space to energy somewhat boils down to the replacement of both previous notions by the latter one, we now think that entropy ought to replace time.

Such a concept of time does not seem to be in opposition to what the other authors think about the genesis of this essential general notion.

Many colleagues agree that some succession of discontinuous (Wundt, James) or continuous (Mach, Waitz, Ward, Ribot, Fouillée) sensations ought to be required.

"It is probable," says Mach, "that the feeling of Time relates to the organic wear, which is in turn necessarily related to the production of consciousness, and that the Time we feel might probably be due to the work of our attention. Indeed, during the Time of watching Something, the fatigue of the relevant organ of consciousness grows incessantly. Hence, the work of our attention increases incessantly.

"Then, the impressions that are attached to a more significant amount of the prevenient work do appear to us as the older ones."

The design of temporal signs (Ward), like the local signs of Lotze, might be easily explained. The temporal sign, that is, a movement of attention from state 1 to state 2 embodying the evolution stretch state 1 $\rightarrow$ state 2 ought to be the entropy variation experienced by the brain in this passage; the correct synonym of the psychological term "temporal sign" should then be the mathematical term "element of transformation," like the one that we have given to the elementary variations of entropy at the beginning of the work on thermodynamics foundations.

One consequence of this conclusion will be that if there are data where time intervenes as the main element or as a primordial condition, then we might in principle give a thermodynamic interpretation to these data, as a first approximation obviously, by expressing the time variable in terms of the entropy.

Such notions as evolution, progress, memory, stability, and tendencies, might, therefore, be approximately considered as thermodynamic concepts based upon the entropy notion or at least upon this notion as the main factor of the notions just mentioned.

Thus, our opinion at this point would be quite the opposite of that of Broadbent, who was one of the rare authors trying to introduce energetics into the field of cerebral physiology.

Indeed, Broadbent chose to apply the principle of energy conservation (in the following very imperfect form, he had suggested that nervous force ought to be only a variety of motion). Specifically, he had admitted some single mass at a "high chemical tension" situated in the nerve centers or, in other words, such a molecular structure that its atomic groups regularly tend to abruptly resume their typical arrangement immediately free from their movements, in contrast to their natural affinities.

And then he adds, "This high chemical tension has nothing in common with instability, a perpetual tendency to any disorganization." It is in this way that he is suddenly taking over the standpoint of an author who adopts kinetic and atomic—and, therefore, basically non-evolutionary—theories.

It is, however, impossible to reject the principle of energy dissipation if one does admit the principle of energy conservation, like Broadbent.

This conclusion is yet another example of the impossibility of tuning atomo-mechanical theories with energetics.

Otherwise, we must be in opposition to the theoretic standpoint that has dominated all the natural sciences since the seminal contributions by Lamarck and Darwin. There should be no contradiction between the biological and physical sciences on the condition that we reject the contradictory atomo-mechanical theories. We cannot repeat the following too often: there should be no such contradiction, because of the principle of Carnot, that is, the principle of entropy increase and with the entropy taken as a factor of thermal evolution. Therefore, space and time, the two a priori forms of our sensitivity (Kant), would have to be replaced in the future by the more scientific notions of energy and entropy.

[**Our immediate comment:** Many sincere thanks to Ameline, who could help us envisage an entirely different (but throughout plausible) perspective in developing the natural sciences as a whole!]

The dualism of general energetics is, therefore, not the same as that of the atomo-mechanical theories. While the latter ones admit to only something that ought to be constant in the universe, this something has two basic elements, matter and force. As per the latter fact, the energetics instead admits one constant element and one variable element.

Nevertheless, let us not forget that we may reduce the dualism of energetics to a monism having one of two alternative conceptions at the bottom of it—the concept of energy and, even better perhaps, the concept of entropy, indicating at the same time a more or less sensitive evolution to a term that we may conjecture from now on.

Whatever we have said about Descartes's theories, let us not forget that he had tried to account for the universe with two constant elements: one being form or matter and the other that he had tried to express mathematically by defining it as the "quantity of motion," that is, the product of mass and velocity. Then, he stated that this product (mv) ought to be constant in the universe. Meanwhile, it was

Leibnitz who could show that the element to be constant is not equal to mv but to mv^2, that is, mass multiplied not by the first power of velocity but by its second power.

The latter is but the livening force (Vis Viva) that, as we have already seen, ought to be the origin of the present energy notion. Even if the resulting mathematical expressions do differ, they have a basic analogy so striking that we must never forget the genius of Descartes, who first attempted to express this conception accurately; still, Descartes's finding was only the first approximation of the truth.

"What do the first steps of all kinds cost?" asked d'Alembert on this subject, "the merit of doing them dispenses one from the merit of making them great"

[**Our immediate comment:** Ameline helps arrive at an essential conclusion: There is no way to conceptually separate the total energy conservation from the entropy increase, that is, the first basic law of thermodynamics from its second basic law, like Clausius had suggested (that was his conceptual error, in effect). Remarkably, the voluntarism of the revolutionary physicists has consisted of their choosing only one way to reconcile the erroneous gap produced by Clausius. We deal with a unique basic law of energetics: the law of energy conservation and transformation. The entropy notion is of immense importance for grasping the real physical sense of the energy transformability. For more details on this crucial point, please refer to the monograph in Ref. [1] of Chapter 1. The following describes a skillful application of the energetics in a complicated field of psychiatry.

Interestingly, Ameline's essay helps also get some insight into why the successful development of the atomo-mechanical theories had required such (at first glance) strange steps as the solemn "killing the time and space," which could be accomplished by Einstein's relativity theories and finally by the works of Prigogine's school.]

A2.2.6 Third Part

A2.2.6.1 The muscle system and energetics

Summary

- Plan of the third part
- Introduction: The thermodynamics of muscle

- Heat and movement from chemical reactions (Liebig)
- Mayer wrongly applying the principle of equivalence and believing that work is producible at the expense of heat.
- Experiences of Béclard, Chauveau, and Laborde (Since heat is a term in energy transformations (Carnot), we must reject the theory of the muscle-heat machine. This fact does not imply that the origin of the forces in living beings is metaphysical. On the contrary, the consequences of energetics are rigorously verifiable.)
- The electro-capillary concept of the muscle

Since time and space, which are the two basic elements of psychological phenomena, can be regarded as the first idea of the more complete and scientific notions of entropy and energy, respectively, the conclusion that emerges legitimately is that the laws of energetics apply to function of the encephalon and the rest of the nervous system. Indeed, we seek to prove the latter point in the present work.

Still, this proof that psychologists might regard as a direct and sufficient one will not be considered so by physiologists. For the latter colleagues, we will first show, by coordinating the recent research works on muscles and nerves, that such a close analogy does unite these two organs, and that the general physiology of the former must apply to that of the latter. Now, it is already well demonstrated that the functioning of muscles is the domain of thermodynamics.

We hope thus to give the experimental and direct physiological arguments that would fill the gap revealed by Schiff in the physiology of the nervous system when he said, "Science does not possess a single direct experimental fact capable of indicating that the transformation of impressions into active perceptions is a phenomenon subject to the general laws of motion."

Finally, a particular study of some pathological mental phenomena (states of degeneracy) constitutes a sort of later verification of our concept.

***

The relationship between energetics and physiology dates to the time when we were concerned with the origin of the energy form known as "heat."

Initially, heat was confused with life, and this opinion, which goes back to the religions of India, is found in Hunter's statement: "The vital heat is not producible by physical or chemical acts, it is a particular principle, it is a vital force."

The first author who gave a clear and explicit idea as to the origin of animal heat was Mayow. He attributed the property of combining with blood to the nitro-air spirit (oxygen). Thus, the chemical theory of animal heat was born, the theory that Lavoisier had to prove by studying the respiratory function definitively.

The exact source of heat formation is of great concern to physiologists, for Lavoisier thought that combustion was going on in lungs.

Nevertheless, Liebig and Matteuci could show that contracting muscular fibers produce carbonic acid.

Then, Becquerel and Breschet, Helmholtz, Valentin, and Heidemann could demonstrate that it is just the contraction of the muscular system that ought to mainly contribute to the formation of animal heat. At the same time, it became clear that the essential chemical reactions taking place in muscles are responsible for the thermal effects.

So, what is the relationship between mechanical work, chemical reactions, and the heat produced? It is just this poser that we must examine in some detail.

According to Liebig, "However closely related the conditions of heat release and the development of mechanical effects may be to the observer, in no way can we regard heat as a source of mechanical effect"; then the source of mechanical energy and heat energy lies in the intramuscular chemical phenomena, and according to Liebig, we have the following transformation:

$$\text{Chemical effects: } \left\{ \begin{array}{l} \Rightarrow \text{Heat} \\ \Rightarrow \text{Movement} \end{array} \right\}$$

Contrary to this opinion, Mayer maintained that mechanical work might come from the heat released by the intramuscular chemical reactions and we arrive at the following transformation scheme:

$$\text{Chemical effects: } \Rightarrow \text{Heat} \Rightarrow \text{Movement}$$

In effect, the principle of equivalence had just become known, and the inverse transformation of mechanical work into heat had become almost transparent for physical facts. Consequently, the muscular work had to be accompanied by some heat absorption, and Hirn could check this fact approximately.

It was precisely at this moment that Béclard undertook his experiments, and then the problem acquired quite another aspect, probably without Béclard's knowing anything about the work notion and ideas of Mayer.

Béclard found that between the three kinds of phenomena, chemical, mechanical, and calorific, there was a close interrelationship, the expression of which had been given definitively by Berthelot in the form of the following theorem:

> Living creatures must accomplish their functioning without the aid of any energy except for that of their food (water and oxygen included). The heat being developed by a living creature during a period of its existence is equal to the heat produced by the chemical metamorphoses during any intrinsic work of its tissues, plus that to digest its food, but minus the heat absorbed during any extrinsic work done by this creature.

The heat released and the useful work (and only this useful work) should be complementary. Indeed, Béclard found that static muscular contraction, that of a muscle supporting a weight, for example, always develops a quantity of heat superior to that of dynamic muscular contraction, that is to say, accompanied by external mechanical effects and useful (positive) work. Hence, there is not necessarily any heat absorption while a muscle is contracted.

However, this seemed to be the case according to the principle of the equivalence of heat and work. The authors who criticized Béclard wanted to see in the cooling of the muscle at the beginning of its contraction a verification of their muscle-heat-machine concept (Herzen, Heidenhaim, etc.).

The consequences of the debate were but entirely unexpected.

The muscle-heat-machine theory was seemingly related to the monistic conception of all the forces acting throughout the universe, including the vital force. Conversely, to deny the transformation of heat into muscular work would be equivalent to adopting some particular immaterial and supernatural essence that could be valid for the activity of the living beings. Such a standpoint would help at least some minds to doubt that the so-called vital force was just a form of energy.

Later, Chauveau, together with Clausius, Helmholtz, and others, could demonstrate that heat can never be directly transformed into muscular work. The latter, just as Liebig thought, had nothing in common with animal heat, except for its origin in the pertinent chemical reactions.

Moreover, a simple calculation of the muscular efficiency shows that either the theory of the muscle-heat machine is false or it is slightly incompatible with the second fundamental law of thermodynamics.

We base such calculation on the formula that expresses the value of the yield we discussed earlier. Indeed, we know that the yield in question should be expressible by the ratio of the work produced to the quantity of heat expended (Q).

In turn, the work produced is equivalent to the difference between the heat quantity released by the hot source (Q) and that absorbed by the cold source (Q'). Therefore, we arrive at the following relationship:

$$R = \frac{\Theta}{Q} = \frac{Q - Q'}{Q}.$$

After replacing the Q and Q' by the pertinent absolute temperatures T and T', which, as we know, characterize the heat sources in question, we shall have

$$R = \frac{T - T'}{T}.$$

However, the human body operates between 37°C and 45°C, as measured by the mercury thermometer, and in terms of the absolute temperature, we arrive at

37°C + 273°C = 310°C and 45°C + 273°C = 318°C.

Hence, we have

$$R = \frac{318 - 310}{318} = \frac{8}{318}.$$

Then, for the ideal outcome, the following ought to be true:

$$R = 0.025.$$

However, we know this well from experiments that the actual yield is approximately equal to 1/5, according to Helmholtz, which could be between 1/3 and 1/5 according to Fick and is equal to 1/5 according to Hirn.

The experiment is thus in absolute contradiction to the theory based on the idea of the muscle-heat machine, and science must abandon the latter concept.

***

We think it is useful to insist on the necessity of abandoning Mayer's hypothesis. To prove this, we will report some results obtained by Chauveau and his pupils.

Their results urge us to apply the Carnot principle and, therefore, to study the performance of a muscle in its actual operation.

Chauveau says, for example, that the muscle that lifts a weight works less and less economically as the muscle shortening lasts longer. Again, one of the best ways for a muscle to work economically while it is shortened while doing work is to be held as close as possible to its maximum length (Pomilian).

Now, we know that according to the Carnot principle, the machine that has the best yield is the one whose components are in the state nearest to equilibrium, so that the system is reversible. In this case, the work done is close to zero. However, the elastic forces should then be close to zero as well, since a rupture is about to occur. Therefore, the elastic forces hardly intervene in the lifting of the weight. They can only do a job close to zero.

In short, here we are obliged to return to Liebig's theory while modifying it in the sense of the theory that he had but abandoned.

Therefore, the Carnot principle does apply to the physiology of the muscular system. If the principles of energetics do not seem to be extended toward this part of biology yet, then it is just because Mayer had incompletely understood the mechanism of the transformation among the different forms of energy. It is merely the interpretation of the first principle of thermodynamics that was incompatible with the experimental physiology, not with the thermodynamics itself.

Most physicists, like the physiologists of Chauveau's school, come to the same conclusion as the author of the essay at hand: namely, that heat appears only at the end of a series of various energy transformations, and that heat bears the character of an excretion.

***

Finally, the preliminary conclusions about the performance of muscles are in full agreement with the muscular contraction theory of d'Arsonval.

D'Arsonval attributes muscular contraction to the variations in the superficial tension of the liquids because of the capillary phenomena. He insists on this idea. He has been able to explain most of the phenomena observed in muscles by physiologists.

On the basis of the same concept, Imbert was similarly able to foretell the experimental laws by which Chauveau could demonstrate the variability of the yield by the animated engine, depending on the circumstances of experimentally measuring the output by the latter.

In a word, d'Arsonval's concept is in full accordance with the principles of thermodynamics.

Note that we may put vasomotor and circulatory phenomena out of action in the release of heat by muscles since it occurs in a long-time dead muscle as well (Laborde, Pompillian). Remarkably, the same is true of the brain.

Patrizzi could demonstrate that variations in the muscular reaction time are independent of the variations in circulatory disorders, as recorded by the method of plethysmography.

A2.2.6.2 Analogy between the muscle system and the nervous system

Summary

- Original analogy
- Physiological analogy
- Contractility
- The muscle fiber and the neuron
- Other examples showing the physiological analogy of muscles and nerves
- Physical analogy
- The electro-capillary theory of the nerve: Lippmann, D'Arsonval, and De Bueck
- Energetics seemingly legitimately applicable to the functioning of the neuron

We must show now that all the above essential conclusions concerning muscular physiology do apply rigorously to neurophysiology, by relying first upon the close analogy that connects both systems in question: muscular and nervous.

First, this analogy already noted by several authors (Richet, Frédérico, etc.) does exist from the phylogenetic point of view, that is, for a progressive evolution from the bottom of the scale, for in animals, the nervous system forms a unique whole with the muscular system.

Thus, in some species of hydra, for example, the muscular fiber continues on to become a sensitive cell on the external surface of the body, whereas in other species, it is separate from the main body while remaining connected to it anatomically and functionally by a thin stream of protoplasm, which ought to be the first draft of the motor nerve. The differentiation accumulates even more: the neuromuscular cell (Kleinenberg) gives rise to three cell types, one of which continues to be connected by a protoplasmic net, on the one hand to the muscle cell and on the other hand to the ectodermal cell, forming thus an intercentral fiber. Finally, the ectodermal cell moves away from the surface of the body, remaining connected by a protoplasmic network, which represents the sensory nerve. Of the three cell types, one becomes a sensory neuron; the other a motor neuron; and the last cell type becomes an ordinary muscle fiber, which is only a neuron adapted to a particular function. Without representing a select type only from the histological point of view, it is a contractile element from a physiological point of view, such as the motor-sensory neurons.

Such contractility is indeed something that seems to be demonstrated today (Pugnat). Under the title "Hypothesis on the Physiology of the Nervous System," Duval took up the theory of amoebic-like dynamics in the terminal arborizations of neurons. This idea goes back to Walther (1868), who observed frog brains during their thawing. Later, Popoff (1876) could show that nervous cells might absorb solid particles of dyestuffs, and he concluded that this ought to imply the protoplasm's capability of contracting itself.

Indeed, several authors, for example, Köllicker and especially Cajal, have also disputed such a standpoint for some time.

Cajal, unlike in the recent experiments by Heger, could observe any variations in the form of neuronal dendrites, although he examined these cells on dead animals in diverse ways.

However, on the neuroglia, he saw changes in the length of the prolongations, changes that he did not even hesitate to relate to different psychic states.

Thus, in sleep, there would be relaxation (Cajal: relajaçión) of neuralgic prolongations, and in cerebral activity, there would be contraction: Hence, as Soury remarks, he wrongly opposed the theory of Duval, since Duval had in view the nerve cells themselves and not the neuroglia cells.

Let us note that the neuroglia have the same ectodermal origin as the neurons and that the nervous elements, if they were not contractile, would be very different from the muscular and neuroglia elements, with which they have so many embryological and phylogenetic analogies.

The origin of these movements would be attributable to the exchanges, combustions, or assimilations in the cellular body and in the nucleus so that the "sleep" of the neurons would be comparable to the state of the leucocytes in asphyxia, and the arrival of oxygen followed by the departure of carbonic acid would "wake up" these leucocytes and trigger the movements (Meynert).

The idea of mobility in the arboreal extensions of the nervous elements was used by Lépine to explain the suddenness of hysterical phenomena: paralysis, anesthesia, and natural or induced sleep.

The general modalities of natural sleep are also discussed in the work by Pupin.

According to both latter authors, sleep ought to be the state of isolation by retraction of the neurons' terminal extensions, and such movements have been observed in the nucleus, in the nucleolus of the nervous cell body, as well as in the distribution of its granulations.

Besides, this is not the first time when the study of sleep conditions could demonstrate the analogy between muscle fibers and nervous cells, since Obersteiner had already noticed that the sleep and rest of the brain are similar to the absence of contraction, which is the rest of the muscle. Moreover, he had related the origin of sleep to the accumulation of some products, which, if not eliminated, do prevent the functioning of the cerebral organ; likewise they produce the fatigue of the working muscle.

We cannot list here all the facts that prove the analogy between the two kinds of tissue in question. However, we will cite several other, even more complex, examples where the resemblance between the muscle fiber and the neuron could still be apparent.

Thus, Gotch and Macdonald recently found that the three tissue types—nervous, voluntary muscle, and cardiac muscle—behave similarly in the presence of galvanic currents of 1/100th of the duration of a second, when studying their excitability variations due to temperature changes.

Finally, Broca and Richet found that, just as the heart presents a periodic refractory phase, for which it does not respond to electrical excitation, nervous centers exhibit periodic inexcitability, which is only of longer duration. Besides, it is still the cooling of the heart or the nervous centers of the animal under experimental conditions that lengthens the duration of the refractory period.

***

In assuming that the muscle ought to be a body with high capillary tension, we remember our having mentioned above that Imbert could manage to reveal the same laws related to this organ's performance as those found by applying the Carnot principle to its study.

Meanwhile, it is precisely this design that considers the functioning of the neuron as being reduced to electrical variations in tension inseparable from variations in its surface tension, which does explain its amoeboid movements used by Duval to construct his theory of neuronal sleep.

There is no point in recalling the histology of the neuron, so let us see what its physiology is like. Since a protoplasmic prolongation of a truly elongated neuron constitutes the peripheral nerves, the physiology of this neuron and that of the nerve proper are often confused.

Countless theories have sought to explain the innermost phenomena of the nervous function, and we can divide them into two main groups.

While some colleagues have admitted the fundamental chemical nature of nerve transmission (Liebig, Ranke, Becquerel, Hermann, Wundt), others have primarily admitted the physical modality and the electrical modality.

For example, Hermann, in 1867, assumed that a substance of excessive chemical instability, in a word, explosive, forms the cylinder-axis, which would be detachable under the influence of the slightest excitation.

However, a fundamental objection can be formulated against this theory; it is because, in a deflagration, the entire chemical energy would be releasable at once, and yet the nerve, even after several frequently repeated transmissions, loses none of its properties.

Indeed, it is impossible to tire a nerve or to exhaust it. So how do we reconcile this fact with the idea of a chemical consumption of some importance, especially when we are experimenting on nerves deprived of blood flow?

Another objection is that while thermal variations accompany any chemical reaction, the temperature of the nerve varies exceptionally weakly. We draw the statement of the problem from de Boeck in his thesis. Such is at least the conclusion of this colleague.

Thus, Wundt's hypothesis, which implies a reconstitution of unstable products, could be rejected.

As for the physical hypotheses, it is the electrical, physical hypotheses that have been the most controversial. In general, the nerve is comparable to a conductor that unites two electrical stations (e.g., according to Frédéricq). Meanwhile, the transmission rates are entirely different, and although Beaunis claims that such a model would not have all the significance it was given because of the high resistance that nerves exhibit when conducting electricity, we may expect that, if so, the nerves should heat up a lot, which is not the case, and we know from Hermann that the actual electrical resistance of the nerve is 12 million times greater than that of mercury.

Then the ligation of a nerve passes an electric current and stops excitation, so here again, we cannot invoke an analogy with an electric telegraph.

According to the Dubois–Reymond hypothesis, a nerve would be composed of bipolar molecules immersed in a low-conductive medium, molecules having a negative pole at each end, and an intermediate negative zone. Unfortunately, it seems to show that when a nerve is at rest no electric current circulates in it, and François-Franck points out that this conception is in contradiction to the differences in potential he has observed between the asymmetrical points of the same fragment of a nerve, although we seem to connect points where the electromotive force should be identical.

There remains the theory that allows the nervous phenomena to depend on the variation in the superficial tension of the nervous elements or neurons; it is the electro-capillary theory.

Becquerel had had an intuition about this theory, but, as we shall see, he related it to a chemical origin:

> The facts set out in this memoir lead to the following consequences: The nervous, bony, and other muscular currents have been observable in living or dead beings. In all these experimental settings, the tissues form closed currents and communicate the interior with the surface. They accomplish this either with a wire or with a nerve isolated from all adjacent tissues. Thus, the processes do have a chemical origin. They do not come from an electrical organization of the muscles and nerves. Therefore, we cannot treat the observed functions as muscular and nervous aspects of this organization.
>
> The electro-capillary currents play the principal part in these same functions; these are the only currents, whose existence is well ascertained so far: Living bodies produce them wherever there are two different liquids separated by a cellular membrane. As Life diminishes, cells enlarge, liquids mingle, electro-capillary currents cease, and the putrefaction begins; At this point stop the researches of the physicist, because all that is due to the cerebral excitement transmitted to the sensory system, which reacts by a reflex action onto the motor nerves, as well as onto the mechanical action of the heart, depends on the physiology and not physics.
>
> . . . The currents are such that the inner wall of vessels and nerves is the locus of the reducing effects and the outer wall of oxidation effects. They act to trigger oxidation in the parts of the gray substance in contact with the two substances. Moreover, they trigger also a reduction in the parts of the white matter near this same contact.

Lippmann, in 1875, could show that not only does the deformation of the free surface of a liquid give rise to mechanical and thermal phenomena but that the capillary energy gives rise to variations in electric potential, and conversely that a variation in electric potential modifies the surface of a liquid.

D'Arsonval, inspired by this idea, could construct an artificial nerve that exhibits some electrical properties like those of the natural nerve.

He says, "I take a glass tube of the diameter from 1 to 2 millimeters and fill it with the drops of mercury alienating with drops of acidulated water. Thus, they are forming a conductor composed of cylinders alternately formed of mercury and acidulated water, constituting as many capillary electrometers of Lippmann as there are mercurial cylinders."

"On several occasions," continues d'Arsonval, "I drew attention to the importance of the phenomenon of Lippmann, that is, the variation of the surface tension, in the field of physiology.

"It was only using this effect that I could explain the negative oscillation of the muscle and the discharge of electric fish. The experience I have just reported will enable me to explain the negative oscillations in the nerves just in the same way.

"As shown by M. Ranvier, the nerve is indeed, a conductor composed of cells placed end to end and composed, like any cell, by an irritable part (protoplasm) and non-irritable parts or liquids, which have separation surfaces, where variations of surface tension can arise, just as in the tube I described above. This irritation can spread from cell to cell and must necessarily trigger an electric wave having the same speed as the entire length of the nerve."

In the same way, it is easy to understand why a simple ligation or crushing of a nerve leaves its electrical conductivity intact but abolishes its nervous conductivity.

Still, d'Arsonval assumed the continuous cylinder-axis, just as it was (Ranvier), and his artificial nerve was a discontinuous cylinder-axis, as Engelmann wrongly stated. That is why de Boeck modified the artificial nerve in the following way:

D'Arsonval's schematic nerve contains good conductors of electricity, that is, both mercury and acidulated water. However, we have no reason to assume that the electrical conductivity of the latter is comparable to that of the successive disks of albuminoidal and greasy materials superimposed in the cylinder-axis.

Therefore, de Boeck sought to modify the schematic nerve of d'Arsonval by associating intrinsically bad conductors.

He used olive oil and a mixture of alcohol and water of the same density as oil.

In a glass tube 2–5 mm in diameter, successive disks of oil and alcoholic water were introduced.

Electrodes in communication with a Thompson galvanometer were plunged into the ends of a tubewell filled and closed on both sides by a gutta-percha membrane. He also introduced a lead wire like that used to construct his electric thermometer. Then, he insulated the wire carefully from an electrical point of view and connected it to the Wheatstone bridge to try and detect any temperature variations that might have occurred in the schematic apparatus. The entire system was carefully surrounded by cotton wool, protecting it from any external heat radiation.

Each contact on the gutta-percha membrane determines an electrical variation in the state of the system, without producing any modification of temperature, however slight it may be.

Hence, the apparatus does represent a schematic model of transmission that has many analogies with the one we have in the healthy nerve:

- The substances that make up this artificial nerve are truly bad conductors.
- The operation is purely physical and does not include any decomposition or reconstitution of new chemical bodies.
- The schematic nerve is indefatigable.
- Any interruption in the continuity of the liquid column stops operation; this interruption corresponds to the ligation of the nerve.
- This electro-capillary system represents electrical variations analogous to those of the living nerve.
- It does not undergo any appreciable variation in temperature during transmission.

In summary, the typical excitations are transmitted in a nerve by some modification of the surface tension in the successive heterogeneous segments that compose it and by correlative variation in their electrical state.

Therefore, as de Boeck points out, the nervous system ought to consist of a series of successive elements, each formed of a cell and its single extension or its multiple prolongations, which put it in simple contact with other elements, without any histological continuity. While it was impossible to admit this hypothesis on the basis of the former models, it has become plausible nowadays, not only owing to the findings of Demoor, but even more so since the publication of the works by Golgi, Cajal, and van Gehuchten.

Now it is natural to draw the following conclusion concerning both nervous and muscular tissues: there is not only a clear evolutionary analogy and a structural analogy between both that follow from the latter but also a particular analogy in their functioning that allows describing them both according to the general laws of energetics.

It is now essential to discuss the direct experiments made on the physiology of the nervous system to find its actual physical conditions. We shall see that in this case, the story is again similar to that of the analogous research on the muscular system.

A2.2.6.3 Energetics and the nervous system

Summary

- The excitation of calorific energy by the brain (Lombard, Schiff); objection due to Mr. Gautier
- Answer from Italian physiologists (Herzen, Tanzi, Mosso, etc.)
- Note from Mr. Richet and Mr. Laborde: Heat excretion in the form of chemicals
- Objection due to Pouchet and its refutation

Lavoisier pioneered the affirmation that psychic phenomena are possessed of some physicochemical nature, by writing these famous lines:

> We might evaluate what is a mechanical aspect in the work of thinking philosophers, in writers' writing, in musicians' composing. These efforts, usually considered as purely moral, do have something intrinsically physical, material, which makes it possible, in this respect, to compare them to what happens if we feel pains.

Since Magendie, Edwards, Müller, Bernard, etc., reiterate this affirmation, experiments have resulted that have unfortunately been misunderstood because of a faulty interpretation of the heat-to-work-and-work-to-heat equivalence principle.

The first experimental data seem to go back to 1866 and are due to Lombard. His results are very remarkable, indeed, for he could reveal the immense importance of the intellectual effort in the production of brainwork.

The most remarkable effects thus produced were those during the composition and those during the arithmetic operations and copying.

For example, if one is used to work like calculating something, to have a sensible temperature evaluation in such cases, the calculations must be either complicated or carried out at a very high speed. On the contrary, those who don't tend to expose their ideas in writing might help in estimating the elevation of their cerebral temperature when they try to describe a simple subject in writing something.

Finally, it is not only the nature of the intellectual work that comes into play, it is also necessary that the duration of this work be significant so that the temperature varies appreciably.

The only objections we can make to Lombard do not relate to his conclusions but rather to his mode of experimentation (François-Franck)—when directly measuring the temperature of the skull's skin, Lombard concluded the temperature of the encephalon. Primarily, we have to refute his way of interpretation, for Lombard attributed the variations in temperature to the different degree of vascularization; remarkably, Schiff could soon demonstrate that this was not so, as one of his conclusions reads:

> A sensible impression of the trunk and limbs of the extremities produces heat in the brain by the fact alone of its transmission to the nerve center and independently of the circulation, which is identical to Laborde's results about muscle.

Broca, Paul Bert, Seppilli, and Maragliano (1879) could arrive at the same result, and here we must mention Gley (I884) and Bianchi together with Bifulco (1885) and Montefusco. All the colleagues had noticed the elevation of the epicranial temperature under the influence of sensations, emotions, and intellectual work. We should note, however, that Conso (1881) even noticed a drop in temperature at the beginning of his experiments.

A pupil of Schiff, Herzen could underline the significance of these results in their entire generality.

"The conscious process," he said, "is the transition phase from a lower brain organization to a higher brain organization."

In short, the marrow can have an individual consciousness, as Schiff was already maintaining since 1858; this means that the marrow, like the brain itself, has functional disintegration phenomena.

Buccola (1881) applied these ideas to cerebral pathology: he thus noticed that in hypnosis, dementia, and the obscuration of consciousness, the functional disintegration decreased in the bark, even reducing to zero. For example, patients coming out of delusional mania usually do not know whether they have had a dream.

Indeed, the repetition of the same sensation during hypnosis was producing a sort of habit in the nervous centers, and Schiff, Lombard, etc., observed that the latter habit diminishes the elevation of the cerebral temperature.

In dementia, there is no possible disintegration.

Likewise, the psychological automatism existing in a dream or a delirium ought to be accompanied by almost no disintegration (Max-Simon).

It seems that mental pathology could thus receive some basis for its rational explanation. Still, there arose an excessively exciting discussion about the nature of consciousness. This discussion did shake the conventional convictions for a moment. Hence, some time is now necessary to understand precisely the meaning of experiments being the starting point for such high and hasty generalizations.

Gautier could show that it is not possible to immediately conclude from the conception of thought that any release of heat should accompany the energy transformation required by intellectual work.

Indeed, work must always be transformable into heat. And, reciprocally, to produce work, it is necessary to spend some heat, and we immediately fall into the same discussion that we have summarized by studying the thermodynamics of muscles—so, the answer would also be identical.

First, as to the very facts objected to by Gautier, we might adhere to the answer of Richet that Conso had noted in 1881 a lowering of the temperature preceding the final elevation.

Still, we may give a valid explanation of the error that the adversaries of the unity of physiological phenomena have fallen into and persist in such a state. According to them the principle of equivalence implies that the heat-to-work transformation ought to be as easy as that of work-to-heat transformation. Meanwhile, now we know well that the Carnot principle shows that the latter is not the case.

We must consider the work of the brain as typically accomplished in a static way (Laborde) but dynamically in several aspects, as well as the muscular work in the experiments of Béclard.

Presently, the real phenomenon observed is this: the static intellectual work, the work without any external manifestation, ought to release less heat than the respective dynamic work accompanied by manifestations external to the nervous system.

Besides, Tanzi and Mosso (1888) observe that alternative elevation and lowering of temperature might still occur during the work performed by the brain.

Tanzi also concludes (1889) that there are reciprocal equivalence and convertibility between psychic energy and other forms of energy, like heat.

Let us mention again the recent work of Boet and Henri (1898), who agree with the preliminary work.

The importance of externalization is very significant, and Laborde has made it clear in the conclusions of his lecture course at the School of Anthropology.

In prehistoric times, he says, human beings had a life that was almost exclusively automatic from a cerebral point of view, but they had an extreme muscular development, and thus, externalized almost all their energy.

In ancient times, on the other hand, the muscular system was more degenerated, while the nervous system experienced a much more considerable development at the same time, as intellectual work replaced muscular work.

An interesting poser is then, why should the equivalence between the energy accepted and the energy transformed disappear, although this equivalence is so noticeable when looking at the beginning of the appearance of the human species?

The mentioned equivalence is a well-demonstrated fact (here we explicitly insist on this), and it must also take place when considering the replacement of the muscular work by the intellectual one.

Some colleagues even consider this in the broadest outline, such as the ingenious reasoning of Laborde. This result is a direct consequence of Béclard's experiments interpreted in a sense consistent with Carnot's ideas.

Nevertheless, nobody could yet undermine the conviction of the opponents of these ideas, and they persist in having a simple relation of parallelism, with no other connection, between the psychic phenomena and the calorific ones that accompany them.

Meanwhile, Carnot treats heat as a kind of excretion. Hence, why should not we admit that the disintegration of the nervous system would produce a heat excretion like it produces an excretion of chemicals? As clearly demonstrated by the research of Byasson, Mairet, Thorion, Stcherback, etc., both excretion types, chemical and physical products (heat), would increase under the influence of intellectual work.

So Liebig's chart

$$\text{Chemical effects:}\begin{Bmatrix}\Rightarrow \text{Heat}\\ \Rightarrow \text{Movement}\end{Bmatrix}$$

might well be modified in the following way, which only somewhat complicates the latter without altering its essence.

$$\text{Chemical effects:}\left\{\begin{matrix}\Rightarrow\begin{Bmatrix}\text{Chemical products.}\\ \text{Physical products (heat).}\end{Bmatrix}\\ \Rightarrow\begin{Bmatrix}\text{Muscular work.}\\ \text{or}\\ \text{Intellectual work.}\end{Bmatrix}\end{matrix}\right\}$$

***

We believe to have thus shown, once again like many other colleagues, that it is true that thoughts can be regarded as a form of energy. Our original suggestion here would be to explicitly rely upon the Carnot principle. For the phenomena that are the inseparable conditions of thinking processes, the two basic principles of thermodynamics rigorously apply: transformability of all energy forms into each other and their successive degradability into the terminal form of heat. Degradability is the true sense of "increasing the entropy more and more," with heading for even greater stability of the systems under study being the actual basic physical aim/result of any energy transformation.

But before we finish this chapter, it is important that we do not pass over an article by Pouchet, entitled "Anatomical Remarks on the Occasion of the Nature of Thought."

This article contains a curious objection that seems to have escaped the sagacity of Mr. Soury and Mr. Golgi.

Indeed, Pouchet remarks that the unconscious (instinct) is more important than the conscious (thought) in the manifestations of life in general, and this seems to him in favor of the opinion supported by Mr. Gautier.

Now, we understand this sentence as follows: "Thought" is only a notion of a chemical work of the brain (Gautier); since the chemical work due to the unconscious is by far the most important, the chemical work due to the conscious (assuming it exists) is only a small portion of the measure of the total chemical work, and then the variations in the chemical work that really belong to the conscious can be compatible with the experimental errors as soon as the total chemical work is measured experimentally.

And, consequently, we cannot say anything about the relation of chemical work to thought until we have established that chemical work due to consciousness is of an amount comparable to chemical work due to the unconscious and is not infinitely small vis-à-vis the latter.

The answer to this argument of mathematical form will not be mathematical but anatomical. After having shown the possibility of explaining muscular phenomena by thermodynamics, we have shown that the phenomena taking place both in neurons and in muscular fibers must obey the same laws.

Now, if the phenomena of Pouchet's unconscious happen in neurons, then where are the phenomena of the conscious located? In neurons, no doubt. It is in the neurons of the superficial layer of the brain, associating the plumes of the pyramidal cells, where the most complicated phenomena of thought take place. The next poser is then, do these neurons undergo thermal variations that can assimilate their physiology to that of the muscle fiber and other neurons of the cerebrospinal system? An affirmative answer is in order because as all the experimenters since Lombard have noticed, it is when the associations of ideas are the most numerous that the elevation of brain temperature is the strongest.

Therefore, it is the disintegration phenomena occurring in neurons that are responsible for a large part of the thermal phenomena observed.

It is in this way that Pouchet's objection might be refuted directly; only the ingenuity of the form remained, which was to be expected from the rest.

To sum up, we conclude together with Lambling that "The application of physico-chemical laws to biological phenomena is unrestricted."

On the following pages, we will try to show the relations that connect them: the law of Carnot on the one hand and the law of Fechner on the other hand.

We hope to combine intimately the two scientific bases, the psycho-chemical sciences and the psychological sciences, as well as the two concepts that correspond to them: entropy and thought, respectively.

A2.2.6.4 Energetics and the nervous system (*Continued*)

Summary

- The duration of psychic acts proving the transformation of cosmic energy into electro-capillary energy in the neuron
- The law of Weber–Fechner: Facts and interpretations
- The real meaning
- Sensations successively increasing the entropy of neurons, their stability, or their progressive inertia
- Duration and entropy in psychic phenomena
- New verification of the analogy of these two notions.

The time has come to give a more direct legitimacy proof of how to apply the energetic laws to the physiology of the central nervous system.

To accomplish this, we are going to study the conditions of sensation and perception. That is, we are interested in psychic phenomena that are of a higher order than the functioning of the muscle and the nerve since the brain plays an incontestable part in it.

We will divide this section into two parts.

In the first, we will briefly outline the support of the monistic and unitary concept of the physical and vital forces and those associated with the nervous system. Thus, first and foremost, we will mention the argument drawn from the fact of the duration of psychic acts, which implies the following conclusion: Any process that requires a specific amount of time can only consist in the transformation of one kind of energy into another. Hence, mental activity must also be the result of some energy transformation.

In the second part, we will look for the meaning of Weber–Fechner's law, which expresses the relations between excitation and perceived sensation.

The first observations of the duration of psychic phenomena go back to astronomers' research studies on the well-known "personal equation," that is to say, on the delay (differing with each individual) that one notes between the hour of the passage of a star at the meridian and the exact time of this passage (Maskelyne [1795], Bessel [1828], Nicolaï, Treviranus [1830]).

For Nicolaï, this fact comes from a non-instantaneous conflict between consciousness and the organ of the senses; but Müller declares that he cannot admit such an interpretation, for he has observed that the reaction time in question is infinitely short and invaluable.

When studying frogs poisoned by vomited nuts, he says that he could not notice any interval of time between touching and convulsion (1840).

Since then, there have been many experiments by Hirsch, Wolff, Donders, Jaeger, Helmholtz, Schiff, Herzen, Buccola, Richet, Wundt, Exner, Beaunis, etc. It would take too long to enter the details of those experiments. Remarkably, the latter could have demonstrated the existence of reaction times (personal equation) with absolute certainty.

Let us now retain only this incontestable fact and seek the interpretation that constitutes the direct proof of the identity that exists between the psychic activity and one of the energy forms, just as requested by Schiff.

Here is an abridgment of Schiff's reasoning:

> The interval of time, which elapses between the sensory impression and the realization of the corresponding act, cannot be inert, inactive; it can serve only to establish continuity between the cause and the effect. Indeed, the suspension of this continuity for only one-millionth of a second would be equivalent to a suspension for the entire eternity.
>
> Now, a movement that has begun must continue; that is just what is always and always observed in Nature. Hence, the duration of the psychic act does surely indicate some modification of an external movement into the internal one taking place in the substratum of the psychic acts in the nervous system.

We would now conclude that there ought to be a transformation between external energy and internal energy, which may, in principle, manifest itself in another form of internal energy, that is, in muscular work, or maybe not.

Therefore, the available research results in psychophysiology teach us unmistakably that time is the condition par excellence of the transformations of some external energy portion into nervous energy first. We know nervous energy to be electro-capillary energy. Finally, nervous energy might be transformable into muscular energy, which in turn is also reducible to electro-capillary energy.

***

The search for the meaning of Weber–Fechner's law will enable us to explain the nature of the nervous force outside of any metaphysical hypothesis.

Colleagues have long claimed that the nervous system functioning ought to be a kind of movement. Here, we have already learned that the straightforward and convincing reason for the latter statement is true (see above): there is a clear analogy between the functioning of the muscle fiber and the neuron. In addition to this, we shall now consider the results of calculations and experimental measurements that will allow us to be even more affirmative than the research on the existence and origin of the reaction time could allow itself.

Interestingly, as for the recent research results, it is in the memoirs published outside the physiology and psychology fields that colleagues have reported the first results concerning excitations and sensations (Euler, Bernoulli, Laplace, Bouvier, Arago, Poisson, Steinheil, etc.).

Let us say right away what the law of Weber is.

Sensations grow in equal amounts when excitations or irritations grow in relatively equal amounts.

Or again in somewhat different wording: The smallest perceptible difference between two excitations of the same nature is always due to a real difference, which increases proportionally to these same excitations (Weber).

Thus, consider, for example, a series of sensations increasing by the same quantity and for simplicity starting from I. This series will then be I, II, III, IV, V, VI, and so on.

Then, let 3 be the increase that we must give to excitation number 1 to have sensation II, and the corresponding excitation will be 3.

According to Weber's law, to have sensation III, it will be necessary to increase the excitation in proportion to 3 so that 3 times 3 = 9.

Further, to have sensation IV after having sensation III, it will be necessary that irritation 9 (which corresponds to the third sensation) should undergo an increase proportional to 9, that is, 3 times 9 = 27, since 3 is the minimum variation in the increase of excitations.

It will, therefore, be the progression 1, 3, 9, and 27 of the irritations that will produce the progression I, II, III, and IV, respectively, of the sensations.

We call the former as geometric progression (GP), whereas the latter is known as arithmetic progression (AP), which grants the following form to Weber's law:

While sensations grow in an AP, the excitations grow in a GP.

Mathematically, if some sequence of integers follows an AP, then the subsequent logarithms of these numbers should follow a GP, and we might conclude that the perceptibility of a sensation increases proportionally to the logarithm of the irritation (Fechner).

Let us now examine the various criticisms concerning the Weber–Fechner law in its substance, as well as in its form.

Among all the criticisms addressed to this psychophysical law by a host of authors [Bernstein (1868), Brentano (1874), Langer (1876), Helmholtz, Aubert, Mach, Classen, Tannery (1875), etc.] and especially Héring (1872–1875) and Delboeuf, we will summarize here only the main points, and only those that are sober:

- The mathematical difficulties raised by the law.
- The modification or abandonment of the psychophysical formula.
- The partial or total disagreement between the experiments and the proposed expression.

Among the reported mathematical difficulties, colleagues claimed that Fechner's formula implies the existence of negative sensations. Here we must note that:

- In physics/chemistry, a negative temperature appears only because the temperature scale has its zero in the reasonable middle indications. We already know for sure that it is enough to take a new starting point at ~270° below the current zero so that all the temperatures become positive again (this is just how we introduce absolute temperatures).
- Likewise, there seem to be no negative sensations for they arise solely because of the imperfect nomenclature used.

As for the formula of the law, we would not like to dwell much on the criticisms that concern it; we only say here that it is a very natural formula describing many ordinary physical laws, which do not seem absurd for such a reason, for example, the law of Mariotte, according to which the work of the pressure necessary to compress a gas is proportional to the logarithm of the ratio of the initial volume to the final volume. Likewise, we arrive at the law that expresses the atmospheric density decrease versus the altitude, the law of the heat propagation along a metallic bar, etc.

Finally, there remain several partial negations (refer to, for example, Hering) concerning Fechner's observations. Still, more recently, colleagues have already published many research results confirming the validity of the latter. Waller's recent experiments (his results on the mode of reaction could be applied both to the nerve and to the muscle) demonstrate that depending on the experimental conditions, one obtains an *S*-shaped curve that agrees with the experimental data by Fechner, on the one hand, and with those by Hering, on the other hand. This fact proves the importance of the experimental conditions and gives us legitimacy to invalidate several criticisms concerning the psychophysical law.

Moreover, as Delboeuf points out, one only needs daily observations to convince oneself of the following facts:

Suppose we add some excitement to already existing excitement. We shall feel the addition if it is stronger than the previous excitement. The amount of this additional strengthening ought to be the same for any addition in the row.

For example, if there is noise around two interlocutors, they both raise their voices to be heard.

When the sun rises, its light renders invisible the moon and all the stars, whose brightness has not varied.

On the other hand, experience has shown that an indefinite increase in the number of performers in vocal or instrumental concerts does not produce action of proportionate intensity on the listener's ears.

All this agrees with the psychophysical fact that the variation in the excitation must increase more and more to produce an appreciable variation in sensation.

In short, everything is already available conceptually. There is no need to look for some mathematical formulae, because they are easy to find, and they need only be further complicated if we wish to render them more and more consistent with the experimental data to come. This single result should suffice to tempt a thermodynamic interpretation of the Weber–Fechner law to find direct proof of the unity between the physical and psychic energy forms.

Here is the interpretation of Fechner's law that we propose:

Let us consider physically-chemically the energy of an excitation that increases the entropy or stability of the peripheral neuron. The brain is then less likely to change under the influence of the excitations that follow. Consequently, the corresponding modification of the neurons responsible for the sensation must be produced by excitations that are more and more intense to remain perceptible.

Herewith we do not invoke the vague notion of fatigue: it is now the precise and scientific notion of stability, of entropy, that, in our opinion, gives the key to the phenomena from which we wanted to draw the psychophysical law.

Conversely, the existence of the psychophysical law is the direct experimental proof that the principle of Carnot applies to the functioning of the nervous system elements, neurons. It also proves that the second law of thermodynamics (degradation of energy) governs the nervous system, as it is the case for the first fundamental law (energy conservation) from the moment the existence of the psychical reaction time could be realizable.

We have therefore reached the end of our task, and we think we have definitively established that the energetics laws must henceforth be taken as a starting point in the study of the physiology of the nervous system, even in its most complicated phenomena, like sensation, perception, and association.

Thus, the transformation of energy is in line with the increase in entropy, that is, in a word, the two conditions of thought.

***

Before finishing this chapter, we must remember that in trying to generalize the concepts of entropy and energy, we would like to reveal two conceptual forms that could help to render the notions of time and space to be more precise.

First, let us note that psychic phenomena do require both a variation in duration and a variation in entropy.

Then, why not melt both these variations into one? Does the Carnot principle not imply the principle of equivalence? Likewise, the notion of personal equation does prove the transformation of physical energy forms into psychic energy, the psychophysical law proving that the increase of entropy in such transformations must imply the duration of psychic phenomena.

Consequently, time and entropy are two notions so intimately united that any distinction between them appears artificial and useless.

Moreover, that time and heat release do vary in parallel to each other should again support this idea. Such a conclusion sounds true because the physicists had to realize that the heat notion is identical to the entropy notion. We have already discussed this above.

Thus, on the one hand, Ribot remarks, "Psychometric research shows every day that the state of consciousness requires a significant time because it is more complex than primitive and acquired automatic acts, whose speed is extreme for they are not entering the consciousness."

On the other hand, we also know that according to Lombard, Schiff, and Mosso, as the nervous centers repeat an act, this act becomes more comfortable and more accessible, whereas the heat released becomes weaker and weaker, while the consciousness decreases until it disappears.

Therefore, time and entropy (heat) do have concomitant variations in psychic phenomena. This fact does legitimize the fusion we propose.

Let us observe, however, that the notions of time and heat are not absolutely and strictly identical with each other and with the entropy notion and even that they cannot be so since entropy is just a virtual stage in the evolution of all these notions, being the evolution that progresses by successive approximations toward a still unknown notion limit.

First, the new notion ought to differ from the earlier ones by the exclusion of certain phenomena hitherto unrecognized as non-existent. A notion more scientific than the other ones should thus remove mysticism and ignorance from our thoughts while causing both doubts and belief.

It is in this way that we believe that the new notions of energy do not have the metaphysical character of the notions employed in the atomo-mechanical theory of heat.

The notions of temperature and calorie fall almost immediately under the category of senses; they are properties as sensitive as color, form, surface roughness, density, etc. It is from temperature and calorie that the entropy notion arises when we divide the latter by the former. Hence, the entropy notion has an elementary mathematical form, and while all the hypothetical and primarily metaphysical explanations of psychic phenomena by movements of atoms and/or molecules are overcomplicated mathematically, they cannot have experimental support and almost immediate observation, unlike the energetic concept that we propose here.[j]

A2.2.7 Thermodynamic Design of Some Mental Situations

Summary

- First attempts of which the principle of equivalence has been the only object
- Reversibility and capsizability of mental phenomena
- Hallucinations and memory
- Definition of degeneration
- The dissolution of heredity and the constitution of an abnormal type
- The energetic conception of the degenerate: Total or partial stopping in the increase of the entropy
- Psychophysics of the primary delusions of degenerates, entropy increase close to zero, either in the revolution of the individual (confusion, hallucinations) or in the evolution of the species (false interpretations)
- Anxiety, depression, excitement, etc., being only secondary affective disorders (the man of genius seeming to break heredity but remaining normal, continuing to increase the entropy of the species)

[j]Ameline's appeal to consider in detail the interrelationship between the time and entropy notions is of immediate philosophical interest. To Ameline's deliberations, we must add the philosophical conjecture by Nils Engelbrektsson, who viewed entropy as a "limiting concept, where the actual limit ought to be imposed from outside" (see Ref. [1] of Chapter 1). To bridge the apparent logical gap between the "outside limit" and the "thermal factor of evolution/stability/sustainability" would definitely be a challenging philosophical poser. On this theme, it would be of interest to recast a glance at the legacies of Henri-Louis Bergson (1859–1941) and Edmund Gustav Albrecht Husserl (1859–1938). Meanwhile, due care should be taken not to drown in the fetishism concerning atomo-mechanical theories, whose fully metaphysical basement somehow did not interfere with them to prove their validity and usefulness.

- Different types of degenerate, characterized by the absence of the entropy increase of his species
- The general evolution of ideas by thesis, antithesis, and synthesis, creating an apparent difficulty in the diagnosis

As already announced, in this last part of our thesis, we will aim to investigate whether the principles of thermodynamics apply not only to the functioning of an isolated neuron (nerve) or even some group of neurons (arc reflex, reaction time, Fechner's law) but also to all the neurons making up the encephalic mass and even the entire nervous system. We would like to clarify whether the mentioned neuron communities might have some operation modus, where the conditions imposed by energetics are realizable.

If we succeed in this test, then it follows, logically and inevitably, that no metaphysical concept, that is, neither atoms, nor molecules in motion, nor vital fluid, nor psychic force, can have any utility in the study of mental phenomena and that the latter could bring only confusion and delusion.

Colleagues have reported some attempts in this direction. These are especially noticeable, for they are mostly general and confined almost exclusively to discussing energy increase or decrease—either in neurons, or in the brain, or even in the entire nervous system.

We have already quoted the work of Pupin on the histological nature of sleep; we will not return to it here.

Bombarda also seeks to explain the phenomena of natural and artificial sleep: in the first case, there would be relaxation, or paralysis of the neuronal extensions, while in the second case, there would be a contraction, or tetanization, of these same extensions.

Soukhanoff goes deeper into the details of the functioning of neurons, but his work is limited to speaking of accumulation, defect, and energy discharge of neurons, which lead to more or less frequent vibrations of and encounters among the neuronal dendrites, depending on the case.

All these authors, in short, consider only the principle of energy conservation, which implies only that the nervous energy comes from other forms of energy and that it can increase or decrease. However, none seeks to apply the Carnot principle, which indicates how some energy variety can transform into another one.

Only Buccola, as we have seen, had sought to consider metabolism (excretions, disassimilation of the nervous system) in trying to rationalize some mental phenomena.

Here we will discuss how the main symptoms observed in cases of delusions might be interpreted from an energetic point of view and thus try to show the thermodynamic significance of these symptoms.

In doing so, we will not look at the state of brain energy but solely at the ways energy might be transformable. In short, we shall try answering the poser, what is the variation in entropy in the main classes of delusions observed generally?

"When I remember playing with a hoop," reports Guyau, "the image I evoke is present, just as present as this paper on which I am currently expressing abstract ideas."

First, to enable this, a fresh sensation must produce a definitive modification in the nervous centers.

Then, this modification must be repeated in the opposite direction, just to restore the illusion of a current phenomenon. It is true, as Guyau remarks, that if one is conscious of feeling the second time, the feeling is noticeably weaker than the first time.

Richet characterizes the memory by the fact that a brief excitation leaves a prolonged reverberation that can be entirely latent and that can be revocable later.

Hence, the poser arises, how can we conceive the thermodynamic significance of memory?

In straightforward terms, to our mind, indeed, the stability of the nervous system ought to increase after each successive excitation, and this stability cannot diminish since no real phenomenon might bring a null variation in the acquired stability. Hence, such an excitation should indelibly modify the nervous system.

As for the evocation of excitement already felt, Guyau reports it himself; the modification ought to be repeatable in the opposite direction.

Thus, the phenomena capable of modifying the nervous system ought to be capsizable and thus can be reproduced in the opposite direction.

Meanwhile, their capsizability is never perfect for in effect there is no right reversibility in the ordinary sense of the latter word at least.

The actual evocation is much weaker: when capsizability is perfect, there is complete reversibility, and the reversed occurrence of the phenomena should be identical to that after the initial excitation. Hence, there would be a firm belief in the reality of the sensation. In other words, it is then a hallucination.

To sum up, if the nervous system were indeed very stable, no excitement could change it. Since it has medium stability in effect, it would still be modified, whereas the phenomena should be capsizable without being truly reversible, with the latter property being possible only if the nervous system stability were almost zero.

Let us now replace the term "stability" with the notion of "entropy." We see that when hallucinations occur, it is because the nervous system or some part of the nervous system in play undergoes only reversible modifications, so that entropy in its processes does change very little, so little that even the actually non-existent/irrelevant excitements seem to be happening.

Thus, it is clear that a malfunction of the nervous system causes relevant phenomena to be reversible. Moreover, this is independent of the nature of the excitations producing reversible modifications in the nervous system; this depends solely on its failure.

When we discussed Fechner's law, we noticed that, for excitation to be perceived differently from the preceding one, it must undergo a proportional increase compared to the preceding one.

The relative size of the necessary increase is a measure of the stability of the neuron and should increase if the neuron is functioning by degrading more energy, that is, increasing entropy.

Well, the nervous system ought to be even more sensitive to the manifestations external to it if it will undergo modifications in an almost balanced state, that is to say, while not producing a considerable increase in entropy.[k]

Likewise, in any other system, it is known that the entropy of the nervous system must typically increase, and if the nervous system's entropy increase appears to be too little or almost zero, then we might speak of its pathological functioning, and the nervous system in question ought to be profoundly abnormal.

***

---

[k]It is just here that Ameline's viewpoint as to entropy could be reasonably combined with Engelbrektsson's "outside limit" description of this notion (see Ref. [1] of Chapter 1). Medically seen, the nervous system has no "immunity" in regard to external manifestations, and its sanity/stability cannot be ensured. The latter seems to be the zest of Ameline's idea: The organism must pose some limits to defend its nervous system against various external manifestations. Insufficiencies of such limits might lead to the pathologies just discussed.

We think that there is a thermodynamic explanation for the state of degeneration as well.

What is degeneration then?

If we go back to the definitions we have already given, we see that the degenerate is a new type that embodies some biological conditions of the hereditary struggle for life, but only in an incomplete way (Magnan, Morel).

Degeneration, says Mr. Morel, is the dissolution of heredity; normal heredity implies the perpetuation of the species, adaptation to the biological conditions imposed by the environment, etc.

Meanwhile, Mr. Magnan insists that the degenerate is not a product of developmental discontinuity, nor an atavistic product, and that the degenerate should be a rather new type. Heredity, which is the most necessary condition for the species, is lacking in the degenerate.

It is a new type with a nervous system that no longer functions normally. Whereas evolution ought to have the thermal factor (entropy), heredity ought to be the *sine qua non* of the evolution; therefore, in the degenerate's nervous system, entropy must occur only very inadequately.

At no time may the nervous system function reversibly (with a zero variation in entropy).

The degenerates, who differ so much from a regressive type, cannot therefore have nervous systems, which are functioning otherwise than in a reversible way, where so little is needed. Otherwise, it would undergo irreversible changes, and then it would be a standard type, but only late vis-à-vis its current environment, and it would not present anything pathological.

The degenerate is not appropriately unstable. While lacking stability, the degenerate should be overly balanced with the environment and sustaining the latter in a very profound way. An unstable system, whose equilibrium could be changed, might pass to a stable state and should become perfectly normal over there. On the contrary, a system being neither in equilibrium nor in disequilibrium with the external forces, should be the toy of any variation, of fluctuations of any type of forces, and in this regard should correspond well to the idea of degeneracy. Neither stability nor instability, which would indicate the possibility of quickly acquiring this stability and of preserving it, but a complete lack of both should be the characteristic of the state of hereditary degeneracy.

Therefore, it should not be disadvantageous—both not to suffer the environment and not to be too much in balance with it.

Nevertheless, it is necessary to possess certain cerebral inertia, which may be called the power of inhibition, to be able to preserve and maintain the characters transmitted hereditarily, and if we say laconically that the degenerate is hereditary, then it means that such a heredity offers significant differences with average inheritance. The degenerate is a pathological hereditary, and such an individual must be comparable not to his or her immediate relatives but to his or her distant ancestors. Degenerates must be comparable to the exceptional variety of the human species, to which they belong. They may thus rightly deserve the name of "degenerate"; degeneration can be more or less rapid and it can appear in an individual, in a family, or in a group of individuals. In this latter case, we may speak of the gradual dissolution of normal heredity, while the pathological inheritance may be similar in some members of the clan considered.

In summary, on the one hand, the degenerate is an entirely new type. On the other hand, as compared to the usual type having an irreversible functioning, which demands a constant increase of its entropy, the new type can thus only work reversibly, that is, with almost no increase in entropy.

Such is the thermodynamic concept of the state of degeneracy. We think we have made it clear enough that we are now trying to give an energetic conception of the principal syndromes and delusions that are characteristic of mental degeneration.

We can already answer the questions of how degeneration takes place, why in its transitions it is not methodical and uniform, and why it happens from one generation to another, sometimes in succession, sometimes instantaneously (Saury).

To our mind, this is due solely to the reversibility of the changes in the nervous system of the degenerate, the overcomplete equilibrium[1] with the environment, which causes transformation of the external energy into internal energy without increasing the entropy, without increasing the stability.

In his studies on mental degeneration Mr. Magnan divides the mental situations of degenerates into three groups:

[1]Here, Ameline means *an indifferent balance*, that is, *neither a stable nor an unstable one.*

- States of lucid madness, with the conservation of consciousness attributable to obsessions, impulses, and inhibitions (madness of doubt, emotional delirium according to Morel et al.).
- States of very advanced imbalance (mania of reasoning and persecuted self-accusers, moral folly).
- Delirious states of various forms and evolution, among which Magnan distinguishes two principal forms—polymorphous delirium and systematized delirium—which the German and even some French authors (Séglas, Chaslin, etc.) designate under the name of paranoia. All the alienists almost unanimously describe three forms of the latter.

Interestingly, the mentioned three forms are each characterized by the predominance of a symptom associated with the other two. The three forms in question are:

- Hallucinatory mental confusion (*verwirrtheit*), in which confusion is the main symptom.
- In opposition to this form, systematic hallucinatory madness (*verrücktheit*) in which the systematization of ideas is apparent. The dominant symptom consists of delusionary interpretations (English authors describe them just as "delusions").
- Hallucinatory delirium (*wahnsinn*), a form in which hallucinations of almost every sense predominate, with systematization being barely possible.

Depending on the nature of the predominant idea, these delusions will have varieties: religious, erotic, hypochondriac, ambitious, etc.

The form of delirium varies, whereas its background should always consist of delusional interpretations based upon hallucinations and/or illusions.

We believe that manic excitements and melancholic depressions do not constitute separate forms within the general field of mental degeneracy and that they mingle around or instead accompany delusions.

Indeed, delusions of the degenerates are above all the fundamental alterations of intelligence. The disorders of affectivity ought to be secondary instead: sadness, anger, anguish, etc., occur just when a change occurs in the state of the nervous system of the patient.

In the case of the typical reputed melancholy, there is no delusional interpretation that motivates the state in which the patient finds himself or herself. There is a distinctive melancholy because the patient recalls having done his or her religious duties poorly, as there are melancholy persecutions as a result of terrifying hallucinations.

Is there not the category of "persecuted self-accusers," who indicate that the distinction between a melancholy and a persecuted person is not so profound as it is commonly said and believed?

Finally, let us consider two types: an individual convinced of having committed a crime without being able to specify which one, who lives in a state of continual anguish, and the "persecuted one," who gets angry and becomes impulsive and violent because he/she must follow some accusing and injurious hallucinations. Between these two, we must place several other remarkable types, one being the scrupulous one, who is heavy-hearted and/or unhappy for having forgotten a word in a prayer. Only then would we place the self-accuser, who hears about calling "a thief, a thief" and immediately starts to seek whether it is not a robbery that he/she has dared to covet a coat in a fashion store, by confessing his or her lust as a crime that demands the scaffold in atonement for his or her alleged turpitude.

Are any of these patients not likely to make suicide attempts, which may perhaps be the most characteristic and the least undeniable stigmata of degeneracy?

Whether the reaction is anguish, sadness, anger, joy, revenge, suicide, or whatever, we cannot use it to characterize a given mental form. An erotic persecuted has alternatives of joy and sadness, of anger, according to the intensity of his or her genital hallucinations, and we would not say that he/she is affected sometimes by mania, sometimes by melancholy.

In short, the content of mental disorders (eroticism, religion, hypochondria, etc.) and the content of affective disorders (sadness, anguish, joy, anger) do not characterize the degenerate patients. Their common pathologic feature ought to be the ease with which ideas and feelings are born and disappear; it is always the lack of stability that we note in these patients; it is the abnormal functioning of the brain, consisting of reversible phenomena, that is to say, accompanied by almost no variation in the entropy.

We have insisted enough on the nature of the hallucination to have the meaning of hallucinatory delirium.

In mental confusion, there is also an alteration of memory in some distinctive way. The only difference from the previous delirium is that it is mainly the intellectual phenomena and not just the sensory phenomena that are in the foreground.

In the systemized delirium, reversibility does not appear evident at first sight; but we must note that in such a patient, the reversible phenomena are not individual as in the sensory delirium or confusion and that they are somewhat related to the general evolution of the human species.

Indeed, what are the proposed explanations for the origin of the false interpretation presented by systematized delusions?

Three theories have been suggested in this case, all three having slight differences between them.

Meynert views a superstitious background as the pathogeny, just the opinion that Marie and Séglas share.

Indeed, superstition has much resemblance to the madness of doubt and phobias. Systematic delirium, on the other hand, appears to be only an exaggeration of a fixed idea; hence, many authors designate the obsession even by the name of "rudimentary paranoia" to insist on the kinship of the two delirious degenerate types.

Aside from such pathogenesis, Semérie shares the opinion of Comte on the evolution of the human mind in three periods, theological, metaphysical, and positivist, and he remarks that in systematized delusions, we find interpretational modes of both periods being before the positivist and scientific stage of the present human spirit.

Remarkably, Tanzi expresses an opinion that looks somewhat analogous to both preceding ones: mental debility removes new acquisitions from the brain, and we find only the old ones.

Strictly speaking, in both these latter opinions, we do not find any indication of reversibility, but the irreversibility of the usual functioning of the nervous system is apparent.

The first opinion looks like the best one from our point of view. Indeed, the syndrome of obsession is schematically composed of a fixed idea with consciousness and anxiety followed by an impulse, with the consequent satisfaction. It is uncertain who would finally make sure of the falseness of his or her apprehension, a phobic, who ends up by conjuring up his persecution, or an impulsive, who decides to satisfy his life; all these patients differ from each other only in the predominance of consciousness, anxiety, and impulse and all seek satisfaction by an act of which they are aware.

Now, we might easily give a thermodynamic explanation. Suppose that neurons of the nervous system achieve the conditions of reversibility. These reversibility conditions ought to be those of equilibrium. Furthermore, we know (principle of Maupertuis) that whenever an equilibrium system is disturbed, it gives rise to a tendency to restore the broken equilibrium in the opposite direction to that of disturbance.

Finally, we also know (Paulhan) that any affective phenomenon is the expression of a profound disorder in the organism. Then, is it not evident that the disturbance of the balance of neurons would cause a tendency to recover this balance, which accounts for the conscious struggle of the patient? These affective phenomena join the disorder produced, and the manifestations of anxiety occur.

All this is since reversibility is possible in the functioning of the nervous system: Physically, emotivism, which is only too sensitive to change in the environment, comes because degenerates are incapable of acquiring entropy in their psychic acts.

Basically, as the clinic shows, both psychically and physically, the entire mental state of the degenerate should have the same explanation of its origin, irrespective of its apparently varied manifestations, because the fixed idea is not only the element of the first and second classes of mental situations inherent in the degenerate but also the psychological element of the reasoning mania and of the "persecuted-persecutors" madness.

There is yet another proof of the fact that degenerates might appear as beings whose manifestations are accomplishable without a sensible production of entropy; it is just what we have already determined as a characteristic feature—namely the ease with which they pass from sensations to acts. Likewise, they are automatic both in their movements and their sensations; "At the maximum of automatism," says Taine, "the hallucination is perfect."

However, we may not connect automatism with a stability increase in such cases, and we have insisted enough on the latter result when we have discussed the experiments by Lombard, Schiff, etc.

Above, we have already underlined our claim that there is a very close relationship between the time and entropy notions. Accordingly, in some patients having the madness of doubt, Sollier could notice that both the memory disorder and the localization in the time are of importance.

All the authors dealing with mental illnesses have noticed the alterations related to the notion of time. In effect, this is much more frequent than reported. In studying the symptoms systematically from the beginning of the disease, one always notices a disturbance of the memory and that the patient finds melancholically that "the clock goes too fast," that he/she ought to be a persecuted person, for whom it took a rather long time to be brought to the asylum, that he/she had to be sent to unhealthy places, etc. To sum up, it is not necessary to analyze in detail all the illusions of the patients noticeable during the period of such a state. Not only does the acquisition of entropy appear to be troubled, so does the notion of time.

Hence, an intimate relationship between both these notions should be quite possible, and once again, we could find here points to further strengthen our conviction regarding this theme.

***

The study of mental pathology types has thus confirmed the indications we have drawn from the study of the healthy nervous system physiology. The two principles of energetics can, therefore, be applied in psychology and must be used to interpret all the purely experimental research results in the field of psychophysics up to the present time.

However, before we finish, we are going to use the Carnot principle to formulate a plausible induction hypothesis on the possible nature of the differences between mental degeneration and genius.

Here we would like to confine ourselves to giving a kind of schematic formula for the latter relations. What characterizes genius is, in the universal opinion, the possession of an organic development whose superiority is manifested by a novelty, an originality, such that the man of genius goes before many other human beings in the evolution of ideas/physical dexterity. The man of genius seems to have broken the ordinary laws of heredity, like the degenerate, but it is not a reason to confuse the genius and the degenerate. While genius embodies a future type of human being (Nordau), a degenerate does not even fit the old type, as colleagues wrongly believed, since the degenerate is inherently abnormal, while the old type is standard. Genius is a typical future type. He/she has evolved faster than the average human being, than the crowd of the latter. He/she embodies an energy transformer that works with a considerable increase in entropy, while the degenerate works by keeping his/her entropy constant.

As compared to the rest of humanity, the latter type is then far too stable [**our immediate remark:** or, to put it a better way, neither stable nor unstable; such type of stability constitutes a blind corner]. Instead, the former has no such tendency to stability [**our immediate remark:** or, to put it in a better way, the stability of genius is dynamic and thus open to progress].

Here is the difference. It seems enormous and yet it is challenging to distinguish the degenerate from the awesome [see our conclusion at the end of Appendix 2].

Why that? The reason seems to be the following:

At the beginning of this work, we have indicated that it is possible to represent evolution of the human mind by an alternating series of discoveries, of ideas whose terms oscillated alternately in two opposite directions, with the oscillation gradually diminishing its amplitude to converge little by little toward a limit currently unknown, though being of unmistakable existence.

It is just the procedure of evolution by thesis, antithesis, and synthesis, as stated by Hegel, and that resembles the famous wording by Heraclite:

"Everything flows; the world is the torch that turns on and off."

Suppose that at present, we support a definite thesis on any subject. Then, the old opinion would be the antithesis regarding the current thesis.

Therefore, "a genius being ahead of the evolution" would represent a synthesis regarding the sustained opinion that ought to approach the corresponding antithesis and it will appear to be climbing down in the world of ideas, absolutely like the degenerate is representing the illusion of an atavistic return.

Both conclusions seem to be downsizing, and yet they are not.

If a genius goes two steps ahead of the present day, we may consider him/her much too declined from average, much too eclectic, since this would be closer to the intermediate limit between opposing opinions. Hence, his/her real originality would be at least unrecognized—if even not just confused with the most vulgar banality.

To sum up, such seems to be the psychophysical conception of the degenerate and the man of genius. The difference between both these mental states becomes precise and manifest when we examine the problem in the light of the fundamental principles of energetics.

A2.2.8 Summary and Conclusions

Thermodynamics is a part of experimental physics that studies the conditions of the transformation of mechanical work into heat and back. The first condition is that there is equivalence between the heat produced and the work expended (Joule–Mayer principle). The second is that it is easier to turn mechanical work into heat than heat into mechanical work (the Carnot–Clausius principle).

Energetics (Rankine) aims to study physical phenomena using the results obtained by generalizing the principles of thermodynamics, which then become, respectively:

- The principle of energy conservation. The forms of energy may be transformable into each other without modifying the quantity of total energy.
- The principle of the degradation or dissipation of energy (or the principle of entropy increase).

Heat is the most stable of all forms of energy. Various forms of energy turn into heat, which also appears as a kind of excretion product of the mutual metabolism of other varieties of energy.

Hence, an increase in entropy renders a phenomenon possible (irreversible).

No phenomena are known to occur without variation in entropy (reversible).

In other words, in the universe,

- The amount of energy remains constant.
- The amount of entropy cannot decrease.

As a conclusion, in the first part of our work, we expressed the opinion that since energy conservation and entropy increase ought to be two essential conditions of physical phenomena, both space and time ought to be underlying conditions of psychic phenomena. Then, we can consider space and energy on the one hand and time and entropy on the other as two notion groups both being equivalent to a respective specific condition. First, to demonstrate the merits of such a viewpoint, it is necessary to prove that the principles of energetics apply to mental phenomena. This story is the subject of the second part of the present work.

The proofs we suggest here are of two kinds, indirect evidence and direct evidence.

Indirect evidence:

- The phylogeny shows a profound analogy of the constitution between the muscular and nervous systems.
- The physiology of the muscular fiber is almost identical to that of the neuron.
- The same mechanical design accounts for the functions of both the above organs; they are both electro-capillary systems.
- Finally, heat produced by both is a product of excretion. In other words, both degrade energy and increase the entropy during their operation.

Direct evidence shows that not only does the neuron transform energy (the study of the reaction time), but also it can degrade the energy, according to Weber–Fechner's law.

In our opinion, the latter so-called law simply means that any prior excitation of the nervous system ought to increase its stability, that is, its entropy, which renders it less sensitive to the excitations that follow.

Finally, in the last part of our thesis, we investigate whether the preceding conclusions, which relate only to the healthy nervous system, could be verified when it comes to the nervous system's pathology.

By confining ourselves to the study of the mental situations of the degenerate hereditary, here is what we believe we have demonstrated:

- Usually, stability or entropy tends to regularly undergo a mean increase in the individual and his/her offspring (irreversibility):
- When this increase becomes almost zero (reversibility), the stability of the nervous system is almost nil. Thus, we may characterize degenerates by a rupture, a dissolution of the heredity laws. At the same time, we may ascribe a lack of mental stability to them.

  This state of reversibility in the mental phenomena can be observed in the following alterations, depending on the individual:

  - o The personal psychic acquisitions by bringing trouble in the memory:
    - ▪ For example, sensations: hallucinations (hallucinatory delirium)
    - ▪ For example, associations: confusion (mental confusion)
  - o The ability to acquire ideas and sensations: delusional interpretations (systematic madness).
  - o Finally, the possibility of the reversibility of mental phenomena, which also gives rise to affective disorders that could be:
    - ▪ Systematized: obsessions.
    - ▪ Nonsystematic: mania, melancholy, etc.
- On the contrary, if the increase in the entropy is very significant, instead of being null or average, one has a mental state capable of acquiring extreme stability, which characterizes the man of genius.

The difficulty in distinguishing the geniality of degeneration is only apparent and comes from the fact that in both cases, it has a rupture of heredity, the preservation of which is the prerogative of banality.

Such, in short, is the psychophysical essay we undertook by using the purely experimental notions of energy and entropy instead of metaphysical notions borrowed either from pure psychology or from atomo-mechanical theories.

Bibliography

Concernant la première et la deuxième partie [for Sections A2.2.6.1 and A2.2.6.2]

D'Alembert. Discours préliminaire sur l'Encyclopédie. Genève, 1777, p. 41.

Autonne. Un nouveau livre sur l'atomisme: La critique de M. Hannequin sur l'hypothèse des atomes. Revue générale des sciences, Juillet 1896.

Berthelot. Sur la chaleur animale. Gaz. Méd., 1865.

— Sur les principes généraux de la thermochimie. Annales de chimie et de physique, 1870.

Berthollet. Essais de statique chimique, 1803.

Bouty. Cours de thermodynamique, 1891 (inédit).

Broadbent. Brain origin. Brain, 1895 (été et automne).

Brochard. Sur les arguments de Zénon d'Élée. Revue de métaphysique et de morale, 1893.

Carnot. Réflexion sur la puissance motrice du feu et les machines propres à la développer, 1824.

Clausius. Mémoires sur la théorie mécanique de la chaleur, traduction Folie.

Couturat. Réponse à M. Léchalas. Revue de métaphysique et de morale, 1893, n° 3.

Delboeuf. Prolégomènes philosophiques sur la géométrie. Liège, 1860, et Revue philosophique (passim).

Dunan. Théorie psychologique de l'espace, 1895.

Duhem. Introduction à la mécanique chimique. Gand, 1893.

Evellin. Le mouvement et les partisans des indivisibles. Revue de métaphysique et de morale, 1893.

Féré. Sensation et mouvement, 1887.

Fouillée. Évolutionnisme évolutionnaire des idées forces, 1890.

Guyau. Genèse de l'idée de temps, 1890.

Hirn. Nouvelle réfutation générale des théories cinétiques, 1886.

Haeckel. Le monisme, traduction de Lapouge, 1897.

Lord Kelvin (Sir W. Thomson). Popular Lectures and Addresses (constitution de la matière).

Léchalas. Les géométries non euclidiennes et le principe de similitude. Revue de métaphysique et de morale, 1893, n°2.

— Étude sur l'espace et le temps, 1896.

— Sur les arguments de Zénon d'Elée. Revue de métaphysique et de morale, 1893, n° 2.

Le Chatelier. Sadi Carnot et la science de l'énergie. Revue générale des sciences, Juillet 1892.

— Le troisième principe de l'énergétique. C. R. Acad. des sciences, Juin 1893.

Leduc. La science de l'énergie et la médecine. Gazette médicale de Nantes, Novembre 1894.

Mouret. L'entropie, sa mesure, ses variations. Revue générale des sciences, Octobre et Novembre 1895.

— Le facteur thermique de l'évolution. Revue générale des sciences, Décembre 1895.

Ostwald. La déroute de l'atomisme contemporain. Revue générale des sciences, Novembre 1890.

Poincaré. Cours de thermodynamique.

— Les idées de Hertz en mécanique. Revue générale des sciences, Septembre 1897.

— Mécanisme et expérience. Revue de métaphysique et de morale, 1893.

Ribot. Psychologie allemande contemporaine, 1892.

— Évolution des idées générales, 1897.

Robin. L'évolution de la mécanique chimique et ses tendances actuelles. Revue générale des sciences, Mars 1898.

Stallo. La matière et la physique moderne, 1891.

Tait. Conférences sur quelques progrès de la physique, traduit par Krouchkoff, 1886.

Ward. Art, Psychology, in Encyclopedia Britannica.

Wundt. Éléments de psychologie physiologique, 1886.

N. B. *Enfin les traités de thermodynamique de Blondlot, Briot, Hirn, Lippmann, Moutier, Verdet, Zeuner, etc.*

[N. B. The last but not the least, important references are the monographs on thermodynamics by Blondlot, Briot, Hirn, Lippmann, Moutier, Verdet, Zeuner, etc.]

Concernant la troisième et la quatrième partie [for Sections A2.2.6.3 and A2.2.6.4)]

Ardigo. La science expérimentale de la pensée. Revue scientifique, Avril 1889.

D'Arsonval. Sur les causes des courants électriques d'origine animale. Société Biologie, Juin 1885.

—Sur un phénomène physique analogue à la conductibilité nerveuse. Société de biologie, Avril 1886.

Beaunis. Nouveaux éléments de physiologie humaine, 3E édit., 1888.

Béclard. Traité élémentaire de physiologie, 7E édit., 1884.

Becquerel. De la cause des courants musculaires, nerveux, osseux et autres. C. R., Janvier 1870.

— Mémoire sur la production des courants électrocapillaire dans le cerveau. C. R., Février 1870.

Cl. Bernard. Leçons sur le système nerveux, 1858.

Binet et Henri. La fatigue intellectuelle, 1898.

De Boeck. Contribution à l'étude de la physiologie du nerf. Thèse, Bruxelles, 1893.

Bombarda. Les neurones, l'hypnose, l'inhibition. Revue neurologique, Juin 1897.

Byasson. Essai sur la relation qui existe entre l'activité cérébrale et la composition des urines. Thèse, Paris, 1868.

Chaslin. De la confusion mentale primitive, 1896.

Chauveau. Du travail physiologique et de son équivalence. Revue scientifique, 1888.

— L'élasticité active du muscle et l'énergie consacrée à sa création. C. R., Juillet 1890.

— Le travail intellectuel et l'énergie qu'il représente, 1891.

— Rapports de la dépense énergétique du muscle avec le degré de raccourcissement qu'il affecte en travaillant. C. R., Acad. sciences, Juillet 1896.

Dagonet. Nouvelles recherches sur les éléments nerveux, 1893.

Delboeuf. Éléments de psycho-physique générale et spéciale, 1883.

— Examen critique de la loi psychologique, 1883.

Demoor. Contribution à l'étude de la fibre nerveuse cérébrospinale. Institut Solvay, Bruxelles, 1891.

Dumas. Les états intellectuels dans la mélancolie, 1890.

Duval. Hypothèse sur la physiologie des centres nerveux. Soc. Biologie, Février 1895.

Féré. Dégénérescence et criminalité, 1888.

—La famille névropathique, 1894.

François-Franck. Physiologie des nerfs. Dict. encycl. des sc. méd.

Frédericq et Nuel. Éléments de physiologie humaine, 3E édit. Gand, 1894.

Gautier. L'origine de l'énergie chez les êtres vivants. Revue scientifique, Décembre 1886.

— La pensée. Revue scientifique, Janvier 1887.

— Les manifestations de la vie dérivent-elles toutes des forces naturelles? Revue générale des sciences, Avril 1897.

Gavarret. Les phénomènes physiques de la vie, 1869.

Van Gehuchten. Anatomie du système nerveux de l'homme. Louvain, 1897.

Gley. De l'influence du travail intellectuel sur la température générale. Soc. Biologie, Avril 1884.

— Sur la question de la variation des urines pendant le travail cérébral. Arch. de physiologie, 1894.

Gotch et Macdonald. Température et excitabilité. Journal of physiology, XX.

Hack-Tuke. Le corps et l'esprit. Trad. Parant, 1886.

Heger. Préparations microscopiques du cerveau d'animaux endormis et éveillés. Bull. Acad. médicale de Belgique, Novembre 1897.

Hermann. Éléments de physiologie, traduit par Onimus, 1869.

Herzen. Le cerveau et l'activité centrale au point de vue psychophysiologique, 1887.

— L'activité musculaire et l'équivalence des forces. Revue scientifique, 1887.

— Sur le refroidissement du muscle actif. Revue scientifique, 1888.

Hirn. La thermodynamique des êtres vivants. Revue scientifique, Mai et Juin 1887.

Imbert. Le mécanisme de la contraction musculaire déduit de la considération des forces de tension superficielle. Arch. de physiologie, 1897.

Keraval. Délires désignés sous le nom de paranoïa. Arch. neurologie, 1894 et 1895.

Laborde. Des modifications de la température liée à la contraction musculaire. Soc. Biologie, Juin 1886.

— L'échauffement du muscle en travail est indépendant de la circulation. Soc. Biologie, Mai 1887.

— Cours de l'Ecole d'anthropologie, 1897 et 1898 (Inédit).

Lambling. Des origines de la chaleur et de la force chez les êtres vivants. Thèse, agrégation, Paris, 1886.

Legrain. Les délires des dégénérés. Thèse, Paris, 1886.

Lépine. Théorie mécanique de la paralysie hystérique, du somnambulisme et du sommeil naturel. Soc. Biol., Février 1895.

Liebig. Traité de chimie, trad. Gerhardt, 1840–1844.

Lippmann. Thèse, docteur ès sciences, Paris, 1876.

Magendie. Leçons de physiologie, 1836.

Magnan et Legrain. Les dégénérés, 1895.

Mairet. De la nutrition du système nerveux. Arch. de physiologie, 1885.

Marie. Sur quelques symptômes des délires systématisés, 1892.

Matteuci. Cours d'électrophysiologie, 1868.

Max-Simon. Les maladies de l'esprit, 1892.

Mayer. Mémoire sur le mouvement organique, trad. Perard, 1872.

Mendelson et Muller-Lyer. Recherches cliniques de psycho-physique. Arch. de neurologie, 1887 et 1890.

Morel. Traité des dégénérescences de l'espèce humaine, 1857.

Mosso. La fatigue intellectuelle et physique (traduit par Langlois, 2E édit., 1896).

— Sur la température du cerveau. Archiv Ital. de Biologie, Milan, 1894.

Müller. Physiologie du système nerveux, trad. Jourdan, 1840.

Nordau. Psychophysiologie du génie et du talent, 1897.

Obersteiner. Sur la théorie du sommeil. Allgemeine Zeitschrift für Psychiatrie, 1872.

Patrizzi. Le temps de réaction simple étudié dans ses rapports avec les courbes pléthysmographiques cérébrales. Rivista sperimentale di Frenatria, 1897.

Paulhan. Physiologie de l'esprit, 4E édit.

— Les phénomènes affectifs, 1887.

Pompilian. La contraction musculaire et les transformations de l'énergie. Thèse, Paris, 1897.

Popoff. Sur les troubles cérébraux, etc. Arch. de Virchow, 1876.

Pouchet. Remarque anatomique à l'occasion de la nature de la pensée. Revue scientifique, Février 1887.

Pugnat. Modifications histologiques des cellules à l'état de fatigue. C. R., Novembre 1897.

Pupin. Le neurone et les hypothèses histologiques sur son mode de fonctionnement. Thèse, Paris, 1896.

Ranvier. Traité technique d'histologie, 1889.

Ramon Y Cajal. Les nouvelles idées sur la structure du système nerveux (trad. Azoulay, 1895).

Ribot. Psychologie allemande contemporaine, 4E édit., 1892.

— Les maladies de la personnalité, 1895.

— Hérédité psychologique, 1887.

Richet. Physiologie des nerfs et des muscles, 1882.

— Le travail psychique et la force chimique. Revue scientifique, Décembre 1886.

— La pensée et le travail chimique. Revue scientifique, Janvier 1887.

— La chaleur animale, 1889. Essai de psychologie générale, 1889.

Richet et Broca. Période réfractaire dans les centres nerveux. C. R., Janvier 1897.

Saury. Étude clinique sur la folie héréditaire, 1886.

Schiff. Recherches sur réchauffement des nerfs et des centres nerveux à la suite des irritations sensorielles et sensitives. Arch. de physiologie, 1869–1870.

Séglas. Revue critique sur la paranoïa. Arch. neurologie, 1887.

Semérie. Symptômes intellectuels de la folie. Thèse, Paris. 1867.

Soluer. Les troubles de la mémoire, 1892.

Soukhanoff. La théorie du neurone en rapport avec l'explication de quelques phénomènes psychiques. Arch. de neurologie, Mai et Juillet 1897.

Soury. Les fonctions du cerveau. Arch. de neurologie, 1890 et 1891.

Soury. La théorie des neurones. Arch. de neurologie, 1897.

Stcherback. Contribution à l'étude de l'influence de l'activité cérébrale sur les échanges d'acide phosphorique et d'azote. Arch. de physiologie, Mai 1893.

Taine. De l'Intelligence, tome Ier, 7E édit., 1895.

Tanzi. Les oscillations de la température du cerveau, sous l'influence des émotions. Centralblatt für Physiologie, 1888.

Tapie. Travail et chaleur musculaire. Thèse, Agrég. Paris, 1886.

Thorion. Influence du travail intellectuel sur les variations de quelques éléments de l'urine. Thèse, Nancy, 1893.

Wallet. Points relatifs à la loi de Weber-Fechner. Brain, 1895 (été et automne).

Wundt. Physique médicale, traduit par Monnoyer, 1870.

— Eléments de psychologie physiologique, traduit par Rouvier, 1886.

N. B. *En outre pour la quatrième partie, nous nous sommes reportés aux ouvrages de pathologie mentale classiques en France et à l'étranger et surtout aux nombreuses publications de M. le Magnan et aux si remarquables leçons cliniques encore en partie inédites de M. le Pr. Joffroy.*

[N. B. Besides, to write the fourth chapter (A2.2.6.4), we have also made use of the classical books on mental pathology, as published in France and abroad. This especially concerns the numerous publications by Mr. le Magnan and the remarkable clinical lessons by Prof. Joffroy, which are still partly unpublished.]

A2.3 Our General Conclusion

Any attentive reader would exclaim immediately after looking at the first page of the above essay,

"What is this, and what is its purpose?"

The authors have picked up a story published about 120 years ago, which is a master thesis of a bachelor physicist willing to work as a doctor of medicine. How could it be helpful, and what are the actual benefits of taking the time to translate the story and to offer it to the broadest readership?

To our mind, the benefits are twofold.

First, let us consider the purely human value of the story at hand.

A bachelor physicist graduated from Sorbonne decides to attain his master's degree in the medical field, to become not just a healthcare professional but also a professional in a genuinely complicated field of psychiatry. Marius Ameline is not just an average human case, as he is.

After being a disciple of such undoubtedly competent peers as Edmond Bouty (1846–1922) and Jonas Ferdinand Gabriel Lippmann (1845–1921), he has chosen thermodynamics as the field of his primary interest. Then, this is already of direct relevance to our main story here.

At this very point, we immediately come to the professional value of the story told us by Ameline. His main idea was to consider the physical sense of various mental disorders from a thermodynamic standpoint.

This way, we might place Ameline's work onto the field of macroscopic biophysics. On the one hand, he could perform a rather broad and very detailed literature review in the field of physiology of the nervous system. On the other hand, before analyzing the latter, he had presented an introduction to thermodynamics.

Remarkably, just at the time of his studies, the well-known revolution in physics was proceeding, being at its very peak. Therefore, Ameline had two choices: the equilibrium thermodynamics (nowadays a mainstream) and energetics (nowadays considered actual historical trash—for further details, refer to Ref. [1] of Chapter 1).

Ameline had chosen the latter standpoint to employ in his studies and could clearly explain why his decision had come precisely this way.

Ameline's explanation does represent a separate professional value. His depicting the equilibrium statistical thermodynamics as a purely metaphysical branch would cause a shrug among the modern readership. His referring to such colleagues as Mach,

Ostwald, and Stallo would promote further shrugging, which would finally disembogue into yawning. Indeed, to refer to the atomistic representation of the matter as to the metaphysics ought to be nothing more than just a "Dark Heresy," isn't it?

A further apparent awkwardness ought to be announcing that "to increase the entropy of a system is to increase the stability it already has." Finally, the holy "reversibility," which is in effect the basic notion of the good old equilibrium thermodynamics, turns out to be the physical basis of mental pathology. Was everything okay with this author, who presented himself as a bachelor physicist? Was everything okay with his then mentors at Sorbonne?

To decide who is right, let us have a short look at the mechanical definition of equilibrium, which was, is, and will be taught at secondary schools, long before children decide whether to apply for a studentship at Sorbonne or to go to elsewhere.

A2.3.1 The Balance of Bodies: Types of Body Balance

[What follows is our commented English translation from the Russian educational site
http://ru.solverbook.com/spravochnik/.]

Definition

Equilibrium is such a state of the body (body system) that it does not receive any acceleration under the action of applied forces.

The types of body balance are:

- Stable equilibrium
- Unstable equilibrium
- Indifferent balance

Definition

Stable equilibrium is a form of equilibrium in which a body, derived from an equilibrium position and left to itself, returns to its former position [**our comment:** by the way, this is just the essence of Maupertuis' principle duly considered by Ameline].

With a small displacement of the body in any direction from the initial position, this happens if the resultant forces acting on the body become nonzero and are directed toward the equilibrium position.

For example, consider a ball lying at the bottom of a spherically concave surface (Fig. A2.5a).

Definition

Unstable equilibrium is a form of equilibrium in which a body, being displaced from the equilibrium position and left alone to move by itself, will deviate even more from the initial equilibrium position.

In such a case, if the resultant of the forces applied to the body is at least somehow different from zero and directed away from the initial equilibrium position, even an infinitesimally small displacement from its equilibrium position would matter. An example would be a ball at the top of a convex spherical surface (Fig. 2.5b).

Definition

The scholastic definition sounds as follows: An indifferent balance is equilibrium in which a body *is displaced* from a position (state) of equilibrium and *left to itself* does not change its position (state).

In this case, at small displacements of the body from the initial position, the resultant of the forces applied to the body remains equal to zero, for example, a ball lying on a flat surface (Fig. 2.5c):

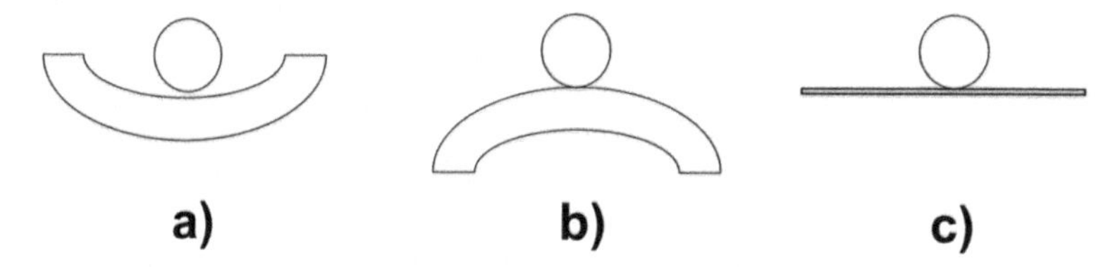

Figure A2.5 Different types of body balance on a support: (a) stable balance; (b) unstable equilibrium; (c) indifferent balance.

Static and dynamic body balance

If the body does not receive acceleration as a result of the action of forces, it can be at rest or move uniformly in a straight line. Therefore, we can talk about static and dynamic equilibrium.

Definition

Static equilibrium is equilibrium when the body remains at rest even under the action of applied forces.

Dynamic equilibrium is equilibrium when the body does not change its movement due to the action of forces.

In a state of static equilibrium is a lantern suspended from a cable on any building structure. As an example of dynamic equilibrium, we can consider a wheel that rolls on a flat surface in the absence of friction forces.

A2.3.2 Our Immediate Comment

The above story about the *indifferent balance* does contain an essential paradox: which kind of realistic forces would *first displace* the system out of its equilibrium and *then leave* the system *to itself*?

Therefore, after speaking about indifferent balance, the scholastic treatment must immediately introduce the notions of static and dynamic body balance. Then, we conceive indifferent balance as a dynamic body balance, which must boil down to uniform motion (motion having zero acceleration). The latter is also metaphysics, and Ameline tells us in detail why it is like this. If we remain within the metaphysical framework, we continue to insist that there is no indifferent balance, because it is still a true dynamical balance, for "the body *does not change its movement* due to the action of forces." Children ask us, but what are the examples of true dynamical balance? The scholastic answer is remarkable: "We can consider a wheel that rolls on a flat surface in the absence of friction forces . . ." People, people, hey, stop! The friction forces we are trying to get rid of here are but ubiquitous.

The scholastic answer would be, *Sure, they are, when we view macroscopic objects, but basically, microscopically, there is no friction, no energy, no time, no space, no impulse: For everything is relative and uncertain. It appears that we have no other way apart from being conceptually mixed to a firm paste with the equilibrium thermodynamics. No smoothing of the mixture might be allowable.*

Highlighted here is just the point the stubborn traditional scholastics all over the world are trying to persuade us about.

In view of this, Ameline's work ought to be of tremendous importance. Why?

The scholastics we have been discussing above comes from the "physicalisation" of the ingenious cyclic-engine model by Carnot and has nothing to do with the actual physics, chemistry, biology, etc., because the physicalized version of the Carnot cycle is a reversible perpetuum mobile near some mystic equilibrium (for the further details, refer to the book in Ref. [1] of Chapter 1 and the references therein). The mentioned "physicalisation" was inescapable solely for providing us with the mathematically ingenious, genuinely valid, and useful physical theory, but without going into the actual physical depths.

Instead, Ameline is discussing the Carnot principle and not its "physicalized" version. He demonstrates that in effect, there is a true indifferent balance; it is just a detectable pathology. There is a very small but still essential difference between "reversibility" and "capsizability." The former corresponds to zero entropy change and therefore represents pure metaphysics, whereas the latter perfectly mimics the former but never brings the entropy to its exact zero because it is throughout impossible to reach zero entropy physically. Instead, the capsizability brings the entropy to the arbitrarily close vicinity of its zero value. Therefore, it has to do with real phenomena.

Indeed, " . . . the word Entropy means a tendency to stability. Entropy indicates a movement; it stands for evolution. Entropy is not stability, for a stable state should be immobile. It is not instability either because of an unstable state, as it might be immobile as well." This is what we learn from the above work. From Newton's mechanics we know that any realistic tendency encounters obstacles, hindrances, counteractions, and resistances.

To reach some stability—either static or dynamic—the system has to have resources enough to overcome the latter. This is why the philosophical sense of the entropy notion is just the dialectic principle of the unity and struggle of opposites, that is, the unity and struggle between actions and reactions.

That entropy is in fact correspondent to the unity and struggle of opposites does not render it dispensable, if we wish to draw the full picture of the "tendency to stability" under study, whatever the nature of the phenomenon involved might be. If we look at the story with the eyes of Ameline, the medical doctor, we recognize that any "bringing entropy to the close vicinity of zero" means but "bringing the resistance to the vicinity of zero." If we are now speaking about some living organism, then we know well that its resistance to some external or internal effects refers to its immunity. Bringing the immunity to the vicinity of zero means opening the door to pathologies.

This is just how an apparently counter-revolutionary work by Ameline might greatly assist scientific research activities in diverse branches of knowledge to remain afloat.

Several excellent posers arise in connection to the above exciting story:

- Could Ameline manage any continuation of his work?
- Could Ameline's work initiative excite some interest, cause some resonance?

The answer to the first poser is throughout positive: yes, Ameline was intensively working on the theme after the publication of his above-mentioned master thesis.

There have been several publications by him, mostly because of the relevant conferences in France. Before World War I, he could publish two papers on the related subject in the French journal *Journal de Psychologie Normale et Pathologique*, in the years 1908 and 1913 [3, 4]. His later publications (mostly in the 1920s) are highly specialized but still not so relevant to our story, which indicates that he was not anymore dwelling on the theme into its depth.

Now, the answer to our second poser brings a sort of disappointment. Ameline's work [4] describing the results of how he was applying his exciting approach could find at least a couple of attentive readers—even abroad, with respect to France [5, 6].

As to his work [3] describing in full detail the very method outlined in this master thesis, it could impress some of his French colleagues at the very least [7], but it looks like being completely forgotten.

The story does really look very easy to forget. Indeed, if we might successfully go ahead with treating the entropy notion *implicitly*, why should we ever try viewing it *explicitly*? Well, engineers do so, and they do get reasonable results. Why is it like this? It looks like a difficult poser indeed. To bridge a logical gap between the *implicit* and *explicit* entropy is but to simultaneously fulfill two orders: "Stay over there" and "Come here." Should we, physicists, go this way? This is just the sense of the work by Tolman [8] and the works alike. This is the just the conventional standpoint.

In his master thesis, Ameline does reveal instead that *treating the entropy notion explicitly* might bring serious benefits, at least in the field of biophysics. Therefore, we might consider him one of the actual actors in the field of "different thermodynamics" (see Ref. [1] of Chapter 1).

Should we then forget Ameline's work? There were colleagues who had answered the poser in the negative. Well, so who were they? Was Ameline's idea hanging in the air somehow? Fortunately, the answer to this poser is positive.

The first and foremost colleague to be mentioned was a Hungarian-born Russian physician, pathologist, and physical biologist Erwin Bauer (1890–1938) [9], who introduced the *principle of the permanent inequilibrium of living matter*, aka *Bauer principle*, aimed at duly explaining life phenomena in physical terms. His actual fate is tragic. His education and the first working experience took place in Austria. In 1925 he had accepted an invitation to work in the USSR and then moved to Leningrad. There was but no more way to come back to Austria in the late thirties of the past century because of his Jewish origin. As a result, in 1937 his wife and he were brutally arrested and then just shot down by Stalinists in 1938, without any credible indictments, like crowds of other victims in the USSR.

Another outstanding colleague was Karl Trincher (1910–1996). His fate was largely similar to that of Bauer, although Trincher could have somewhat better fortune. We present here our authorized and commented English translation of his concise biography published in Ref. [10]:

Austrian victims of Stalin (until 1945)

Karl Trincher (Trinczer) (1910–1996)

Russian name: Тринчер (Тринхер, Тринцер) Карл Сигмундович

Born: *March 17, 1910, Brno, Austrian Empire*

Occupation: *Medical doctor, biologist*

Last place of residence in Austria: *Vienna*

Arrival in Russia/Soviet Union: *July 1, 1934*

Residences in the Soviet Union: *Moscow, Irtyš (Sverdlovskaja obl.), Miass* (Čeljabinskaja *obl.), Kupan' (Jaroslavskaja obl.), Semipalatinsk*

Arrested: *September 10, 1941, Moscow*

Indictment: *Espionage, socially dangerous element*

Judgment: *January 30, 1943, Special Advisory Board (OSO), 5 years in prison*

Rehabilitated: *May 9, 1956, military tribunal of the Moscow Military District*

Motive for emigration: *Others* [**Our comment**: In effect, Trincher's first emigration, namely from Austria to the USSR, was to escape from the Nazis, whereas his second emigration, from the USSR to Austria, was to escape from the pressure due to the Soviet careerist group profiting from Stalinists' repressions. Trincher has presented a detailed picture of his both motives in his autobiography in German, *Mut zur Wissenschaft*.]

Destiny: *Died a natural death on October 10, 1996, Vienna, Austria*

Short biography: Karl Trinczer (Trincher) was born in 1910 in Brno to a Jewish family. He studied medicine for eight semesters at the University of Vienna, during which time he joined the Communist Student Association. In 1927, Trinczer met his future wife, Gertrude Rutgers, in Vienna. She was the daughter of the engineer Sebald Justinus Rutgers, who had cofounded the Dutch Communist Party and was living and working in Russia from 1918 to 1938, attending the first Comintern Congress in Moscow in 1919, while being personally acquainted with Lenin. In 1927, Rutgers visited Vienna for medical treatment. Trinczer traveled with Gertrude Rutgers in 1934 on a tourist visa to his future parents-in-law in Moscow, where, after a year of additional studying, he was able to pass the missing muster and receive the diploma as a medical doctor. The authorities wanted to send Trinczer to the countryside for three years, but he managed to get a job at the All-Union Institute for Experimental Medicine in Moscow. Trinczer specialized in physical chemistry and examined the electrical conductivity of organs at high- and low-frequency voltages. In 1936 he received Soviet citizenship. After having been shown the door by the institute without any actual reason in 1937, Trinczer could get an appointment as a biochemist at a Moscow hospital and in parallel manage to defend his first doctoral thesis (the Soviet equivalent of a PhD) in biological sciences. In 1940 he was allowed to return to the Institute for Experimental Medicine.

On September 10 or 11, 1941, Trinczer was arrested at his workplace on suspicion of espionage and taken to Butyrka Prison in Moscow. After evacuation of the Moscow prisons in the early winter of 1941/1942, he was transferred to the Čistopol' prison. During his stay over there he was delivered an SOĖ (in Russian Социально опасный элемент [socially dangerous element]) indictment. Along with being considered one such, that is, allegedly belonging to homeless people, prostitutes, vagrants, etc., Trinczer was also accused of having kept in touch with his relatives abroad and with foreigners who had been arrested in the USSR (including Felix

Frankl from Perchtoldsdorf near Vienna). Finally, on January 30, 1943, Trinczer was sentenced to five years in prison and deported to a camp near Kazan' and later transferred to Ivdel' (Sverdlovsk region), where he was mostly used as a doctor; he mainly had to treat criminals and work temporarily as a forestry worker. Trinczer was released due to the amnesty of February 16, 1946, but was not yet allowed to return to Moscow. After a short time in a hospital in Ivdel' and then in Kupan' (Jaroslavl area), he was exiled to Sasovo in the Ryazan' area, where his wife worked as a pediatrician.

In 1946 the family was together again and returned to Moscow without permission, where it turned out that strangers [**Our comment**: in fact, those *strangers* were taking part in Trinczer's arrest in 1941] had seized their apartment. First, Trinczer restarted working near Moscow in 1947, and then for some time he acted as a professor of physical chemistry at the University of Semipalatinsk. After his final rehabilitation in 1956, he could be fully active professionally but had to emigrate with his family from the USSR as a result—because his Jewish background and Gulag past were noticeably affecting his academic career. Moreover, in 1962 and 1968 he unsuccessfully tried to defend his second doctoral thesis in biology. Finally, it was not until 1979 that he and his wife had escaped from the USSR and immigrated to Vienna, where he could teach as a guest lecturer at the Physiological Institute at the University of Vienna. His two sons stayed in the USSR. Besides, in the late thirties of the past century Trinczer's parents could escape from Austria and immigrate to Australia together with their elder daughter Maria, whereas their younger daughter Anna did in parallel immigrate to New York together with her husband.

Trinczer's detailed memoirs, as well as a commented full list of his publications, were published in 1995 under the title *Mut zur Wissenschaft* (Ludwigsburg 1995; Courage to Do Scientific Research. My Life as a Medical Doctor, Researcher and Discoverer of the "Physics of Life": An Autobiography).

Note: See also Gertrude Trincher-Rutgers memoirs, *Das Haus in Miass. Odyssee einer Kinderärztin* (Wien 1993; The House in Miass: Odyssey of a Pediatrician).

Source: RGASPI, GARF, ÖStA.

Bauer's legacy is truly difficult to put forward, for he could publish only one monograph in German, where he could just announce his seminal ideas [11]. His work to follow had been published in Russian, whereas Hungarian colleagues tried to advertise it (mostly in Hungarian). It would be an important biophysical project to republish the commented English translations of Bauer's legacy.

Trincher's legacy is widely known: Aside from his original Russian publications, there are numerous English translations and original German publications (see, for example, just a representative selection of them, namely Refs. [12–22]).

Taken together, both Bauer's and Trincher's results do represent a seminal contribution, not only to macroscopic biophysics, but also to the molecular branch of this science. We see that in fact there were at least a number of colleagues who could successfully work on the ideas presented by Ameline. Therefore, nothing dissuades us from pursuing studies in this direction.

A2.4 How to Employ the Ideas of Energetics: A Methodological Reiteration

Keeping in mind Bauer's and Trincher's results, we have decided to republish here (Sections A2.4.1–A2.4.9) the complete English translation of Ref. [3].

A2.4.1 How to Make a Mechanical Theory of Mental Phenomena

The notion of "scientific theory" as applied to the study of any phenomenon is now familiar and no longer has the absolute aspect that scientists once lent to it. We widely accept nowadays that even without damaging the scope of an "explanation," it is entirely possible to attribute only a utility value to the latter, either in revealing the economy of thought embodied by such an explanation or, again, in duly formulating verifiable predictions about the phenomenon studied and explained according to the theory proposed (this is just where the truly scientific character of any theory resides). The most demonstrative example of a scientific theory that fulfills this duplicate role of summarizing, forecasting, and then saving a truly immense quantity of "thoughts" each related to a fact is continually given by the theory of universal attraction; indeed, all the past or future evolution of the most important astronomical phenomena might fit just into a few lines.

Usually, the provisional character of scientific theories legitimizes the formulation of hypotheses concerning the laws or principles of which the whole constitutes the theory. However, on the other hand, it should be noted that these "hypotheses" have borrowed only a little of the definitive aspect from the scientific theories, which the latter always adopt. So, the term "hypothesis" is nowadays held in higher esteem than in the "old days." We admit the very necessity of hypotheses, and even more than that, we do proclaim it herewith. Still, we must always be careful to call them "scientific hypotheses" only on the condition of economizing and predicting at least as much as those that they claim to replace. Let us add here that the hypotheses, and consequently the theories, are only momentarily provisional. Indeed, if it is a fact that seems to be well acquired, it is because the overall scientific progress ought to be produced just by a series of successive oscillations. Consequently, we may divide the theories into two high groups constituting two poles between which the oscillations of scientific thought in a given field would take place, though they might be different from each other. Nevertheless, under expressional varieties, it is sometimes not that easy to recognize the two ultimate extremes that contain the most exact explanation of the phenomenon under study.

We shall see, therefore, that since the scientific explanation varies mainly by the language employed, the means used to construct a scientific theory cannot be very different from each other. Most often, we seek a "mechanical" explanation of the phenomenon under consideration. Meanwhile, the word "mechanical," mainly as applied to psychology, requires several detailed comments. While the construction of conventional classical mechanics was scarcely modifiable, only "mechanical" explanations for the phenomena of the so-called inanimate or inorganic world were justifiable. At this very moment, the famous assertion came and seemed gratuitous: any animal organism ought to be just a machine. Meanwhile, it must be recognizable that for a rather long time, the conventional classical mechanics could not legitimize such a way of conceiving things. Here is why.

What characterizes most clearly the phenomena that were first systematized by the classical-mechanical principles is their obedience to the well-known basic law: the equivalence between action and reaction. As it is manifest, the latter law necessarily implies the impossibility to distinguish the "cause" (action) from the "effect" (reaction). As we know nowadays, this rigorous equivalence between cause and effect is first of all the characteristic feature of equilibrium states or, in other words, the states that are indifferent regarding the "orientation" in one or another time direction. Finally, according to the currently conventional notions concerning the nature of "cause" and "effect," this indifference, this identity, renders the classical-mechanical principles absolutely incapable of conclusively resolving the anteriority problem, which determines the distinction between both basic notions in question.

Therefore, if the action is always indistinct from the reaction, there is not only indifference between the two kinds of "forces" acting on the system and thus characterizing the phenomenon. Mechanically seen, there is also an indifference of the phenomenon regarding the course of the reaction time; indeed, classical mechanics is incapable of clearly distinguishing a phenomenon from its opposite. It cannot determine, for example, whether it is the car that pushes the horse or whether it is the horse that pulls the car; moreover, it is thus not clear whether it is the car that gets into motion first by starting to push the horse (and being thus the actual cause of the movement) or whether it is the horse that first starts to pull the car (and is thus the actual cause). As a result, classical mechanics concerns the "reversible" phenomena only (where the direct phenomenon is always indistinct from the inverse one). Furthermore, such phenomena persist in an equilibrium state. In other words, physicists designate such phenomena by saying that they are "reversible" or "in equilibrium."

On the other hand, it is evident that biological phenomena are never indifferent to their orientation in time. It is also evident that even the most inferior organisms do not begin by dying to live afterward, that their existence has a definite "meaning." Moreover, it is equally apparent that higher beings are also not indifferent to the course of time, that they have the notion of duration, they are possessed of a lifetime, they do distinguish between causes and effects, etc. As a result, we cannot consider biological phenomena as being "reversible"; they are not justifiable using the conventional principles of classical mechanics and are thus sheer "irreversible."

***

You have noticed that we say "classical" mechanics. Indeed, it is next to impossible to "explain" the irreversible phenomena through the principles of mechanics as it existed at the beginning of the nineteenth century. A new branch of physics had to deal with this category of phenomena: this was the role of energetics.

For some time, theorists were wholly unsuccessful in making the laws of mechanics serve the explanation of irreversible phenomena. Then, the opposition of energetics to mechanics was growing, and even the word "bankruptcy" was uttered. Finally, it appeared that the mere generalization of classical mechanics was sufficient to lead to energetics, thus constituting a modern mechanics suitable for explaining irreversible phenomena and, by simplification, even for explaining reversible phenomena. "Modern" mechanics, thanks to energetics, thus becomes suitable for explaining biological phenomena, or at least the strongest objection to its applicability is discarded. Let us now show how we might achieve this without destroying the great and universal principle of equality of action and reaction.

In modern mechanics, the latter principle is called the principle of "energy conservation." This means that if we measure action and reaction according to the amount of energy involved, they are strictly equal. We may bring no objection to this law. Nevertheless, if the "quantity" of energy never varies, there is the "quality" of energy that is variable, and this variation in quality always goes in the same direction: the new quality or, as we say, the new "form" of energy is always less and less "usable." There is no expenditure of energy but a decrease in the power of energy. For example, the two forms of energy are work and heat; the energy form called "work" is superior to the energy form called "heat"; it means that one can transform all the given quantity of work entirely into heat, but the inverse transformation can never give as much work as compared to the one used while producing heat. There will always be some portion of the heat that can never be converted into work; it is not that it has vanished without trace but that a part of this heat must remain in the state of heat, practically, that is to say, of less useful heat. Heat gets degraded and dissipated, that is to say, in this case, "heat" ought to turn to "cold." Hence, we arrive at the following relations, according to the direction of the transformation:

- (Heat ← → Work): Work = Heat.
- (Heat ← → Work): Heat = Work and cold.

However, scientifically seen, "quality" has virtually no existence if it is not measurable. To measure the quality of the energy forms employing their utility, physicists have designated a brand-new word so that there should never be any possible confusion. This term is "entropy," which means "the energy that has been used or the energy that is not usable anymore." It is then easy to understand that this "energy used" increases with each subsequent transformation and that we may thus ultimately enhance the principle of equality between an action and a reaction, which has become that of conservation of energy, by a second principle concerning the utility of energy, and thus formulating the principles of modern energy mechanics (applied to the universe as a whole):

- In all transformations in the universe, the quantity of energy remains constant.
- In all transformations in the universe, the quantity of entropy constantly increases.

This second principle is necessitated by the existence of irreversible phenomena, for there is only one result of all the realistic phenomena: the quantity of entropy would increase, that is, the used energy ought to increase at the expense of the usable energy.[m]

Of course, this second principle is not just applicable to biological phenomena, for it was just the study of purely physical

[m]Just at this point we are truly obliged to Dr. Ameline for helping us with the analysis of the actual achievements by Carnot, Clapeyron, and Clausius. Ameline does show in detail actual backgrounds of how thermodynamics was created, what its growing pains were, so to speak. On the one hand, Isaac Newton had introduced his basic third law: Any action causes a pertinent reaction. On the other hand, Hamiltonian mechanical theory is sheer incapable of properly representing this law, because it proclaims the equivalence between actions and reactions, and therefore, the vision of the basic difference between causes and effects is inescapably blurred. Still, it was a truly tremendous success of the revolutionary physicists to render the basically metaphysical Hamiltonian theory properly functioning for the purposes of realistic physics nonetheless. Nowadays we may evaluate Ameline's deliberations as a unique appeal to properly combine the successful, valuable "metaphysical theories" with the realistic physics, chemistry, and biology—to ensure further progress in the relevant applied sciences.

phenomena that one imposed on scientists, with steam engines being the main reason. As soon as colleagues could demonstrate the principle of work-and-heat equivalence (when transforming the first to the second), that was a stimulus to apply it immediately to steam engines, without noticing that these embody the inverse transformation of heat into work. Initially, it was surprising not to find the equivalent of the heat expended in the work collected; first, it was thought that friction was the cause of this loss of heat. Nevertheless, soon it became clear that the heat lost by the vapor in the condenser or the atmosphere furnishes the complement of the work produced by the heat expended. That was not the then known "work-to-heat" relation anymore but some new relation governing the theory of thermal machines.

Remarkably, the history of muscular physiology exhibits the same phases as the history of steam engine research, with the only additional complication that daring to apply purely physical laws to vital phenomena seemed to overthrow certain beliefs.

As in the theory of steam engines, first, it was supposed that the work produced in the muscle ought to be at the expense of heat.

That was seemingly the conclusion of the mechanical theory of heat (as they were calling it then). When colleagues could ascertain that muscular work had no source in the heat of the pertinent organ, soon did the vitalists triumph over it, insisting on the idea of the "animal-machine." This real misunderstanding lasted a rather long time, and the echo went beyond physiology laboratories. Animal heat has its origin in chemical phenomena (combustions), and it was manifest that the muscular work expended was at the expense of these combustions. But was the caloric energy the obligatory intermediary between the chemical energy and the mechanical energy? That was the question to be solved. The next idea suggested that heat is not just a necessary intermediary as shown by the following transformations:

$$\text{Combustions} \longrightarrow \text{Heat} \longrightarrow \text{Work}$$

Instead, it has become evident that work and heat are produced simultaneously, as follows:

$$\text{Combustions} \Rightarrow \begin{Bmatrix} \text{Heat} \\ \text{Work} \end{Bmatrix}$$

Thus, the embarrassing objection fell, for all the additional expense of chemical energy necessitated by muscular work was found entirely in this work added to the heat released. Moreover, we respect the classification of energy forms according to their degree of utility; the "chemical" form is more fully usable than the "mechanical" form, and, as we have seen, this is more usable than the "caloric" form. Hence, the first principle of energetics (total energy conservation) is applicable, as well as the second one (energy utility).

The muscular system is not a thermal machine, as the old physiologists were misinterpreting the principle of the equivalence of work and heat by turning it over, just as physicists had wrongly done so. The muscular system ought to be a chemical machine or, if one prefers not to be too precise, it is a machine employing the energy form superior both to work and to heat.

Mechanical energetics could thus widen its application field little by little, and next, it had to include the extraordinary phenomena in the nervous system to reach its complete development.

The first indication in this sense resulted from the demonstration of the non-instantaneousness of psychic acts. The existence of a "reaction time," a certain lapse of time between irritation and its perception, could demonstrate an undeniable analogy between the physical and psychological phenomena; psychophysiology has thus appeared. It is impossible to interpret otherwise the length of time used for transformations of organs. Indeed, the spiritual soul concept is opposed to the duration of the phenomena in which it presumably takes part. Hence, the soul could play only a parallel role in the psychic phenomena and, again, it should be valid only for those admitting its existence. Nevertheless, the necessity of such a concept was no longer persuasive for those who would like to penetrate the functioning of the brain and the dependencies of this organ.

Undoubtedly, the existence of a "reaction time" thus demonstrates that some specific energy transformations accompany psychic processes. To conclusively build up the energetic theory of psychological phenomena, it was necessary to show that the principle concerning the utility of the energy forms does find in psychophysiology an almost immediate application. It is the famous "psychophysical" law that offers the object of this application.

We believe that we were the first to point out this remarkable conclusion, as early as 1898. We suggested then that the facts known as psychophysical law do prove the applicability of the second principle of thermodynamics (energetics) to the functioning of the nervous system, in that the latter acquires its energy from non-nervous energy sources and then degrades the former. Since then, other authors have arrived at the same conclusion. Now, we know what the fundamental law of psychophysics consists of: In the first line, it is just its form that alone has had to endure severe criticism. Still, its fundamental basis has remained incontestable, namely, the existence of a definite "threshold." It is just this threshold that the value of irritation must reach to give the subject the impression of a new sensation. If the value of the excitation remains below this "threshold," the sensation does not change. Moreover, we see something more of utmost importance: this value of the threshold increases with the value of the irritation. Therefore, "sensitivity," that is to say, the capability of distinguishing two irritations, gets less and less as the intensity of these irritations increases. In other words, the modifications produced in the nervous apparatus are less and less considerable when the excitation increases. Simply speaking, the nervous system tends to a stable state under the influence of increasing irritation.

Meanwhile, the second principle of energetics also indicates that the universe tends to a stable state, which is the one where the utility of energy will be dissipated entirely. Then, the universe will not tend to change. This fact is just what we might summarize by saying that the universe is "irreversible."[n] Similarly, the modifications of the nervous system that lead the latter to a stable state might be considered irreversible ones. Thus, the notion of the "threshold" is compatible with the energetic representation of "irreversibility." It is this way of argumentation that could experience further development. It is following this way that we might sensibly refine the precision of our proof compared to all the preceding considerations that could, in principle, support our essential finding, which definitively

[n]Here is Ameline's reaction to the then emotional whirlpool around the overhasty idea of Clausius to forcefully separate energy conservation from energy transformation—and then to apply the resulting logical construction to our universe as a whole, without but telling us how to prove the validity of such a global conclusion. Nevertheless, the incapability of Hamiltonian mechanics to distinguish among causes and effects looks much more disturbing. It is just the combination of both idealizations that helped produce valid, useful physical theories from a metaphysical humbug. Ameline's work is a stark demonstration of such a standpoint.

legitimizes the application of the general principles of physics to psychology.

Nevertheless, this is not the only purpose of the present work; we have presented here the above deliberations only to show that modern mechanics, that is, energetics, could claim its capability of explaining the phenomena even at the highest biological level.

As we have said at the outset, it would not be enough to legitimize the construction of a scientific theory by limiting it just to demonstrations of some vague possibilities; we must instead prove that our theory does adapt to the details of the facts already known and, further, that it permits forecasting novel facts. Therefore, we must immediately note that we do not pretend having already suggested such a theory here. Nevertheless, we could have drawn certain conclusions during the development of our premises, and we dare to hope that these conclusions may perhaps become new facts, provided, of course, that the premises can withstand criticism, though being currently still in their infancy.

Moreover, no predictions worthy of such a name might be possible without any kind of calculations or mathematical developments. In what follows, we shall try to remain quite elementary, as far as the subject implies, but we must apologize in advance to our readership for speaking here the "algebraic language" from time to time.

A2.4.2

It would not be useless to remind the readership of the processes generally used by scientists to construct some scientific theory describing a fact or several facts. This conclusion would help us reveal and better define the meaning and scope of the theory that follows.

Let us suppose a realistic phenomenon; no one pays attention to it until someone encounters it. What happens? Do we need this formerly quite non-existent phenomenon "apprehended" by one individual rather than by others?

They say that this is due to associations of ideas provoked in the brain; it is true, but it would even be more exact to add that these associations of ideas did pre-exist in the impressed brain. We may even say that it was a prior "mental" experience that made the following physical experience possible; the role of preconceived ideas—or, say, words, "theories"—ought to be undeniable a priori.

It is only then that systematic experiments and/or observations start playing proper (but still not definitive) roles in refining the scientific knowledge of the phenomenon under study. We often observe that this second stage of the scientific research activity is the last legitimately scientific one and that the subsequent stages bear a distinctly "metaphysical" character (sometimes even rendering them harmful to the overall scientific research progress). We guess this frequently comes from attributing much too much importance to the pedagogical side of the scientific research process. Of course, when a branch of our knowledge is still in its embryonic state, then it is too early to perform a synthesis worthy of such a name. The same is true of teaching; it is impossible to teach everything at once; it is quite natural that we confine ourselves to the first "apprehensions" of scientists, but then it is instead a pedagogical problem and not a philosophical one. It is only later that questions of scientific methodology can be fully understood, while they are clearly of prime importance for professional/amateur researchers who are busy discovering some new areas and with their study and do not need to work on a topic indicated by a master school teacher.

One of the methods of research on some topic, while constituting a new scientific theory, is comparing the phenomenon studied with others already known and seeking similar features, if possible. Successive groping is usually necessary to find the desired analogy. If the resemblance between the two (the known and the unknown) phenomena is fruitful when drawing verifiable conclusions about the unknown one, we might retain the analogy as our guide; but we are usually obliged to change the phenomenon used for the comparison when we have already drawn everything that we could.

This way of working approach is especially inherent to English scholars, while some of them even claim to understand physics differently than to reduce it to a series of "mechanisms" without establishing obligatory links between the rest. Let us now give a basic example of this very seductive method of building-up a novel scientific theory.

We know that a spirally wound wire, forming a sort of cylinder and being traversed by an electric current, is what physicists call "a solenoid." Physicists know that this solenoid behaves like a magnet: when in two parallel solenoids the electric current is in the same sense, these solenoids repel each other; on the contrary, they attract each other when the sense of the current is not the same in both. Let us consider a mechanical device analogous to the system of the two solenoids. Let us imagine two cylinders suspended vertically side by side, each at the axis of two small horizontal pulleys, whereas a flexible cable portion in all directions forms the connection between the pulleys and cylinders. Then, if we force the pulleys into a rotational movement, the cylinders will rotate at the same time as the pulleys and at the same speed; but as the axis of rotation is flexible, the cylinders can deviate from the vertical without stopping. Now, using appropriately arranged belts, let us force the cylinders into a sharp rotational motion. If the direction of rotation is the same for both cylinders, we will see them moving away from each other, repelling each other, because both solenoids conduct electric currents of the same meaning. On the contrary, if rotation movements opposite to each other animate the two cylinders, we shall see them approaching each other, because two solenoids whose currents do not have the same meaning attract each other. As soon as the analogy between this mechanical device and the solenoids becomes meaningful, the experimentation and the calculation concerning the cylinders will be able to replace the experimentation and the computation concerning the solenoids; of course, this is fine as an approach to a scientific research activity, since it will be necessary to verify each induction made by employing the guide mechanism. This simple example shows the spirit of the method, which has been successfully used in a host of cases to help suggest a scientific theory to predict certain peculiarities of a phenomenon still under study starting from the examination of the peculiarities of a known mechanism, of which an intimate analogy with the phenomenon has been previously proven, and/or vice versa.

Since calculations are the only genuinely scientific mode of prediction, it is apparent that it is not only the "sensible" analogies we are seeking but that there are analogies between the respective "equations" describing the two phenomena that would make it possible to reduce the already known phenomenon to an unknown one.

Naturally, it is just the identity of the equations that constitute the analogy. This remark indicates a second process that we can also use to construct scientific theories. We might borrow the methods of comparing equations from physics, chemistry, and mechanics, or any of the most particular branches of these various sciences, the details of the source do not matter; as soon as we realize the possibility of predicting, then the primary function of science can be exercised, and that is enough. This move ought to be the very first step toward the general synthesis that requires looking for more profound and much more complete similarities.

***

Limiting our ambitions, in this work, we will consider a very modest attempt at conclusively describing some mental phenomena according to the equations that may be suitable for expressing the evolution of these phenomena from some unique and essential point of view.

Therefore, one of the "variables," that is, one of the standpoints that we place for studying the mental phenomena in human beings, will be "time."

For simplicity, let us now suppose that we may fairly judge any psychological phenomenon by one of its characteristics under which it is most convenient to recognize and using which it is generally known—say, the "dominant symptom" of the phenomenon in question.

We will thus get the second "variable," the second viewpoint for our study. Hence, the equations we shall arrive at will consequently be in two variables, and then, as we know, maybe represented by some plane curves, the curves we can draw on a plane.

Before presenting our results, we believe it would be worth recalling the works of the same kind that have dealt with psychophysical law; besides, this will offer us an example where the similarity of the equations may be wholly based already upon the analogy of the two phenomena themselves. This situation arises when biological phenomena get compared to chemical processes.

Moreover, some scientists tend to recognize the physiological facts only as consequences of the relevant chemical processes, which would "explain" the former. To our mind, there is no reason why physical agents, such as light and heat, should have no direct influence on the organism apart from any common chemical reactions. At present, the right to use the intangible nature of these physical agents in scientific explanations seems to be undeniable, presumably since the matter itself also ought to be a form of energy; such a representation seems entirely sustainable and plausible, at the very least. Be that as it may, we could mathematically extend the analogy between specific psychological facts and some pertinent chemical transformations to the physiological field. In some instances, not only the same equation may well describe two categories of phenomena, but also the latter both might obey a mechanism of the same nature, assuming that chemical transformations are the main (if not the only) reason for physiological acts taking place in the human brain, for example.

To make the exposition of our example clearer, we must recall the quite basic definition of "logarithms."

Indeed, from math, we know that a "geometric progression," also known as GP, is such a series of numbers such that any one of its members *divided by* the one immediately preceding it in the series should result in the same "quotient" for all the progression in question. On the other hand, an "arithmetic progression," also known as AP, is such a sequence of numbers that any one of its members *subtracted from* the one immediately preceding it in the series should result in the same "rest" for all the progression in question. Let us now place the numbers of both progression kinds one by one, for example, as follows:

$$\text{GP: } 10;\ 100;\ 1000;\ 10{,}000;\ 100{,}000;\ \ldots$$

and

$$\text{AP: } 1, 2, 3, 4, 5, \ldots$$

We notice here that logarithms of a GP form an AP. Thus 3, 5, and so on ought to be the logarithms of 1000; 100,000, and so on, respectively. Hence, we can write the first progression as

$$\text{GP: } 10^1, 10^2, 10^3, 10^4, 10^5, \ldots$$

We see immediately that we have, for example,

$$10^3 = 1000;\ \text{then, at the same time } 3 = \log_{10}(1000);$$

$$10^5 = 100{,}000,\ \text{with } 5 = \log_{10}(100{,}000),\ \text{etc.}$$

If we now generalize the last two identities by replacing the numbers by letters, we will get an algebraic relation defining the logarithms in general.

For this purpose, let's replace

with a the number 10,

with x the numeral that indicates the power, which is the "exponent," of the "base" of the logarithmic system, that is, the power of a,

and with y the number that is part of the GP.

Then we can form an equation that reads

$$x = \log_a y,$$

with x being in fact the logarithm of y with the base a.

Hence, after recasting the numerical example given above, we get the equation

$$y = a^x,$$

with y being, in fact, the power x of a.

Of course, instead of x or y, we can put numbers whose product, quotient, etc., are y or x; for example, if

$$x = b \cdot z,$$

we shall get

$$b \cdot z = \log_a y, \text{ or } y = a^{b \times z}, \text{ and likewise, if}$$

$$x = m - n,$$

we shall get

$$m - n = \log_a y, y = a^{m-n},$$

and so on, even though the expressions for x or y might be more complicated.

Let us also say that the "base" of a logarithmic system can be any number. These are reasons of convenience that guide the choice. The simplest system of logarithms first discovered at the same time as the logarithms themselves is that called logarithms "Napierian," named after Napier or Neper, who invented them. Because of their intimate relationship with the curve called "hyperbola," they are still called "hyperbolic" logarithms. Instead of being, as in the example we have just given, calculated on the "base" (10), they are calculated on the basis of an endless number designated by mathematicians as "number e," where

$$e = 2.7182818284590452\ldots$$

This number is simpler for the algebraic formulae, but in practical calculations, we take 10 to be the decimal cause number.

A2.4.3

Let us now revert to our subject, or strictly speaking, let us compare a chemical fact with a psychophysiological fact. The chemical fact will be provided by chemical mechanics; it always refers to the transformation of one substance into another. Let us consider two gases filling up the same volume and capable of chemically transforming into each other. They can coexist with the following two possible results: either the entire first substance would finally get destroyed, or the second one will tend to retransform into the first one. In the latter case, a kind of "equilibrium state" would be established, leaving both de- and recomposition only halfway. If we now neglect this latter case, then the law that governs the former case is known as the law of Guldberg and Waage.

It purports that the quantity of the new substance formed increases in GP, while the duration of the transformation grows in AP.

Therefore, we may take the duration of the transformation as the logarithm of the quantity of substance transformed at this moment; by calling x the quantity of substance and t the corresponding duration of the phenomenon, we will arrive at the following relation:

$$t = \log x, \text{ or } x = e^t,$$

using natural (Napierian) logarithms. As is evident, we may choose a unit of time bigger or smaller than some conventional timescales, in which case, the number expressing the time will be proportionally higher or lower than this scale. If we denote the latter proportional scale (m), we shall arrive at the following relationship:

$$m{\cdot}t = \log x, \text{ or } x = e^{m \cdot t}. \quad \text{(A2.1)}$$

This equation ought to be entirely appropriate in physiology, especially in the case of the assimilation products, which get destroyed to produce the relevant excretion products. To apply the reasoning on which the law mentioned above gets based, we must assume, as is done elsewhere, that the assimilation products can accumulate and destroy themselves without being immediately replaced. This point is crucial because the Guldberg–Waage law assumes a particular stock of the substance remaining constant during the study of the duration of the transformation. We can admit without much error that the renewal of the stock of assimilation products is not more frequent than the number of heartbeats, about once a second.

On the other hand, we know well that the duration of typical physiological reactions is not much more than the estimate mentioned above and is about half a second, if we consider simple acts, of course. Therefore, even if we do not admit the transformation of "physiological reserves" to have been already carried out, we might still assume that even during some simple psychic acts, the disappearance of the assimilation products follows the law of Guldberg and Waage.

Let us now look at the psychological phenomenon; it concerns the facts referred to as the Weber–Fechner law.

As we know, this law says that the sensation grows along with the logarithm of the relevant excitement irritation. According to this, by calling the "sensation" s and the irritation i and correctly translating the words "varies as" into the algebraic "is proportional to," with the coefficient of proportionality being (n), we can write the equation as

$$n \cdot s = \log i, \text{ or } i = e^{n \cdot s}, \tag{A2.2}$$

using the natural logarithms everywhere.

The similarity of Eqs. A2.1 and A2.2, together with the preconceived idea that vital phenomena are due to chemical reactions, has naturally led to the idea of "trying to explain" Weber–Fechner's law by that of Guldberg and Waage. Here is how:

Let us assume that an external excitation of value i acts on a sensitive organ. In applying the law of Guldberg and Waage, we assume a chemical disassimilation process (x) of unknown duration whose magnitude is rigorously proportional to the excitation. Then, for example, by calling p the ratio or coefficient of proportionality, we get

$$i = p \cdot x,$$

that is, the quantity of assimilation products that will destroy each other and, by hypothesis, stand for the actual value of the excitation. This disassimilation should then take some time given by the law of Guldberg and Waage, and using the formula in Eq. A2.1, we will have

$$m \cdot t = \log i, \text{ or } m \cdot t = \log(p \cdot x).$$

Let us now assume that the phenomenon called "sensation," which is vague and inherently indeterminate as it is, might still be strictly proportional to the time necessary for the chemical decomposition of the relevant assimilation products. Meanwhile, this time must not be confused with the activity time of the external excitation to the organism. This fact makes us introduce some value k as the coefficient of proportionality between the duration of the chemical process and what we call the "magnitude of sensation." We have first

$$t = k \cdot s;$$

if we now replace t in the penultimate equation by this value, then it will be recast as

$$m \cdot k \cdot s = \log(p \cdot x).$$

Next, let us consider the product of m and k and let n be the result, that is to say,

$$m \cdot k = n;$$

so, we may now write

$$n \cdot s = \log(p \cdot x).$$

Finally, since $p \cdot x$ is equal to i,

$$n \cdot s = \log i \ (quod\_erat\_demonstrandum).$$

The above is but the Weber–Fechner law as given by Eq. A2.2. We may thus explain the logarithmic form of the psychophysical law.

This idea is not the least crucial point of this theory since philosophers have sometimes rejected Fechner's law for the sole reason that, according to its expression, there was no "proportionality" between "effect and cause." This reasoning, bizarre and which can only be explained by the forgetting of the most basic mathematical definitions, is also not unique to philosophers, for economists and even politicians have held it on several occasions, also per their specialty.

Nevertheless, the previous theory uses rigorously proportional laws twice and logarithmic law only once.

Moreover, it is easy to show that a simple relation of proportionality can be transformed into a more complicated relation, similar to a logarithmic one, without transforming the phenomenon described by these laws regarding any of its essential features.

Let us consider the well-known physical law, namely the law of Mariotte, according to which the product of pressure and volume correspondent to the former at any given temperature ought to be constant. If p is the pressure, v the volume, and k a constant quantity, the law reads

$$p \cdot v = k;$$

Now, after dividing both sides of this equation by v, we get

$$p = \frac{k}{v} = k \cdot \left(\frac{1}{v}\right).$$

The quantity $1/v$ is the inverse of the volume. Giving it a name, it will be, if you will, the "density" of the gas considered. If we call this density d, we will have

$$p = k \cdot d.$$

Let us represent this law graphically, that is, draw two straight lines intersecting at a point 0, which is called "origin."

On each of these lines, we have lengths equal to the values of p or v from point 0. One of the lines will be for the pressures; it will be the "axis" of p. The other line will be for volumes and will be the "axis" of v. By leading parallels to the axes starting from some points on these axes which correspond to a state of the gas, we will get a point at the intersection of these parallels, a second point for another state, and so on. The resulting set of points will form a line.

If instead of p and v, we take p and d, we will get another line. In the latter case, we conclude that the equation of the line thus constructed is

$$p = k \cdot d; \tag{A2.3}$$

it is a law of simple proportionality. The line that represents this law is straight. This rule is the only relation that critics of the psychophysical law would allow, claiming that "proportionality" is only possible in the relations of the brain with the outside world—an opinion that seems to us judged just by formulating it.

Let us move on to the ordinary equation of Mariotte's law. It reads

$$p \cdot v = k; \tag{A2.4}$$

the line that represents it is a "hyperbole."

What is a hyperbole? Imagine two identical cones welded together at their top points and representing the fully symmetric extension of each other concerning their top points. It is the mathematical object called "diabolo" (two abolos) that represents the image of what the geometers call a double cone, a cone with two layers. Let us cut any of these cones strictly perpendicular to its axis. The resulting section is a "circle," and it is a unique surface of any of the welded cones. Now, if the section were oblique, with the "tablecloths" always remaining only on one half of the diabolo, this section would be an "ellipse." Finally, with a single kerf cut simultaneously through the two cones of the diabolo, the section will be a "hyperbole." Let us now place the diabolo thus sawed well in front of us: The "profile" of the cone will represent the two "asymptotes" of the hyperbola.

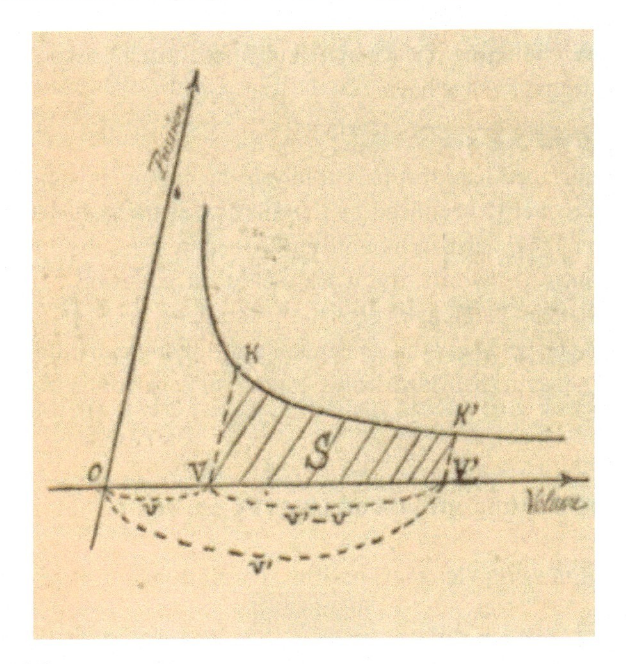

Figure A2.6

Now, if we count volume and pressure on each asymptote taken as an axis, the line representing the formula of Mariotte will be the hyperbola.

So, we see that when we change the way of defining the state of the same gas, namely, by replacing the volume by the density, the line describing the gas states becomes hyperbolic.

Nevertheless, that is not yet the whole story. Let us choose a point corresponding to the state of the gas in the middle of the pressure-volume curve, which we will designate by K (refer to Fig. A2.6); if you wish, it will correspond to the point K. This point defines the points on each of the axes. Hence, the relevant abscissa point V defines a volume V. Let us now consider another state K′, or a point K′, on the curve and an abscissa point V′ corresponding to a volume V'. For example, V′ is higher than V. Then, the straight lines KV and K′V′ cut out a surface between the curve and the abscissa. The area of this surface is in very close mathematical relation with the logarithm of V' and can be used to determine it by setting $V = 1$. It is for this reason that there exist the so-called hyperbolic logarithms (also called "natural logarithms").

Let us denote this surface by S, and since it can be taken equal to the logarithm of V', we have

$$S = \ln(V'). \quad \text{(A2.5)}$$

This surface S has an apparent physical meaning: It is the "work," that is, the energy produced by the dilation of the gas. Meanwhile, the state of this gas can be entirely determined by the "pressure" as well, in other words, the work produced/collected or, instead, expended when bringing the gas to that state.

Moreover, we know that instead of multiplying two numbers, we have only to sum their logarithms; taking the logarithms of Eq. A2.4, we have:

$$\log(p \cdot V) \equiv \log p + \log V \equiv \log k,$$

that is, the logarithm of a product is equal to the sum of the logarithms of the factors, as we have just said.

Let us now replace $\log(V)$ with the area that we have called S. Moreover, $\log k$ is a constant number equal to H; hence the previous relation could be recast as

$$\log(p \cdot V) \equiv \log p + S \equiv H,$$

from where

$$\log p \equiv H - S,$$

which means that a logarithmic law describes the state of the gas.

Thus, we see that the mathematical formula that determines a phenomenon depends on the means used to measure this phenomenon. In the case we have just examined, if the measures are:

- Pressure and density, the line is a straight line

- Pressure and volume, the line is a hyperbole
- Pressure and energy, the line is a logarithmic one

Let us now compare the above with the formula of Weber–Fechner. As we see, irritations and sensations define the state of the brain. Likewise, external pressure and its work on the gas define the gas state (or the work of the gas on the source of the external pressure). Taking the theory of gases as a "mechanistic model," we can, therefore, hypothesize that the sensation s is a logarithm of something still undetermined by physiology and that we call X; in such a theory we would arrive at the relationship

$$s \equiv \log X.$$

Then, always following the theory of gases, we might conclude that the psychophysical law becomes similar to that of Mariotte in its conventional mathematical form and thus consider the product of the irritation amount i with the variable X having some constant value, that is, we will arrive at the formula

$$i \cdot X = K \text{ (}K\text{ is a constant).}$$

In continuing to follow the parallel between our psychological theory and that of the gases, let us imagine a new psychic phenomenon related to the variable X the way density is related to volume in physics, that is, something that is inverse to the latter, namely $(1/X)$. The latter formula will thus take the form of Eq. A2.3 of Mariotte's law. Then, by giving the vague name "impression" to X, we might conclude without any anticipation that $1/X$ ought to be the "density of impression," to arrive at the following relation between irritation and impression:

$$i \equiv K \cdot \frac{1}{X}.$$

Alternatively, using our ordinary everyday language, we conclude that the density of an impression is proportional to the irritation. This way, we have arrived at the mathematical expression for the law, which looks plausible, for it reflects the proportionality between the effect and the cause.

We might recognize that this problem of "choosing the pure form" does not destroy psychophysics any more than Mariotte's law destroys physics because we replace the variable volume by the variable work or energy concerning this volume. Seemingly, it is enough to introduce or "define," if you wish, the notion of "sensation" as the energy set free in the brain by an "impression" resulting from "irritation" or some excitation coming from outside so that the "law of proportionality" becomes a "logarithmic" law.

The above is just how psychophysics might be "explained" in two ways: on the one hand, modeled by an underlying chemical mechanism and, on the other hand, modeled by the pertinent physical analogy. Moreover, these two explanations do not contradict each other.

When we compress a gas (like in a pneumatic air lighter), we know that a part of the energy expended gets used to produce heat, which finally gets dispersed in the atmosphere; when one expands a gas, it cools first, then warms up at the expense of the outside air, whereas heat is still lost. There is energy dissipation in both cases; this produces entropy.

Now, let us go back to the law of Guldberg and Waage: since all the initial amount of the first chemical compound gets destroyed to form that which is second, in this experiment, there would be no tendency to an equilibrium state. Consequently, as we have said at the beginning, such a phenomenon would never be capsizable or even truly reversible—in the sense that entropy production accompanies it by energy dissipation (so that heat is released, that is to say, lost).

This "entropy release" ought to be equivalent to the "wear and tear" of the universe, or instead, of its parts—and, in our brains, this might cause what we have just called a "sensation"; consequently, it would be a feeling of the wear of the organism, which is well known.

Meanwhile, we must admit that the above is only a hypothesis, but it is not gratuitous. To put it in more detail, above we introduced an undefined concept of "sensation," and now we must prove the chemical-theoretical assumption behind the psychophysical law that this "sensation" is proportional to the "duration" of the disassimilation (i.e., the decomposition of the related assimilation products). Further, both time and entropy notions possess the same property: that of being able only to increase. Moreover, already several times since 1898, colleagues have proposed measuring time, and even defining it, through entropy. Therefore, the chemical explanation of the psychophysical law agrees with this idea about the true nature of the time-entropy interrelationship.

A2.4.4

In the preceding section, we have suggested three theoretical parallels to analyze all the facts making up what is called "psychophysics."

The first, which we believe in having been provocative, directly links Weber–Fechner's law to thermodynamics through the entropy principle, without any need to consider a physical or chemical phenomenon.

Various authors have recently exposed the second theoretical parallel based upon the Guldberg–Waage law.

The third one has shown that the same law can be put either into the form of the simple proportionality or as a hyperbolic. Finally, we may even express it as a logarithmic formula, without changing the essential nature of the law.

In our eyes, all three do have a severe defect: all of them do not go further than the phenomenon we are trying to explain theoretically. They seem to lack the relevant fertility, and they have not provided us with any effective means of further penetrating the psychic phenomena. Then, while waiting for such means that may only have occasionally escaped the attention of the researchers, we may perhaps find it more convenient to overcome the difficulty by confining our task to looking for some handy mechanical equations, which could turn out to be appropriate for various mental processes mainly considered in their evolution.

In following the example of physicists, we will look for an equation to adequately describe the relationship between the duration of the phenomenon under study and the most appropriate characteristic sign of the latter that duly accounts for its actual evolution. We may describe this equation using the principles of mechanics.

In doing so, we will obtain a simple "mechanical description" of this psychological phenomenon; we will have "depicted" this phenomenon in mechanical language, with the ideas and words we borrowed from this science. In short, our pretension is genuinely modest, and modest also will be our result. This study will start with the most straightforward hypotheses, as far as our schemata are following the summary of the pertinent experimental observations for certain psychic phenomena.

The most straightforward mechanical ideas that will serve us to describe and to relate the physiology of the brain will be as follows:

Psychic acts will depend on a certain number of "forces," with some of them favoring and others preventing the action in question; the latter are called "antagonistic" forces.

Naturally, we attach the + sign to the favorable forces and the – sign to those that are antagonistic. The resultant of all these forces will be the sum of the favorable forces minus the sum of the antagonists. For example, in a general notation, where

X' is a favorable force,

X'' is an antagonistic force, and

X is their resultant, we may write

$X = X' - X''$.

Let us denote as E the effect of the forces, that is, the intensity of the resulting phenomenon, and then as t the duration during which this phenomenon evolves; here, we will take only these two "variables," as physicists would denote them. Hence, our resulting equations cannot pretend to give a complete or definitive representation of the phenomena to which they will apply. Such equations would solely constitute the very first approximation to properly building "cerebral mechanics."

Suppose now that after having arrived at a moment t, the time has become just a little bit "different" from the latter value. In other words, it has increased by an infinitely small quantity called a "differential," which we denote as dt, that is to say, a differential of t. In parallel, the effect of the forces would undergo some small variation as well in that it will become a little bit "different" from what it was at the moment t, that is to say, E, and this variation will be "differential of E," and we shall denote it as dE, accordingly.

The most straightforward mechanical principle that will allow us to combine these quantities should sound as follows:

The variation in the effect should be proportional to the resulting intensity of the favorable forces and the variation in the duration of this result's validity.

On the other hand, if the effect decreases, the variation of effect will have the – sign, and if the effect increases, its variation will have the + sign.

Let us write, therefore, following the principle above, that dE is proportional to the resultant X and to the duration of action of this force that corresponds to the variation of effect dE; it will come to

$$dE = X \cdot dt, \text{ or}$$

$$\frac{dE}{dt} = X .$$

Moreover, if X is the resultant of the two forces X' and X'', this one being antagonistic, we will have

$$(X' - X'') = \frac{dE}{dt} .$$

This almost obvious equation is fundamental; in general, there should also be a coefficient of proportionality (J), and then the equation becomes

$$(X' - X'') = J \cdot \frac{dE}{dt}, \text{ or}$$

$$\frac{X' - X''}{J} = \frac{\mathrm{dE}}{\mathrm{dt}} .$$

That is, multiplying (dE/dt) by J is the same as dividing ($X' - X''$) by J.

The coefficient J, which weakens the action of the acting forces, might, therefore, be interpreted as something like the "body's mass"—or as a "body's inertia"—if the body resists the motion that the force acting on the body tends to cause.

The ratio dE/dt is then analogous to the "acceleration," whereas E should then be analogous to the "speed" as defined by mechanics.

Such is the model of the mechanical explanation that we will use to build up a theory of various psychological phenomena.

Now, it remains only to study the evolution of such a mental manifestation as one wants. We must look for the formula to be given to all the necessary kinds of forces X' and X'' to arrive at an equation that would be consistent with what the experimental observations reveal about the evolution of the mental manifestations under study.

Thus, first, we shall proceed to study several conventional psychological processes from the standpoint of their evolution.

After that, we shall discuss the equations that seem to agree with the observed progress of the processes under study. Finally, we will arrive at the formulae for the forces the balance of which should lead to such equations.

A2.4.5 The Senses: Theory of the Consecutive Images

These are images that persist and even reappear several times after an external excitation has ceased to influence the organ in question.

This phenomenon reveals the existence of a kind of "inertia" in the sense organs, similar to the inertia of a massive body. The latter explains the fact that the body's movement might persist even after the action of the inciting force is over. In this case, what we are discussing is that applying to some kind of memory asset makes it possible to say that the impressions coming from outside might persist even when the irritation has disappeared. The exact knowledge of this phenomenon is given to us by direct observations, which have the advantage of being immune from any contestation.

Above all, it is owing to the three senses—hearing, tact/touching perception, and eyesight—that the experiments are demonstrative and usable for constructing our theory, and we shall successively examine them in detail.

Hearing: All that we know about the consecutive images, in this case, boils down to the following: After a specified interval of time, the first impression having disappeared, it is possible to rehear, in a way, the sound already heard, which appears to be much weaker and more difficult to distinguish from outside noise. However, according to observers, the number of returns of the image that follows varies. The intensity of these returns becomes weaker and weaker, as the first consecutive image already is vis-à-vis the ordinary image.

If we wanted to graph the intensity of the auditory impression during the whole duration of the phenomenon, by limiting ourselves to the three images—ordinary, consecutive no. 1, and consecutive no. 2—the curve of the phenomenon will show three prominent areas, between which this curve will become the straight line that indicates the nullity of the impact.

In other words, the curve of the process will exhibit oscillations of less and less amplitude, that is, damped oscillations, with a tendency to return to the rest of the organ. These oscillations will come two or three times, with the third being scarcely perceptible.

Tact/Touching perception: About tact, the existence of consecutive images has been the origin of two different theories in terms of the number of images found.

First, colleagues have found a single consecutive image, but recently another observer noted a second one, which boils down to three images because of a single tactile irritation. If we would represent these images graphically, like those in the audition case, then we get a sinusoidal line, as a result, whose "turns" are less and less ample, that is to say, more and more "amortized." The curve of these "conscious" images will make at least three oscillations around the straight line indicating the rest of the organ, that is, the zero impact.

Still, of these tactile images, different physiological explanations have been given, depending upon whether the observer in aces counted two or even three oscillations. When there are only two, we assume that we must consider the two pathways of sensibility possessed by the spinal cord: propagation to the brain would take a different amount of time for each, hence the duplicate perception.

On the other hand, the occurrence of the two consecutive images might also have another explanation; indeed, we come back to the fact that the cutaneous excitation produces a vasoconstriction, which is followed by a vasodilatation, to pretend that it is the perception of these two vascular phenomena that leads to the observation of the two consecutive images.

We will allow ourselves to cast a doubt on the validity of the above theories if we admit that the images in question might remain subconscious and unconscious, which is very likely. Moreover, the fact that the early observers could first count only one or two images (especially for the case of eyesight, as we shall see a bit later here) indicates that attention has been paid to the phenomenon by assigning it to the conscious image. In contrast, until its observation, it could well remain subconscious and even unconscious.

Our way of looking is independent of the number of consecutive conscious images, because mathematically, a curve with damped oscillations always remains sinusoidal, whatever its extent.

Eyesight: It is the vision that gives us the most significant number of these consecutive images. Here again, these images are partly unconscious, and this is just more or less what the authors report. However, we must remark here that at the beginning of the observations only "positive" parts of the consecutive images were reported, that is to say, only the clear/bright parts of the object could be found in the image, and later "negative" images appeared, that is to say, where the blacks replace the clear/bright parts of the object in this image.

Nevertheless, colleagues found later that a rapid negative and consequently fleeting image appeared between the ordinary and the consecutive positive ones, with its short duration being probably the cause of its transience to the attention of the observers.

Be that as it may, colleagues later also pointed out that following the second (ordinary) negative image, two further consecutive images, a definite no. 2, and a negative no. 3, appear. The latter both are longer and less intense than the previous ones, so we have been able to trace the total phenomenon, from the positive image that everyone calls "the image proper" up to the fifth consecutive image (in all six images). Hence, we get a curve composed of six "damped" oscillations of increasing duration, with these oscillations being around an average position giving the state of rest for the eye.

There is still no reason to stop drawing the oscillations after observing six of them, especially since some authors report only four of them; we might conclude that unconscious or subconscious images probably do exist, though too weak to be perceived.

Moreover, in doing so, let us apply a principle vital to any scientific theory: the principle of "continuity"; then, the number of damped oscillations cannot be limited.

The other phenomena can be of long duration—up to 15 seconds. For excitations less than an intense one, which hardly exceeds all the just-perceptible excitations, the form of the evolution law for the intensity of the impressions is not so complicated; at least the current precision of the experiments leads to a simple mathematical relationship only.

***

Let us now try to calculate this relation according to the theories previously suggested to quantify the Weber–Fechner law but placing ourselves here with other standpoints.

Above (refer to Section A2.4.3) we have denoted an external irritation by the symbol i, whereas x stands for the chemical process to which the irritation gives birth in the body, with p being a coefficient of proportionality:

$$i = p \cdot x.$$

Moreover, as we have found that

$$x = e^{m \cdot t},$$

where m is another coefficient of proportionality and t the duration of the chemical transformation. We will have

$$i = p \cdot e^{m \cdot t}.$$

On the other hand, we called the ratio $1/x$ the "density of the impression" (refer to Section A2.4.3) and put the relation

$$i \equiv \frac{K}{x}.$$

Hence, conversely

$$x = \frac{K}{i}.$$

Furthermore, substituting the proper value for i, we arrive at the following formula:

$$x = \frac{K}{p \cdot e^{m \cdot t}}.$$

Let us note here the ratio of the quotients K and p to call it k; then

$$x = \frac{k}{e^{m \cdot t}}.$$

Now, we may assume that the value of the "sensation" s is proportional to x:

$$s = l \cdot x,$$

where l is just a constant quantity. Then for the relationship we have sought, we get

$$s = \frac{l \cdot k}{e^{m \cdot t}}.$$

Finally, after denoting the product $h = l \cdot k$ we arrive at the relation

$$s = \frac{h}{e^{m \cdot t}}.$$

It is a logarithmic law, as it is easy to see, for there is a power of the number e, the base of the natural logarithms.

This formula is well known to physiologists; it indicates that after excitation, the sensation does not reach its full value immediately but instead should take a specific amount of time to perceive an excitation. For example, if we look at a screen on which a weak image gets painted due to X-rays, we notice that first, we perceive an illuminated surface only, and after a brief amount of time, the image becomes fully visible. In the same way, the intensity of a sound emitted by a tuning fork is not immediately perceived in its full intensity; the visibility of flashlights also requires some time to arrive at its maximum.

The time it takes for a sensation just to be perceived should be a small fraction of the total time that can pass before this same sensation gets fully understood, including the time to produce all its consecutive images.

We can, therefore, legitimately conclude that the curve that will represent the entire evolution of a sensation will begin as a logarithmic curve and change after a particular amount of time into a curve with damped oscillations (Fig. A2.7). This is how, at least in short, will the phenomenon that we have just studied be mechanically described in broad outline only.

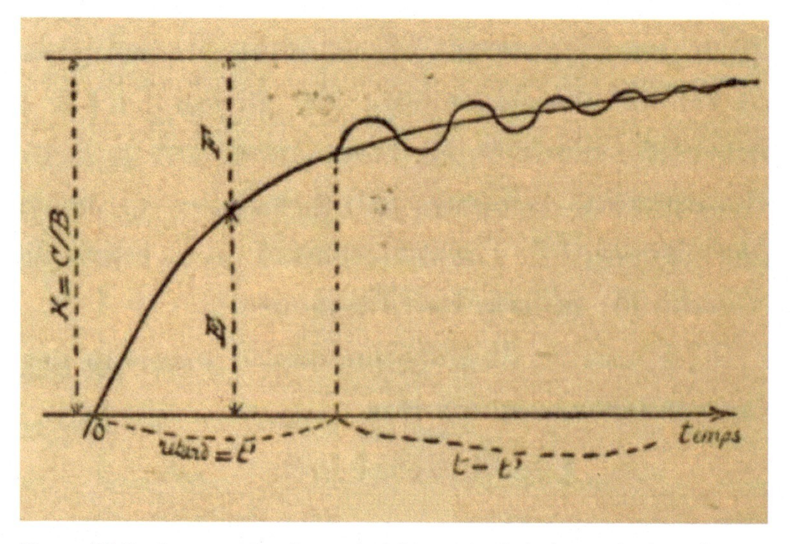

Figure A2.7 An example of a typical "sensation" evolution (ordinate). As a consequence, the abscissa is time, with its first stretch depicting delay = t' and $t - t'$.

A2.4.6 Demential Law by Paul Janet

As a typical subject of the application of our theory, we shall now take a psychological phenomenon that results on its own from the evolution of the cerebral organ during the whole life. This phenomenon is global in that it summarizes all the psychic facts and consists of the variations of their resultant over a prolonged amount of time.

Here is what it is about: In 1877, Paul Janet sought the cause of a fact observable among the people having exceeded a certain age. Indeed, for many people, there comes a time when the years seem to flow more and more quickly as they get older.

Janet reports this as follows: "For example, let us take one year as a measurement unit; objectively, this is just a calendar year. Meanwhile, for a ten-year-old child, it represents one-tenth of the child's lifetime solely, but for a fifty-years-old adult that same year will be no more than one-fiftieth of his/her existence: Hence, in going from a child to an adult, the same year will be recognized to be shorter in the proportion of 50 to 10, that is to say, five times shorter. To sum up, a fifty-years-old adult is living five times faster than a ten-year-old child."

Without stopping at the proportion of 1/50, which is given only to illustrate the idea, Janet draws his conclusion: to give us always the same impression of the time elapsed, it is necessary to consider that the real time, which is the "exterior" time, is in a sempiternal opposition to the impression of time, which is an "interior" time we feel. Then, we might say, as our next step, it is necessary to work out the correct picture of the "excitation" time we feel. Indeed, the latter might always vary (in fact, it increases) by quantity in the same ratio or proportion as the real (i.e., "outside") time to the already elapsed lifetime of the observer (i.e., "inside") time.

Interestingly, the above statement reminds of the actual statement of Weber's law: the external excitation must be in constant relation with its increase, or vice versa, to give the same apparent impression.

Without applying to differential and integral calculus, a simple estimation does immediately show that the latter statement is none other than that of the logarithmic law. Here is how to demonstrate this:

Let us go back to the pure GP that we have used to define logarithms.

It is

$$10; 100; 1000; 10{,}000; 100{,}000; \ldots$$

Let us write the differences between the consecutive terms; these are

90; 900; 9000; 90,000; . . .

Further, we try to identify the ratios between each of the above numbers and the smaller number of the relevant two used to produce it. We shall then arrive at the following result:

90/10; 900/100; 9000/1000; 90,000/10,000; . . .

which can alternatively be written as 9, 9, 9, 9, . . .

Hence, we always get the same ratio.

Now let us assume that the smaller of the two numbers used to find the differences represents the appropriate excitation level. Therefore, the differences should represent the augmentations given for the relevant initial excitations to reach the next excitation level. This excitation will produce the same degree of sensation irrespective of the number of steps completed, that is, the same increase in the sensation's progression. Mathematically, if we formulate a progression out of terms having the same increase or difference, it must be an AP (*quod erat demonstrandum*).

Conclusion: A logarithmic curve can represent the whole entirety of the realistic psychological evolution processes in any individual.

Remarkably, in the simple logarithmic equation of sensation the term "power of number (e)" is but in denominator, because for the same duration, the increases in sensation become smaller and smaller, because, in turn, the "perception" of time becomes less and less sensitive. As a result, we should place this term in the logarithmic denominator, while letting a be the perception of time at a given age t.

Finally, we arrive at

$$a = \frac{1}{e^{Bt}},$$

where B is a coefficient that we call the "dementedness coefficient," as we call the above equation the "dementedness law."

The higher this coefficient B, the faster the intellectual weakening of the individual. Thus, somebody might arrive at complete dementia at an early age (idiocy); others might acquire this at adolescence and still others even later. Some reach such a state at the beginning of their old age (senility), whereas there are individuals who never arrive at this station up to their deaths, if one may say, in preserving the entirety of hardly weakened faculties until their advanced age.

Therefore, the classification of the dementia states presented above acquires a purely quantitative form.

We can also render it more complete and more adequate in terms of the clinical facts, but this is not the object of this present work.

A2.4.7 Psychoses

We will be brief in this section, and yet this is where our theory seems to bring us the most clarity, but the subject is too special to go in for details here.

Meanwhile, two major classes of psychoses remain if we dismiss the group of pure "dementia": organic ones (those with some anatomical lesions of the brain) and only physiological ones (toxic ones, for example).

These two classes are those in which the "predisposition" is, respectively, either strong or weak; lastly, in the latter group the predisposition can be hereditary or acquired. Often, to take together the repeated madness ictuses, chronicity replaces heredity, as has already been duly recognized.

We have given in brief the evolution of pure dementia of the one fully or almost without delirium. Let us now see what can serve as a general character to other forms of psychosis.

First, let us bring together the psychoses that evolve on the same individual because we can consider the hereditary diseases as having begun in a generation and continuing to evolve in the following ones—a consideration that is not new.

Then two main varieties appear, one formed by follies with many hallucinations of acute or chronic form and another form where hallucinations are solely accessory, which is called intermittent madness, with melancholy or manic or mixed ictus.

The intermittent or even remittent (the one with less marked intermittences) character is observable in all forms of psychosis.

This fact is especially observable if we do not submit the insane to the means of constraint; the intermissions are then particularly sharp in accentuated patients. Even their existence is due to the success of treatment methods. The latter could be either living in a hospital or at liberty, but within a foreign family. Indeed, even if the pathological mental state does not change from time to time, especially at the very intimate intervals when it comes to acute states, it would be sheer impossible to leave a mad person alone for a moment, and it would be above anyone's strengths to exercise continual supervision without any means of restraint.

On the other hand, we may even scientifically define any madness as "dementia." Thus, we will be able to describe the evolution of psychoses by a "dementia" curve, that is to say, a logarithmic curve with some oscillations superimposed on the latter. These oscillations should be "amortized" because the symptoms become less and less marked, that is, more and more attenuated, as the weakening of the faculties progresses, with the chronicity of the disease developing and, at the same time, with the duration of the disease getting prolonged.

These oscillations will not be simple ones. Instead, in their turn, they will be deformed by other oscillations, which should also be damped if they are hallucinations (persistent sensations would reappear and get mixed with the sensations of the moment), with each variety of hallucination delivering its curve. If we might employ the acoustic analogy here, the resulting curve describing all these superimposed phenomena will be like that produced by assigning the whole set of vibrations to a fundamental/principal sound tone plus the curves of its higher harmonics.

The acute forms will be represented by a curve whose principal oscillations will soon get damped after the first oscillatory outburst of a significant amplitude. To adequately describe such cases, it is necessary to give only ad hoc values to the coefficients of the equation describing such a curve.

Finally, the hereditary psychoses will be representable by curves analogous to the preceding ones. Still, we will suppose that several generations share the curve, just as some alienists suppose that several generations do share the illness, but it can still concentrate all its evolution on a single individual.

Nevertheless, and this is the very point at which our theory prescribes attention, it will be necessary to consider the "improvements" of pathological inheritance. Indeed, only very rarely do alienists note the good hereditary antecedents of patients; meanwhile, we ourselves feel obliged to point out such cases. It is only the study of our equations that could help us reveal examples of "hereditary improvement" in some cases of episodic syndromes (phobias, impulses, manias) that occur in hereditary patients with marked pathological antecedents, as well as in individuals with superior intelligence. Similarly, individuals of the latter category often have alternatives for cerebral hyperactivity and cerebral inertia that show up as indeed attenuated periodic intervals. It is in such cases as well that our theory could allow us to conclusively distract them from the group of regular periodic intervals to which one may conventionally ascribe them and to recognize them as examples of "improved pathological inheritance."

The above is just a sketch of "prediction" that may grant a specific "scientific" character to our theory, but only if our conclusions could be subsequently verifiable.

Let us now turn to the actual equation of the psychological phenomena that we have just schematized.

A2.4.8 Mechanical Representation of Psychic Phenomena

The fundamental principle of cerebral mechanics is, as we have said, the same as that of ordinary mechanics: it is the definition of "force" that is its basis. We define force at the same time as "mass" in stating that their ratio is "the acceleration" that takes a body of mass equal to J under the influence of a force equal to X, for example.

Let $\mathbf{E}$ be the velocity attained by the body at the moment t. Acceleration is by definition, as we know, an increase in this velocity during the unit of time. Now, if the speed varies—becomes a little different from what it was—we write that this variation is $d\mathbf{E}$, that is to say, "differential" of $\mathbf{E}$; but the time—the duration—of the motion is then increased by dt, that is to say, differential of t, and acceleration is the ratio $d\mathbf{E}/dt$. Let us write, then, that force is equal to the product of this acceleration and the mass in motion, since, according to the preliminary definition, we have $d\mathbf{E}/dt$. Let us write, then, that force is equal to the product of this acceleration and the mass in motion, since, according to the preliminary definition, we have

$$X/J = d\mathbf{E}/dt, \qquad \text{(A2.6)}$$

so that we may arrive at the formula

$$X = J(d\mathbf{E}/dt).$$

Moreover, if we assume that the mass of the body is equal to unity, we only have the following:

$$X = d\mathbf{E}/dt, \tag{A2.7}$$

which simplifies the formula because J is useless for the moment.

Instead of a single force, several of them may be acting on the body to produce the phenomenon studied; we will then distinguish between the following two categories of these forces:

The forces that favor the production of the effect should have the + sign.

The forces antagonistic to the phenomenon should have the – sign.

We will sum up the forces of each kind, and the difference between the two sums will be the resultant of all the acting forces. Therefore, if X' designates the favorable forces and X'' the antagonists, we shall have

$$X = X' - X'' \tag{A2.8}$$

and consequently

$$X' - X'' = d\mathbf{E}/dt. \tag{A2.9}$$

Now, we will call **E** the effect of all the forces contributing to the production of a phenomenon. Hence, the fundamental principle expressed by Eq. A2.7 could sound as follows: for a given time, extremely or infinitely small, the immensely or infinitely small increase (or decrease) of the effect is proportional to the resultant of the acting forces as well as to that given time, and if one wishes to take the coefficient J into account or, that is to say, to state Eq. A2.6, we must add "and conversely proportional to the inertia of the organ."

Moreover, Eq. A2.6 expresses precisely the principle of inertia.

The preceding equations describe a phenomenon using the dominant symptom of this phenomenon. In other words, we employ the sign of its characteristic manifestation or, at least, what appears to us as such. Finally, we use the time notion by looking at the evolution of the chosen symptom.

In Eq. A2.6, it is solely necessary to introduce the actual sum of forces on which the phenomenon ought to be dependent. Then, the completion of the calculation involves only mathematical difficulties, the solution of which sometimes imposes certain additional conditions to fully determine the mathematical expression for the "forces" to seek.

Such are the elements of the mechanical language in their remarkable simplicity utilizing which we will try to analyze several psychological processes whose evolution we have described in ordinary language.

According to these principles, let us seek the equations of the three psychological phenomena: pure dementia, sensations, and psychoses.

A2.4.8.1 Mechanism of dementia

We will start by considering the insanity law by Paul Janet: the accurate perception of the time elapsed becomes less and less sensitive with age and ends up being practically inappreciable in a very advanced old age, as we know.

Therefore, there is a rather rapid diminution of the time perception, which occurs almost abruptly. The latter observation is obviously enough to reject a law of simple proportionality, for such a law would apply only to a perception that is always regular and without variation.

On the other hand, as the time perception decreases with age, it varies in the direction opposite to increasing time so that if we call this perception degree a, we cannot regard it as merely equal to $1/t$. Of course, t being raised to the power of some number would surely grow much faster than t. Then, to get the proper expression for the denominator of the fraction a, we would choose the number e, and to render its powers rigorously equal to a, we shall take the product $B{\cdot}t$ instead of t. Finally, we would arrive at the following formula:

$$a = \frac{1}{e^{B \cdot t}}.$$

Now, the intellectual weakening, the dementia, increases as the perception of time decreases. Since the relevant psychic phenomena have only a literary, only a qualitative definition, nothing prevents us from considering the weakening of the pertinent faculties as measurable by the "complementary" quantity of a, that is to say, from considering the relationship between the intellectual weakening E and a in the following plausible form:

$$E = K - a,$$

which shows as a result that when a increases, E decreases, and vice versa.

It is the most straightforward relationship that can be assumed, and that is just what made us choose it at first glance.

Now, replacing a by its time-dependent value

$$E = K - \frac{1}{e^{B \cdot t}}.$$

Let us call C the product of B and K:

$$C = B \cdot K.$$

Then we will have, after taking this into account,

$$E = \frac{C}{B} - \frac{1}{e^{B \cdot t}}. \tag{A2.10}$$

Such is the law of pure madness without delirium. It represents the diminution of a quantity C/B under the influence of age, and this quantity can serve to appreciate the mental capacity of a subject.

We also see that the higher the B, the lower is C/B. Hence, we should call B the "dementia coefficient" and then C ought to be the "vitality coefficient" of the brain. We can see that the higher the C, the more time it should take for the quantity C/B that is subtracted from K to reduce the "mental level" or the "mental capacity" that K ought to represent.

The same equation would be found by first putting the equations expressing the fundamental principle and the mathematical expressions of the forces:

$$\left\{ \begin{array}{l} \dfrac{dE}{dt} = X' - X'' \\ \text{with:} \\ X' = C \quad \text{and} \quad X'' = B \cdot E. \end{array} \right\} \tag{A2.11}$$

In performing detailed calculations, Eq. A2.11 leads to Eq. A2.10, which, consequently, makes it possible to conclude that dementia can be due to two causes: a constant force (cerebral vitality) and an antagonistic force proportional to the intellectual impairment in the ratio B, called the dementia coefficient.

A2.4.8.2 Mechanism of sensations

If we consider consecutive images, and since a logarithmic law governs the evolution of sensation, the whole phenomenon will result from the superposition of a series of rapidly damped undulations onto a logarithmic law (refer to Fig. A2.7, together with the qualitative discussion of the topic).

As we have seen, at the beginning of a sensation, the law giving s using time t is

$$s = \frac{h}{e^{m \cdot t}}.$$

Still, to take the consecutive images into account, we also need to damp decreasing oscillations with some proper kind of the "sinusoidal" law of variation. Mathematically, a sinusoidal law is of the form "M·sine (b·t)" or "N·cosine (b·t)," or a sum of these two expressions.

(It is easy to know what a "sine" or a "cosine" is. It is enough to look at a watch. One chooses two perpendicular diameters, for example, IX–III and XII–VI and monitors both needles of the watch. Then, the distance from the tip of one of the needles to one of these diameters gives the sine of the angle it makes with the diameter (say, IX–III), whereas the cosine is then the distance from the tip of this same needle to the other diameter (XII–VI), that is, it is always the same angle. We see at once that when the needle turns, these distances vary depending on the length of the radius of the dial. Sometimes from one side and sometimes from the other side of these diameters, they are sometimes positive and sometimes negative, and this changes periodically.)

Since the oscillations ought to be damped, they must diminish when the time increases; then, we will multiply the previous sinusoidal expressions, as already explained, by a factor having t at the denominator. Since only a few consecutive images might be conscious, the damping must be genuinely rapid. We will take a power $a \cdot t$ of the number e, with the parameter a adjusting the damping velocity (whereas parameters M and b should adjust their amplitude and frequency, respectively).

To sum up, let s' be the portion of the phenomenon that presents the undulations. Then we will present it as the following law:

$$s' = \frac{\left(M \cdot \sin(b \cdot t)\right)}{e^{a \cdot t}}.$$

Now, it only remains to add s to s' to have the result.

Still, to enable our equation to properly describe the dementia states, we must modify it a little bit. Indeed, we will look at $s + s'$ as analogous to the perception of time; this analogy should indeed be real since the words "sensation" and "perception" ordinarily have the same meaning. Remarkably, as we have already evaluated the intellectual weakening by cerebral wear due to the perception of duration, we should likewise consider the effect of external agents in the form of cerebral wear produced by sensations. So, just as we have written down already

$$E = K - a,$$

we will then write down

$$E = K - (s + s')$$

and also

$$C = B \cdot K. \tag{A2.12}$$

Finally, by replacing all these quantities with their respective values, we will get

$$E = \frac{C}{B} - \frac{h}{e^{m \cdot t}} - \frac{M \cdot \sin(b \cdot t)}{e^{a \cdot t}}. \tag{A2.13}$$

The above will be just the equation sought. We should not forget that the letters in Eqs. A2.12 and A2.13 do not have absolutely the same meaning since they do not concern the same realistic facts. In effect, they do have, to some extent, similar meanings for K, B, and C but the same meanings for e and t.

A2.4.8.3 Mechanism of psychoses

We do not have much to say about the general mechanism of psychoses. Given that their fundamental character remains clinically linked to the dementia process, to which multiple periodic processes add dying out with the chronicity of the disease, it will be enough for us to replace the single periodic term of the preceding equation with several periodic terms exhibiting a gradual amortization.

Indeed, after choosing S for some damped periodical term, and $S^{I}, S^{II}, S^{III}, S^{IV}, \ldots$ for the other ones, we will get the following general equation for psychoses:

$$E = \frac{C}{B} - \frac{k}{e^{n \cdot t}} \pm \left(S \pm S^{I} \pm S^{II} \pm S^{III} \pm S^{IV} \pm \ldots\right), \qquad \text{(A2.14)}$$

with the ± sign indicating that the combination of the periodic terms will be made in such a way as to accurately account for the differences and varieties of the pathological states, just as the relevant case under study may dictate.

Now, suppose we have written down the system of preliminary equations

$$\left\{ \begin{array}{l} \qquad \dfrac{dF}{dt} = X' - X'' \\ \text{with:} \\ \qquad X' = C' \quad \text{and} \quad X'' = B \cdot F(t - t') \end{array} \right\}, \qquad \text{(A2.15)}$$

in which the force X'' is an antagonistic force, which does not come into action until after a certain time t or after the "delay" time t, while X' is a constant force. The calculation leads to the relation

$$F = \frac{C'}{B} - \frac{k}{e^{n \cdot t}} + \left(S + S^{I} + S^{II} + S^{III} + S^{IV} + \ldots\right)$$

when we introduce the following simplifying condition:

$$B \cdot t' = 1/e.$$

As we have already explained twice, let us now evaluate the wear produced by the action of the forces acting on the brain; we will then have to write down

$$E = K - F.$$

Furthermore, we will finally get

$$E = K - \frac{C'}{B} - \frac{k}{e^{n \cdot t}} - \left(S + S^{I} + S^{II} + S^{III} + S^{IV} + \ldots\right).$$

Alternatively, by bringing together $(K - C'/B)$ in a single term (C/B), we get

$$E = \frac{C}{B} - \frac{k}{e^{n \cdot t}} - \left(S + S^{I} + S^{II} + S^{III} + S^{IV} + \ldots\right),$$

that is, just what we have called Eq. A2.14.

To sum up, the mechanical law explains both the psychological process (still referred to by the ill-defined name of "sensation") and other mental phenomena, like psychoses, whose detailed study seems to be so tricky at first sight because of their complexity and whose submission to any calculation looks sheer impossible for the same reason.

Recall that the image obtained at the development will not be of increasing intensity, depending on the length of the printing time in the light of a photographic plate. It may be more apparent than that for a time of less significant illumination and/or for a longer illumination time; the alternations represent damped oscillations, and we may explain them by a mechanism identical to that in Eq. A2.15; three of such oscillations regarding the prolonged photochemical action could have been observable, which is comparable to the consecutive images of the senses.

A2.4.8.4 Consequences

The first consequence that emerges from the above discussion is that the oscillations in the psychic states can come with retardation from a force acting on the effect already begun, that is to say, on a previous state due to a constant force, which ought to be an acquired state provided with a certain momentum and, therefore, less sensitive to a subsequent modification. Most of the known pathological factors are, by nature, delayed. As a result, we may explain the pathogenesis of intermittent phenomena in such a straightforward way. It is indeed easy to see that the difference between Eqs. A2.12 and A2.13 or A2.14 comes from the sinusoidal terms, while the difference between the systems in Eqs. A2.11 and A2.15 comes from the $(t - t')$ factor and hence these differences ought to be somehow interrelated.

The above is not the only possible mode of explanation, but it is the simplest of them all. Moreover, most of the equations of mechanics are reducible to expressions containing logarithmic and sinusoidal terms in a variable number. The result is due to two convincing reasons: first, in the practice of differential and integral calculus, these terms are straightforward and convenient to handle; second, we understand that we can very roughly replace an oscillatory phenomenon by the mean value around which the oscillations occur, provided that the latter are not very considerable.

The number of forces acting with retardation might increase in the system in Eq. A2.15, with their resultant being a delayed force as well; thus, nothing essential will change in our final equation. This conclusion is right because, let us repeat, the known infections and intoxications even in their acute states should always have a long incubation period before the noticeable pathological accidents break out.

A second consequence is that Eq. A2.14 can be used to summarize the evolution of the most psychic processes.

Indeed, it is a unique feature of the general pathology that the morbid states are solely an exaggeration, in the broadest sense of the word, of this or that normal phenomenon: hence, it suffices to reduce the number of sinusoidal terms to find Eq. A2.13. Meanwhile, instead of purely and merely suppressing them, it is mathematically enough to suppose them existing in the normal state, but in some considerably diminished form or, if one wishes, being "unconscious" because of their low values, assuming all of them usually weak or, in trying to embrace all of the psychological processes during the whole life, even assuming at last that, in the standard states, we might consider them negligible according to the importance of their resultants as compared to each other. Then, Eq. A2.14, after being reduced to Eq. A2.13, should finally boil down to Eq. A2.12.

It can thus be said to sum up, with a high degree of accuracy, that both normal and morbid mental phenomena must be attributable to two main groups of causes or forces: one being constant and the second group being proportional to the effect thus produced, which is antagonistic regarding the first group and lagging the latter.

A2.4.8.5 Influence of the cerebral inertia coefficient

In our equations, it is the coefficient that plays a role analogous to the mass of a moving body and that we have called J, or inertia coefficient of the brain subjected to the action of various forces.

Let us show the role of this coefficient; for this purpose, the fundamental equation that defines any force, and any mass, is as we know

$$X = J \cdot dE/dt$$

or

$$X/J = dE/dt,$$

which proves that, on multiplying the second term by J, the result must be the same as on dividing the first term by J; hence, the latter coefficient weakens the forces acting on the brain.

From considering the cerebral inertia, several consequences follow:

- Let us suppose that this inertia coefficient can vary itself and its changes come in the banalest way: if the process is periodic at regular intervals, it will be able to transform the action of nonperiodic forces into a periodic effect. The result is visible because the quotient of X if it is not periodic by J, as it is, should remain to be so.

  Thus, either the simple dementia law or the transformations of the organ that it governs are not periodic; but suppose cerebral resistance is diminished at regular intervals by any event of external or internal origin, like, for example, menstruation or weekly or daily alcohol intoxication. Then, the coefficient J will decrease periodically.

  The effect will also be periodic, which goes without saying.

  Meanwhile, let us suppose that to fluctuations and oscillations involved, something known as a "beat" in acoustics happens. Both processes having about the same periodicity, reinforcement should result in both modes with about the same frequencies. From our standpoint, the two oscillatory processes will reinforce each other to produce an exaggeration of both almost normal facts, thus leading to a morbid phenomenon.

  If the periods of both oscillations are very close, the beats should be frequent; but if the periods are somewhat different, these beats will be very distant and will perhaps even be observable only once.

  The above is the perfect image of all the intermediaries between psychoses with frequent ictuses. Besides, the image is valid for those with unique or truly rare ictuses. The principal conclusion is that a periodic cause can give rise to a single acute psychosis if the "periodicity" of the subject differs enormously from that of the pathologic cause.

  Following this way, we might explain at least one of all the plausible production modes of acute psychoses. Sure, an acute variation in J (an infection or an acute intoxication) should give rise to an acute effect, and this could result in an acute psychosis as well.

Thus mechanically explained might also be attacks of acute delirium occurring in the course of psychoses in evolution, either periodic or not, if we consider the latter ones as somewhat less sensitive regarding periodicity.

The other mechanical theory may even suffice for the explanation of psychoses; it assumes that three forces are at stake in this case. A force with a logarithmic variation, that is, dementia; and another one with a periodic variation, which gives rise to the oscillatory S and the oscillatory S^{I}, S^{II}, S^{III}, S^{IV}, etc., which deform the regularity of the S contour, in a way similar to the way numerous harmonic vibrations contribute acoustically to a participation in the "timber" of a fundamental sound S emitted by this or that musical instrument. Finally, there might be the third periodic force, which would combine itself or its higher harmonics with the second periodic variation of J, depending on the case.

Meanwhile, it is easy to see that this way of representing psychosis says nothing more than the ordinary clinical observations and that in reality, it might explain only "idem by idem," that is, without going in for more depth than just the very observation necessary to explain, for example, like the dormitory virtue of opium. Nevertheless, it is always possible to explain the actual source of the rare or unique ictuses of any illness by adding one periodic force to another one; it should always be something like this.

Let us also remark that if the two forces have the same periodicity, they may compensate each other accurately, with the second canceling the effects of the third; the result will be that the periodic variations of the mental state that still exist will get suppressed. In this case, we may then say that the brain will remain in a state of torpor, which should be even more complete, the more perfect the compensation—the coincidence of the participating periodicities—is. Such could be the case of the cerebral "stupor" states.

If one does need three forces to build up a mechanical theory, it is better to add this required third one to Eq. A2.15), which should thus be more fertile in conclusions, as we will see.

- The variations in J studied previously are either continual or only transient ones; nevertheless, they could be possible only during the life of an individual. Those that we are now going to consider have no subsequent variations: J changes in value once and preserves it. Obviously, this can occur during the lifetime of a particular person but may also indicate a cerebral change between two generations—by the mechanism of transformed heredity.

Here is what we mean by that. We may regard the same family as doing kind of everything, despite the succession of generations that compose it, and this is the definition of the word.

It is evident that by the intervention of a new ascendant, the cerebral inertia coefficient of this family may sometimes undergo a sudden change in essential proportions.

Suppose that the increase in J results from the weakening of one or the other of the forces X' or X''. We see that if the weakening of forces can come from the increase of J, the converse should also be true.

So, let us go back to the simplified equation

$$dE/dt = X' - X''.$$

Moreover, suppose that due to heredity or accidentally during life (cranial trauma) X' is reduced to X'/J; we will then have

$$dE/dt = X'/J - X''.$$

Alternatively, after multiplying both sides by J, we will have

$$J \cdot dE/dt = X' - J \cdot X'',$$

that is, X'' is increased as compared to X'.

Now, according to Eq. A2.15,

$$X'' = B \cdot (t - t') \cdot E.$$

The increase in X'' means the increase in the oscillations, which is pathological. Remarkably, since the inertia of the brain also gets increased in the same proportion, the pathological effect is diminished and will not constitute anything more than just a semidelusional, semipathological state that renders the individual in question a candidate for all the delusions placed on the border to madness instead of being a real mad person, thanks to the improvement of his cerebral inertia coefficient. We may imply that X'' is not insignificant in advance, without which the diminution of X' would increase the inertia of the brain to the extent that the latter would become too inert and that the individual would simply get a congenital intellectual weakness (if the reduction in X' is congenital) or an acquired intellectual weakness (if X' is reduced accidentally).

Meanwhile, in X'', the reinforcement can occur either via B or via $(t - t')$.

If it is via B, that is, via the dementia coefficient, we will have a state with a marked maddening tendency and hence an almost continuous state of semidelirium, something like a reasoning madness or an interpretative madness.

If it is but via $(t - t')$, there will be intermittences in psychosis with no accentuated tendency to dementia. It would then suffice to compare the systems in Eqs. A2.11 and A2.15 to realize that it is just the presence of $(t - t')$ that adds oscillatory terms to Eq. A2.12 in rendering it Eq. A2.13.

To sum up, if it is $(t - t')$ that increases, then intermittencies and/or relapses should also increase.

We might, therefore, distinguish between two possible varieties according to whether these intermittences consist of pathologic ictuses having short- or long-time intervals between each other.

In the former case, we will have the representation of the episodic syndromes called "phobias," manias, impulses, and obsessions, which represent rudimentary delusions, being sometimes aborted, appearing and disappearing, imposing themselves on the brain with somewhat diminished stability, though still susceptible to resistance. At the same time, the resistance of the brain might also be both lessening and increasing. This fact is difficult to rationalize without considering the equations.

In the latter case found in the presence of intermittent disorders, we have no accentuated dementia, a condition that often gets noted as hereditary but that we distinguish into intermittent states independent of the variations in the coefficient J. Again, we must also distinguish here the intermittent disorders owing to the transformed inheritance and those of other origins, for example, due to just a simple predisposition; similar is the case when the delay $(t - t')$ of the action of X'' is essential from the beginning of the action of the latter force.

- Let us now turn to study what should be the consequences of diminishing the force X''. The interpretation should not be complicated. First, we shall consider the equation

$$dE/dt = X' - X''/J.$$

We know that the delaying force produces oscillations in the mental state; then, if it gets weakened, the damping of the oscillations should grow. Likewise, in the above case, if J were of considerable magnitude, the cerebral inertia should then be such that one might, in principle, face a case of marked impairment of the faculties. Nonetheless, instead of any improvement in the mental state, there would be a destruction of every mental state. This is just what happens when an insane person gets interned—the monotonous life of the asylum calms then that person's wild impulses. Still, often, the mental faculties diminish as the delirium gets into subacute states if this is not remedied as far as possible by occupying the sick person at various jobs or better still by returning him/her to a near-normal existence through a trial outing or allowing him/her to live in a foreign family under medical supervision. Remarkably, an insane person calm in an asylum reveals the picture of the awakening of his/her faculties and his/her delirium at the same time.

Further, let us consider a force X'' of still remarkable intensity (the damping of the oscillations it causes will be faster than if it were not diminished by J). We shall thus have a disturbance of shorter duration, that is to say, a more or less acute one. On the other hand, if the entire X'' gets diminished in a general way, then the B would also be weak and the $(t - t')$, that is to say, the delay or the incubation would be weak too (this brevity of the incubation should be related to the acuity of the process).

Thus, we may conclude that acute psychoses should have a weak tendency to dementia. Meanwhile, if $(t - t')$ was excessively small, then B could still acquire a significant value. We might translate this by stating that because an excessively short incubation indicates a hyperacute condition, the tendency to dementia could then get considerably increased. This effect seems to be evident and is just what our equations highlight on their own, thus offering an excellent opportunity to verify their validity.

Of course, if X'' were not very noticeable and J does remain important, then the pathological episode would be minimal, as one would expect when the cerebral vitality X' is increased alongside the resistance or inertia of the brain. Besides, the preceding equation might get recast as follows:

$$J \cdot dE/dt = J \cdot X' - X'',$$

by putting these facts in relief.

- We must then speak of the high values that both J and X' may simultaneously adopt in lending; thus, an exclusively considerable force to the brain that we might use to characterize the mental state is denominated "genius." In such a case, it is probable that the coefficient B to indicate the power of transforming X' into X'' may even be more significant than in an ordinary individual since both forces acting in the brain might approximately be considered those to regulate the disassimilation. Then, the coefficient B, though being also considerable relative to the normal ones, would most probably be offset by even larger values of J and X'.

  We might even catch a glimpse of the fact that a correspondingly more significant value of B is related to an accumulation of several dissimilative and destructive influences due to several pathological causes combined to a state of degeneration that would nonetheless be improved, and even ultracompensated, by the intervention of truly healthy heredity.

  These are, in fact, just the conditions in which the production of "geniuses" has been quite frequently noted: mental accidents coinciding with excessively developed intellectual faculties.

  Let us say that the coefficient B is not necessarily very high, as well as the degenerative state, because men of genius have presented no psychic disorder; Leibniz, from this point of view, is opposed to Descartes, Pascal, Newton, Comte, and other colleagues, who were/are even less prominent but anyway far from being rare.

Nota bene: Equation A2.14 does not precisely follow from the system in Eq. A2.15; this would be the case only if the condition

$$B \cdot t' = \frac{1}{e}$$

is fulfilled; otherwise it would have a small modification to undergo (we have already discussed this above).

The general solution of the equation

$$\frac{\mathrm{d}F}{\mathrm{d}t} = K - B \cdot E(t - t')$$

should in effect be cast as

$$F = \frac{K}{B} + \frac{A'}{e^{x' \cdot t}} + \frac{A''}{e^{x'' \cdot t}} + \left(S^{\mathrm{I}} + S^{\mathrm{II}} + S^{\mathrm{III}} + S^{\mathrm{IV}} + \ldots\right),$$

that is, we arrive at the equation in which

- o x' and x'' are the roots of the equation

$$x + \frac{B'}{e^{x \cdot t'}} = 0.$$

- o $S^{\mathrm{I}}, S^{\mathrm{II}}, S^{\mathrm{III}}, S^{\mathrm{IV}}, \ldots$ are expressions of the form

$$S = h \cdot e^{a \cdot t'} \cdot \sin(b \cdot t + c).$$

  Then, finally, the coefficients a and b should be given by two new equations:

$$a + \left(\frac{B}{e^{x \cdot t'}}\right) \cdot \cos(b \cdot t') = 0$$

  and

$$b + \left(\frac{B}{e^{x \cdot t'}}\right) \cdot \sin(b \cdot t') = 0.$$

- o The quantities A', A'', h, and c are constants, which one must determine according to the studied phenomenon and its peculiarities.

Noteworthy, most of our previous deductions here are not changed. Interestingly, according to the particular condition imposed on B and t' variations, they ought to be in an inverse relationship to each other—namely, when B increases, t' should decrease and vice versa—because their product must always remain equal to 1/2.718, that is, about 0.308.

As a result, we could prove that if B is small and t' is large, the oscillations are essential and the tendency to dementia is weaker; otherwise, if B is large and t' is small, the oscillations are weak and the tendency to dementia is marked.

Therefore, we arrive at the two major categories that have been explicit at each stage of the preceding discussions.

There are two groups of psychoses: the insane and the intermittent ones; we fall back on the classification of certain alienists: early dementia and maniacal depressive madness. Noteworthy, the oscillations described by our equations do indicate the alternatives of calm and delirium and not only those of excitation and depression.

The psychoses unique to the so-called hereditary insane might be represented by our mechanical equations when the time parameter t' has considerable value, but when the t' is rather small, several generations might be describable by our equation.

It is easy to realize that when t' is extensive, the oscillations are almost completely damped and become insensitive. Indeed, when t increases, its difference with t', which is always the same, becomes more and more close to the value of the t' itself. In other words, after a very long time, the initial delay of X'' becomes insignificant and negligible; therefore the oscillations, which owe their existence to this delay, become negligible, for they are getting amortized.

It is essential to underline here that $(t - t')$ does not have the same meaning as during the discussion of J; what we consider here is the final value of this quantity, whereas above we have dealt with its initial value, that is, with that at the very beginning of the force's (X') action.

A2.4.9 Conclusion

Such is a mechanical theory of many psychological phenomena, both normal and morbid ones. It seems to us that it is a theory that possesses a specific undeniable scientific character—some inductions concerning new categories in studying intermittent pathological states, intellectual weaknesses, as well as genius, especially regarding the influence of the genetic improvement. The above ought to reveal a not insignificant fecundity of our standpoint; in any case, provocative questions have been raised that might already have obtained affirmative answers.

Still, to our mind, the vital thing should lie not so much in our results themselves as in the method that allows us to construct a theory, that is, to go to the very first step, to suggest the first approximation toward the desirable goal from so many points of view—toward a true psychophysics, a valid mechanics of brains.

—Marius Ameline

References

1. F. L. Holmes (1962). From elective affinities to chemical equilibria: Berthollet's law of mass action, *Chymia*, **8**, pp. 105–145.
2. E. B. Starikov, K. Braesicke, E. W. Knapp, W. Saenger (2001). Negative solubility coefficient of methylated cyclodextrins in water: a theoretical study, *Chem. Phys. Lett.*, **336**, pp. 504–510.
3. M. Ameline (1908). Comment faire une théorie mécanique des phénomènes mentaux, *Journal de psychologie normale et pathologique*, pp. 398–446.
4. M. Ameline (1913). Psychologie et origine de certains précèdes arithmétiques adoptées par les calculateurs prodiges, *Journal de psychologie normale et pathologique*, **10**, pp. 465–490.
5. W. Brown (1914). General reviews and summaries: the recent literature of mental types, *Psychol. Bull.*, **XI**(Nr. 11), November 15, pp. 397–399.
6. V. H. Rosen (1953). On mathematical "illumination" and the mathematical thought process, *Psychoanal. Study Child*, **8**, pp. 127–154.

7. C. Mangin-Lazarus (1994). *Maurice Dide, Paris 1873–Buchenwald 1944: Un psychiatre et la guerre*, Les Éditions Érès, Toulouse, France.
8. R. C. Tolman, P. C. Fine (1948). On the irreversible production of entropy, *Rev. Mod. Phys.*, **20**, pp. 51–77.
9. http://www.eoht.info/page/Erwin+Bauer
10. http://members.inode.at/777911/html/stalin/trinczer.htm
11. Erwin Bauer (1920). *Die Grundprinzipien der rein naturwissenschaftlichen Biologie und ihre Anwendungen in der Physiologie und Pathologie* (Springer, Berlin).
12. Karl Sigmundovič Trincher (1960). *Teploobrazovatel'naja funkcija i ščeločnost' reakcii legočnoj tkani* (*Akad. Moskva*) (Heat-Generating Function and Alkalinity of Lung Tissues).
13. Karl Sigmundovič Trincher (1964). Biologija i informacija: ėlementy biologičeskoj termodinamiki (*Nauka*, Moskva) (Biology and Information: Elements of Biological Thermodynamics).
14. Karl Sigmundovič Trincher (1965). *Biology and Information* (MA Springer USA, Boston).
15. Karl Sigmundovič Trincher (1967). *Biologie und Information: Elementare Diskussion über die Probleme der biologischen Thermodynamik* (Teubner, Leipzig).
16. Karl Sigmundovič Trincher (1969). *Structurally Bound Water and Biological Molecules* (National Research Council of Canada, Ottawa).
17. Karl Sigmundovič Trincher (1977). Biology and Information: Elements of Biological Thermodynamics (Biologija i informacija, Consultants Bureau, New York).
18. Karl Sigmundovič Trincher (1981). The mathematic-thermodynamic analysis of the anomalies of water and the temperature range of life, *Water Res.*, **15**(4), pp. 433–448.
19. Karl Sigmundovič Trincher (1981). *Die Gesetze der biologischen Thermodynamik* (Urban u. Schwarzenberg, Wien).
20. Karl Sigmundovič Trincher (1981). *Natur und Geist: zur physikalischen Eigenständigkeit des Lebens* (Herder, Wien; Freiburg; Basel).
21. Karl Sigmundovič Trincher (1990). *Wasser: Grundstruktur des Lebens und Denkens* (Herder, Wien).
22. Karl Sigmundovič Trincher (1998). *Die Physik des Lebens. Band 1. Die thermodynamischen Grundlagen der Biologie, Leben, Kunst, Wissenschaft* (Verlag der Synergeia GmbH, Klausen-Leopoldsdorf, Austria).

Appendix 3 to Chapter 1

A Methodological Outlook

With what is discussed in Appendix 2 in mind, we recognize that such ultimately unfinished conceptual building as atomo-mechanics either in Boltzmann's or in Gibbs' representation ought to look like something quite surreal. Likewise, surreal is its derivative, the meanwhile good old tried-and-true realm of quantum mechanics, which we can view "as an extension of the classical probability calculus, allowing for random variables that are not simultaneously measurable" [1], although this is a throughout valid and productive physical theory at this same time. This same pertains to the (in the meantime already conventional) conceptual tandem of the equilibrium thermodynamics and statistical mechanics, which also consequently employ the "magic" probability notion, leading us to a genuinely unusual situation described by Prof. Wicken (1942–2002) as follows [2]:

> The irreversibility of macroscopic processes is explained in both formulations of thermodynamics in a teleological way that appeals to entropic or probabilistic consequences rather than to efficient-causal, antecedent conditions. This explanatory structure of thermodynamics does not imply a teleological orientation to macroscopic processes themselves, but to reflect simply the epistemological limitations of this science, wherein consequences of heat-work asymmetries are either macroscopically measurable (entropy) or calculable (probabilities), while efficient causal relationships are obscure or indeterminable.

Entropy-Enthalpy Compensation: Finding a Methodological Common Denominator through Probability, Statistics, and Physics
Edited by Evgeni B. Starikov, Bengt Nordén, and Shigenori Tanaka

ISBN 978-981-4877-30-5 (Hardcover), 978-1-003-05625-6 (eBook)
www.jennystanford.com

Such a situation results from uncritically following the legacies of Boltzmann and Gibbs, not because both peers were in error, but because they both had no time to bring their seminal works to some exploitable level.

In this connection, there are serious methodological problems articulated by the outstanding physicist Margenau (1901–1997) [3]:

> Methodology might be understood to mean a description of various individual procedures, which have led to the successful solution of specific problems. In studying the subject of physics from this point of view, i.e., with the special emphasis on method, one would naturally turn his attention to the traditional divisions of experimental and theoretical physics, the former with its measuring devices and the latter with its mathematical technique. In no other sense than this, does the term methodology make any direct appeal to the working physicist, and if we would ask him to define his methods, he would probably answer with the description of experimental technique or the methods of setting up and solving differential equations. His answer would tell us **how he solves his problems**, but hardly **how he finds them** and **why he solves them**.

Remarkably, the above wise words were published in 1935, just when the validity and usefulness of both quantum and statistical mechanics were no more in doubt. Still, the difficulty mentioned by Prof. Margenau as concerns the posers "How may we find the physical problems?" and "Why should we solve them at all?" in the conceptual realm of quantum and statistical mechanics persists till today, to our sincere regret. To our mind, Prof. Margenau could provide us with the right direction for our trains of thoughts [4]:

> **Do masses, electrons, atoms, magnetic field strengths, etc., exist?** Nothing is more surprising indeed than the fact that in these days of minute quantitative analysis of relativistic thought, most of us still expect an answer to this question in terms of yes and no. The physicist frowns upon questions of the sort: "is this object green?" or "what time is it on a distant star?" He knows for sure that there are many different shades of green. He knows for sure that the time depends on the state of motion of the star. Almost every term that has come under scientific scrutiny has lost its initially absolute significance and acquired a range of meaning; of

> which even the boundaries are often variable. The word to be has escaped this process.
>
> A glance at the various usage of a term may persuade us that some confusion "exists" in this connection. The chair exists because I can see it; the United States exist; fairies exist in my imagination. I have been using the word, I think, in a manner sanctioned by refined discourse in all these instances. If there is no more uniformity in the standard usage of the word, **we had better drop the question as to the existence of the constructs of explanation.**

Meanwhile, there remains but one more poser, as to the expressed conceptual duality of the theories in question, namely, their intimate adherence to the classical probability calculus (first, to its frequentist branch), which has no immediately recognizable physical basis. Nowadays, we might assume that this is at least partially due to the sheer absence of clarity in such an essential point as the real physical sense of the entropy notion (see Ref. [1] of Chapter 1 and the references therein for more details).

To this end, we would like to cite here the wise words by Prof. Williams (1899–1983) [5]:

> ...The above is not to impugn statistical generalizations effected by the physicist, nor to deny the actuarial usefulness of the mathematics worked out by the proponents of the more sophisticated theories. However, it is to deny that the theory of probability, which is a counterpart to such achievements, implied by them, or is adequate to the demands of sound knowledge, or is true to fact. Especially fallacious, I propose, is the claim of its champions that it is in some unique sense "factual," "scientific," "empirical," or "operational."

The search for an adequate physical methodology has been going on until recently. Prof. Bhatt, the editor of an exciting and instructive book [6], draws the following conclusion equally applicable to the conventional quantum and statistical mechanics (emphasis added):

> Contemporary quantum physics is driven to posit *dual-levels of reality*, the empirical and the trans-empirical, and yet many theoretical scientists do not take this seriously. They still cherish the idea of the ultimate state of matter. But *no account of any reductionism*—behaviorist or naturalistic or *any sort of physicalism*—can be free from logical flaws and such trains of thoughts are, therefore, untenable. There is an incurable limitation

of theoretical reason, and the trans-empirical is not accessible to current positivistic methodologies.

Prof. Wallace shares the above opinion and tries to analyze it in more detail in his book [7]. His analysis leads to the following methodological recommendations [8]:

> . . . Some general conclusions can be drawn:
>
> 1. State spaces, ubiquitous in physics, should not be (blindly) reified.
> 2. Only very general, abstract conclusions about the ontology of quantum (or classical) mechanics as a whole are likely to be possible.
> 3. There is no "fundamental" classical mechanics: no single classical-mechanical theory from which all other empirically relevant classical theories can be derived.
> 4. There might be a "fundamental" quantum mechanics in this sense, but if so, it is not nonrelativistic particle mechanics; rather, it is a quantum field theory, the standard model of particle physics.
> 5. In principle, metaphysicians of quantum theory ought then to be looking at quantum field theory—but they should do so with a clear appreciation that the theory (at least as understood by working physicists) is, by design, largely silent about "fundamental" ontology. If there is a physical theory, which does tell us about fundamental ontology, we do not as yet have it.
> 6. Metaphysicians interested in the ontology of various modifications or supplements to nonrelativistic quantum particle mechanics need to have a methodological story to tell as to what they are doing and why it is worthwhile.

Fascinating to this end are the suggestions to try finding and analyzing the parallels between the modern physical theories and the Buddhist theory of *śūnya* (emptiness/wholeness) published in Ref. [6].

The duality mentioned above or even a kind of multitiered structure of quantum and statistical physics urges us to look in detail for the deficiencies in our total world view by arriving at a poser that might cause serious concerns [9]. Indeed, the "end of science" might, in principle, come, but not due to reaching the "limits of knowledge" (they do not exist at all). Instead, it may come due to fruitful *voluntarism*: Here, we mean the sense of inclination to realize the desired goals without taking into account objective circumstances and possible consequences. If such voluntarism leads to definite success, it necessarily results in a *fetishism*: any success attracts numerous con artists having hardly anything to do with the professional field in question, who strive solely for their profit, who produce fetishes for this purpose.

Probably, this is just why Planck was practically all the rest of his life trying to analyze in detail the philosophical notion of free will [10, 11], after getting a deserved prize for his quite emotional, to a certain extent even irrational, idea of "energy quantum," which could nonetheless lead to the well-known, very revolutionary breakthrough in physics.

Noteworthy to this end is that one of the "champions" critically mentioned in the above-cited work by Prof. Williams [5] was Reichenbach (1891–1953), a noticeable proponent of the revolution in physics (see, for example, Ref. [12]). He could take the criticism very seriously and perform a comprehensive and thorough logical analysis of the frequentist approach to the probability, with particular reference to its applications in physics.

As a result, Prof. Reichenbach could have logically analyzed the actual intrinsic sense of correlations between/among the physical variables/observables containing some random contributions or being throughout random. He had arrived at the seminal idea of the "common cause" by recognizing that not only might correlation mean a direct cause-and-effect relationship between/among the pertinent variables but also there may be some hidden/latent (i.e., not measurable directly but valid at any rate, and realistic) variables correlated to the observables. He came to this conclusion by carefully analyzing the conventional statistical mechanics built up upon Boltzmann's works dealing with the ideal gas (no interparticle correlations) and upon Gibbs' works (dealing with statistically independent subsystems, that is, with the ideal gas as well) [13].

The untimely departure of Reichenbach did not allow him to consider this idea in more detail, although there had been a noticeable resonance among his colleagues, the philosophers of science (see, for example, works in Refs. [14–17]). What ought to be but truly surprising is the sporadic responses on the theme among physical, chemical, and biological theorists.

Meanwhile, the colleagues working on the problem of artificial intellect have picked up the idea of interconnection between the probabilistic/statistical treatment and the hidden/latent variables. They are already productively working with it [18– 20].

Interestingly, outstanding British psychologist Spearman (1863–1945) pioneered the idea of the common causes behind observable correlations back in 1904 [21], that is, well before natural scientists could learn about this from Reichenbach. Importantly, that was Prof. Spearman who could suggest the formal mathematical method to deal with such common causes that could have formed the conceptual basis of the factor analysis of correlations [22].

Nowadays, the factor analysis of correlations is one of the standard methods of multivariate statistical analysis, well known in diverse fields of science and engineering [23]. It is existent in two primary forms: confirmatory factor analysis and exploratory factor analysis. Of these, it is just the exploratory factor analysis that is correspondent to revealing and studying the "trans-empirical" and thus hidden/latent common causes. Meanwhile, the natural sciences slowly started profiting from the employment of such methods as well [24–31].

To our minds, a productive outlook for further development in natural sciences might be a more detailed acquaintance with the modern multivariate statistical analysis. Such an approach—along with using Buddhist methodology consequently and systematically—might help access and conclusively explore the transempirical [6] levels of physical reality.

An attentive reader might well exclaim here that all the above deliberations sound like an apparent stretch, for the actual relationship between thermodynamics and the factor analytical approach ought to be far-fetched. Is the latter approach indeed the right one to investigate entropy-enthalpy compensation?

A careful analysis of the relevant literature brings a definite answer. The colleagues performing a detailed analysis of the concepts inherent in the factor analysis of pairwise correlations were working not only in Great Britain and the United States but also in German and Austrian-Hungarian Empires from the middle of the nineteenth to the starting of the twentieth century. In the first line, we have to refer here to the works by Fechner (1801–1887), a German physicist, philosopher, and experimental psychologist. In Appendix 2, we have already considered suggestions by Dr. Ameline as to the intrinsic interrelationship between Prof. Fechner's psychophysical theory and the "different thermodynamics" (energetics). Next we dwell on the latter topic.

Prof. Fechner was teaching physics at the University of Leipzig but had to quit this post due to a severe workplace accident that practically cost him his eyesight. Nonetheless, he was actively continuing his fruitful work in the field by thoroughly dealing with the physical and philosophical implications introduced by the atomistic structure of matter [32]. This point is of direct relevance to our present topic. Meanwhile, Fechner's further work was connected mostly to psychology. As a result, Prof. Fechner might be considered one of the real founders of a new physical discipline: psychophysics.

Interestingly, Prof. Fechner's train of thoughts could bring him to working out the numerical-mathematical (in effect a statistical) tool to analyze psychophysical systems described by multiple variables. His tool bears the German name *Kollektivmaßlehre* (multivariate statistics, to put it in modern terms). To our sincere regret, Prof. Fechner had time just to recognize the general applicability of such a tool and could start successfully applying it to demonstrate that the relationship between the amount of psychological sensation S and the physical intensity of a stimulus we may describe via the formula $S = k \cdot \ln(I)$, which psychologists call the Weber–Fechner law.

Remarkably, the Weber–Fechner's formula closely resembles the Boltzmann–Planck formula for entropy. This fact is of direct relevance to our present topic, as well.

Fechner's ideas as to the *Kollektivmaßlehre* have been thoroughly analyzed by a number of his German and Austrian colleagues, who were investigating the parallels between Fechner's ideas and the work by his British colleagues Spearman and Pearson (1857–1936) (the well-known authors and proponents of the generalized factor analytical approach) [33–38].

The apparent parallel between the mathematical expressions for the Weber–Fechner law and the well-known Boltzmann–Planck expression for entropy requires separate consideration. Indeed, in Appendix 2 we presented suggestions by a French alienist

Dr. Ameline as to considering the Weber–Fechner law in the context of the Carnot principle. Remarkably, Prof. Fechner's followers all around the world could recognize with time that the logarithmic formula in question might only be a valid approximation for some particular cases. In contrast, the relevant general approach ought to have somewhat different mathematical expressions. Of immediate interest to our present topic is the result by an outstanding German psychologist Lysinski (1889–1982), who suggested the corrected expression for the Weber–Fechner law in the following handy form [39]: $S = \frac{\ln(1+a\cdot I)}{\ln(b)}$, where a and b are constants to be determined by fitting this formula to pertinent experimental data.

We know now from Ref. [1] of Chapter 1 that Prof. Lysinski's expression closely resembles Linhart's expression for entropy, and we may formally mathematically derive it using Linhart's unique approach. Moreover, the foundations of Weber–Fechner's law might also be analyzed in terms of energetics, which is the actual conceptual root of thermodynamics. A French psychiatrist, a physicist by his bachelor diploma, Ameline, could have carried out this work and published his findings in French. We have presented in Appendix 2 the English translations of his two publications on the theme.

To sum up, our suggestion to apply exploratory factor analysis in connection with the concept of entropy-enthalpy compensation (EEC) in (bio)-physical chemistry has quite a severe basis.

Should we completely discard or thoroughly modify equilibrium thermodynamics, statistical, and quantum mechanics?

The problem is not the main topic of the present communication, but we have approached this poser rather carefully, for we employ here the good old, tried-and-true methods of statistical mechanics to reach conclusions beyond the conventional equilibrium thermodynamics. Here we just try to look at the possible directions of answering it.

To our mind, equilibrium thermodynamics in its conventional form has to be discarded and substituted by energetics, for it is the latter that fills the entropy notion with a clear and distinct fundamental physical sense (more details on this theme in Ref. [1] of Chapter 1). Above, we have noticed that statistical mechanics tries somehow to cope with the notion of probability, for this looks like the only way to clarify the entropy notion of the equilibrium thermodynamics. On the other hand, the adepts of this line of thoughts are merely considering the notorious "second basic law" of thermodynamics a statistical rule and using the mathematical form of the latter to build up the theory of quantum mechanics. Conceptually, this project was to fail in its very essence, although the actual result has been quite the opposite. It is the work by Linhart (see Ref. [1] of Chapter 1) that helps at least start thinking over the reasons for such a truly unexpected result. Nevertheless, the probabilistic/statistical motives intrinsically unite statistical and quantum mechanics. Has anyone upon Earth tried analyzing this interconnection in detail? If so, what have been the results?

The answer is affirmative and leading to the works by Prof. Simon Ratnowsky (1884–1945).

Figure A3.1 Ratnowsky, approximately in 1914; his portrayal is from https://de.wikipedia.org/wiki/Simon_Ratnowsky.

Here is the full list of his publications:

- (1910) Guye, C.-E., Ratnowsky, S., *Sur la variation de l'inertie de l'électron en fonction de la vitesse dans les rayons cathodiques et sur le principe de relativité*, Comptes rendus hebdomadaires des séances de l'académie des sciences, v. 150, pp. 326–329

- (1911) Guye, C.-E., Ratnowsky, S., *Détermination expérimentale de la variation d'inertie des corpuscules cathodiques en fonction de la vitesse*, Archives des sciences physiques et naturelles, Période 4, tome trente et unième, pp. 293–321
- (1912) Ratnowsky, S., Die Zustandsgleichung einatomiger fester Körper und die Quantentheorie, *Annalen der Physik* 343, pp. 637–648
- (1913) Ratnowsky, S., *Experimenteller Nachweis der Existenz fertiger elektrischer Dipole in flüssigen Dielektricis*, Friedrich Vieweg & Sohn, Braunschweig
- (1913) Ratnowsky, S., *Theorie der festen* Körper, Verhandlungen der Deutschen Physikalischen Gesellschaft 15, pp. 75–91
- (1913) Ratnowsky, S., *Experimenteller Nachweis der Existenz fertiger elektrischer Dipole in flüssigen Dielektricis,* Verhandlungen der Deutschen Physikalischen Gesellschaft 15, pp. 495–517
- (1914) Ratnowsky, S., *Zur Theorie des Schmelzvorganges*, Verhandlungen der Deutschen Physikalischen Gesellschaft 16, pp. 1033–1042
- (1915) Ratnowsky, S., *Die Ableitung der Planck-Einsteinschen Energieformel ohne Zuhilfenahme der Quantenhypothese*, Verhandlungen der Deutschen Physikalischen Gesellschaft 17, pp. 64–68
- (1916) Ratnowsky, S., *Die Entropiegleichung fester Körper und gase und das universelle Wirkungsquantum*, Verhandlungen der Deutschen Physikalischen Gesellschaft 18, pp. 263–277
- (1916) Ratnowsky, S., *L'entropie des solides et des gaz et le quantum universel d'action*, Archives des sciences physiques et naturelles, Société de physique et d'histoire naturelle de Genève, Période 4, tome quarante et unième, p. 502
- (1916) Ratnowsky, S., Über die absolute Entropie starrer rotierender Gebilde, Mitteilungen der Physikalischen Gesellschaft zu Zürich, Nummer 18, pp. 126–133
- (1918) Ratnowsky, S., *Zur Theorie molekularer und inneratomarer Vorgänge, Annalen der Physik* 361, pp. 529–568
- (1921) Guye, C.-E., Ratnowsky, S., Lavanchy, C., *Vérification expérimentale de la formule de Lorentz-Einstein faite au laboratoire de physique de l'Université de Genève*, Mémoires de la Société de Physique

et d'Histoire Naturelle de Genève 39, fasc. 6 (1921), pp. 274–364 + pl. 4–6

Why his work may be interesting to us

First of all, he was participating in experimentally checking the validity of Einstein's relativity theory. There were severe problems with the latter—see, for example, the work by Proctor [40]. Prof. Guye and his apprentices could successfully reject the negative note the latter colleague had given to Einstein's theory. Remarkably, most of Ratnowsky's works deal with theoretical physics and, specifically, with a thorough analysis of the physical foundations of statistical and quantum physics. What was he looking for, and what could he find? The answers would be of immediate relevance to our present theme.

To start answering both posers, we present here the French abstract to his whole work published in *Archives des sciences physiques et naturelles, Genève*, in 1916, followed immediately by our English translation of it.

S. Ratnowsky (Zürich): *L'entropie des solides et des gaz et le quantum universel d'action*

L'auteur pose l'hypothèse suivante : « Un système matériel, même en l'absence de toute énergie calorifique (donc au zéro absolu), représente un réservoir d'énergie auquel il peut puiser comme à une source étrangère » et montre qu'elle conduit, avec l'aide de la Mécanique statistique de Gibbs, à des expressions formellement identiques à celles que donne la théorie des quanta pour l'énergie et l'entropie.

De plus, il en découle avec nécessité que le rapport entre la fréquence ν *et la valeur limite de l'énergie propre* ε_0 *d'une liberté, caractéristique du système donné, est une constante universelle* ($\varepsilon_0 = h\nu$)*, ce qui établit la concordance avec la théorie des quanta.*

De même, l'expression de l'entropie des gaz telle qu'elle a été donnée par Tétrode, à partir de l'entropie des solides de Planck, en découle pour ainsi dire de soi-même, sans avoir besoin de faire appel au quanta. En résumé, l'auteur, par une voie tout à fait différente, parvient aux résultats connus quant aux formules donnant la tension de vapeur des solides et la constante d'entropie des gaz.

Here is our English translation of the above:

S. Ratnowsky (Zurich): The entropy of solids and gases and the universal quantum of action

The author posits the following hypothesis: "Even in the absence of any heat energy (therefore, at the absolute zero of temperature), any material system does represent a reservoir of energy, which we may think of as an external energy source," and with the help of Gibbs statistical mechanics, he shows that this leads to expressions formally identical to those given by quantum theory for energy and entropy.

From the above, it necessarily follows that the ratio between the frequency ν and the limit value of the inherent energy ε_0 of an arbitrary degree of freedom, characteristic of the given system, ought to be a universal constant h (i.e., $\varepsilon_0 = h\nu$), which establishes accordance with quantum theory.

Similarly, the entropy expression for gases as given by Tetrode, starting from the entropy of the solids by Planck, derives from it, so to speak by itself, that is, without needing to appeal to quantum theory. In summary, the author follows an entirely different route to reach the known results as to the formulas giving the vapor pressure of the solids and the entropy constant of the gases.

[**Our immediate comment:** The above means that Prof. Ratnowsky could formulate the actual primary conceptual point of departure for the whole quantum theory, as well as show that all the basic formulas of the latter are mathematically derivable from this sole point in a straightforward manner, without any other assumption. To derive these formulas, he used Gibbs' approach to statistical mechanics. Noteworthy to this end, Prof. Gibbs had his lifetime to treat statistically independent systems in full detail. But he had no more time to study correlations.]

To estimate the actual significance of Ratnowsky's result, let us recall that assuming "the absence of any heat energy in the system under study and therefore placing the latter at the zero of absolute temperature" is expressly not physical, for zero absolute temperature is never reachable in realistic processes.

The poser is then: Why statistical and quantum mechanics being in effect some handy chapters of the probability theory do nonetheless adequately describe realistic systems?

Let us first try to clarify who Ratnowsky was in effect, for he was and still remains widely unknown, irrespectively of his provoking results.

Fortunately, there is a recent monograph by Meyenn about Erwin Schrödinger's time at the Physical Institute of the University of Zürich [41], where he became a tenured professor for theoretical physics well before his revolutionary breakthrough and exactly when Ratnowsky was also actively working there. There were contacts between them both, and Prof. Meyenn devotes several pages to describe the actual nature and modalities of their encounter. The next paragraph is our translation into English of the biographical note in Ref. [41]:

> Simon Ratnowsky (1884–1945) was born in Rostov-on-Don and moved to Bern in 1903 with his parents. There he had received his first scientific education in his parents' house. After studying in Bern, Nancy, and Geneva, he received his doctorate in 1910 with a study on the inertia of deflected cathode rays, under the guidance of Charles-Eugène Guye (1866–1942). He then continued his studies in Zürich and dealt there—in addition to experimental work in the physical laboratory of the Swiss Federal Institute of Technology—with the equation of state of solids. In 1912 he became Alfred Kleiner's (1849–1916) assistant at the Physical Institute of the University of Zürich. A year later, on the recommendation of Max von Laue, who gained a very high opinion of Ratnowsky's abilities, he applied for the *venia legendi* (in Germany, Austria, and Switzerland, the official right to hold university lectures). Such a position had finally been granted to him at the beginning of 1916, though it required periodical renewal.

Our research shows that Ratnowsky was indeed an offspring of a wealthy Jewish mosaic family living and working in the Russian city of Rostov-on-Don from the end of the nineteenth to the beginning of the twentieth century. Meanwhile, partial escape of that family in the direction of Switzerland was, in fact, the best decision. From 1905 to 1920, the Don area, together with the rest of the Russian empire, was thoroughly obsessed with revolutions and the bloody civil war, which among other tragic events were characterized by wild anti-Semitic pogroms. After that, the Leninists and Stalinists brought overall terror, while during World War II, the area of the Don capital became the scene of fierce fighting four times; as a result, German troops could occupy the city twice (everywhere German troops performed systematic massacre of Jewish inhabitants).

Nevertheless, between 1910 and 1924, the University of Zürich had difficulties of quite a different kind: From 1909 to 1911, Einstein was occupying the chair for theoretical physics and from 1912 to 1914 Laue. Both peers had left Zürich so that there was a pause in teaching for several years. Edgar Meyer (1879–1960), who was teaching experimental physics and held the post of physical institute director from 1916 to 1949, tried to solve the problem either by filling the vacant associate professorship for theoretical physics anew or by performing a more comprehensive refurbishment of that professorship. With no prospect of regaining Einstein from Berlin, several younger physicists were considered, of whom four were eventually shortlisted in November 1919: Paul Scherrer (1890–1969), Paul Sophus Epstein (1883–1966), Franz Tank (1890–1981), and Simon Ratnowsky (1884–1945).

Among them, the Privatdozent and then assistant professor in Göttingen, Scherrer, who was initially on the list, was to be given priority as a native Swiss. Peter Debye and director of the Göttingen Observatory Johannes Hartmann had agreed that he was a very talented young man whose previous scientific work entitled to high hopes and who proved to be an excellent lecturer at the same time. However, after Scherrer accepted a call to ETH Zürich, his candidacy had to be dropped.

Tank was also finally dropped, but due to insufficient performance, though being a native of Zürich. He had also moved to ETH after that and was successfully working and teaching in the electrical engineering field.

Epstein had excellent references. Born in Warsaw in 1883, he grew up in Russia and studied physics with Peter Lebedev in Moscow. In 1911, he came to Arnold Sommerfeld in Munich, where he earned his doctorate with a theoretical contribution to the theory of diffraction on flat screens in 1914. Subsequently, he wanted to habilitate in Zürich to gain admission as a lecturer at the university. As a result, Epstein's teaching qualifications for the advanced lectures in theoretical physics were beyond doubt, but the university circles could not yet furnish the proof that Epstein would also be able to hold the introductory lectures equally well.

Consequently, he could get his *venia legendi* in the tenure-track form, like Ratnowsky. Still, in the summer of 1921, when the Zürich Department of Education was discussing the appointment of a theoretical physicist, the objection had already been raised against Epstein that "despite his first-class scientific ability, he was considered too repugnant—to the extent that working together with students and colleagues would hardly yield a fruitful result." Finally, he moved to the United States (Californian Institute of Technology) and could find there his place "Amid the Stars" (Linhart could meet him over there; see Ref. [1] of Chapter 1).

Ratnowsky's performance and subsequent fate were but somewhat different. In an application dated March 16, 1920, which the then dean Alfred Wolfer (1854–1931) directed to the Education Department of the Canton of Zürich to staff the extraordinary professorship for theoretical physics, still vacant at the time since Laue's departure, he characterized Ratnowsky in the following words:

> Ratnowsky is scientifically less highly rated, although he is a keen thinker, who goes his own—though sometimes not truly happy—ways. Einstein calls him the right researcher. In terms of his teaching, he is also recommended by Einstein as one of the three candidates, Scherrer, Tank, and Ratnowsky, who could be considered alone.
>
> Presently, Ratnowsky has somewhat depressing living conditions, which forced him to take over sideway jobs—lessons at the local canton school and in private schools—which inevitably had to limit his scientific activity.

To this end, we must mention that Ratnowsky was the father of two children, Nora Anderegg-Ratnowsky (1908–2000), outstanding Swiss painter and decorator, and Raoul Ratnowsky (1912–1999), outstanding Swiss statuary.

Meanwhile, at the beginning of 1915, Ratnowsky had published a paper in which he tried to derive the Planck–Einstein energy formula by introducing a different assumption about the potential energy of a solid without the help of the quantum hypothesis. The report of March 2, 1920, concerning Ratnowsky among the four candidates in the list for the appointment of associate professor in theoretical physics, reads as follows: "[T]his work should be considered a failure." Moreover, this same year, a paper came out in *Zeitschrift für Physik* [42] written by Herweg (1879–1936), a professor in the University of Halle and then in the Technical University of Hannover, who was a specialist in high-frequency physics, telegraphy, and telephony but never in quantum theory or physical chemistry. In that paper, Herweg tried to persuade the readership that Ratnowsky's 1913 paper on preformed dipoles in dielectric fluids ought to be pure nonsense. A possible explanation of that affront could be that Herweg, being already an emeritus, has nonetheless signed the notorious *Bekenntnis der Deutschen Professoren zu Adolf Hitler.*

Nevertheless, despite his successful teaching (from the summer semester of 1919 to the winter semester of 1924–1925, the deanery has compiled the list of attendants at his lectures) and numerous publications on the molecular theory of solids and liquids, Ratnowsky had unsuccessfully sought a permanent position with the Education Directorate of the Canton of Zürich. After he was given a teaching assignment for the last time in the summer semester of 1922, his situation became more and more acute, so now Schrödinger had to deal with the matter. 1922 was the year he joined the Physical Institute of the Zürich University as a tenured lecturer in theoretical physics. He was stubbornly trying to help Ratnowsky. While Schrödinger read about the theory of light as well as about the atomic structure and the periodic system of elements in the summer semester of 1923, Ratnowsky held lectures on atomic structure and spectral lines. The motives and details of Schrödinger's struggle for Ratnowsky are truly remarkable. Professor Meyenn has provided us with the correspondence between Schrödinger and Sommerfeld concerning this case. Here we would like to publish our English translations of the relevant paragraphs in Ref. [41], for they appear to be genuinely insightful regarding our present discussion:

Schrödinger to Sommerfeld

Zürich, November 10, 1924

Dear Esteemed Professor!

I intended to introduce Mr. Ratnowsky for the coming semester to a small teaching position in theoretical physics. Edgar Meyer, with whom I spoke today, tells me now that this would be pointless given the "devastating" judgments you and Planck[a] had delivered about Ratnowsky's scientific qualification when you were occasionally questioned about the candidacies for the professorship post here in 1921 (or earlier).

[a]{Karl von Meyenn: and Langevin?}.

Your letters are admittedly filed in the records of that time, although it is somewhat awkward for me to demand their inspection—for everything ended up with my appointment, and thus the records do indeed contain judgments about me. Besides, it does make a difference whether it is an appointment of a tenured professor or solely granting a small teaching assignment—which anyhow sounds unacceptable for Meyer, who is a decided opponent in this case.

Would you be so kind as to write me a few lines in the affair, which I may also show? I do not ask for a "corrected" judgment—I know that scientifically, you do not value Ratnowsky very highly, and precisely so do I. But the "devastating" verdict Meyer attributes to you ought to be a mystification, I think. They say you wrote "A man who wants to do quantum theory without quanta, there is but not a jot of truth in this . . ." or something like that. As I said, if you think with a clear conscience that Ratnowsky is a man who does not deserve a teaching job, then I am the last one who will even lift a finger. Nevertheless, I have the impression that this is an intrigue I cannot quite wrap my head around, but I am not even sure if—despite Meyer's Judaism—anti-Semitism is not involved. However, it is probably something else—though I do not know what exactly. Therefore, I beg your overall judgment for me not to depend on an unrelated gossip all around this.

Now something very unpretentious: the meeting takes place on the evening of Thursday, the 13th.

Only today—just today, and only by accident!—I have learned that one has to have submitted the teaching assignments already for this meeting. Of course, I will make a noise. However, if you could make it possible to answer me right away so that maybe the letter will be here on the 13th, I would be delighted. Please do not be angry with me. Otherwise, if Meyer opposes me and the faculty refuses, I would probably insist on a separate vote and deliver your (and Planck's) letters.

Nevertheless, if you find it unpleasant that I should give the letter directly from my hand, I shall not do so, but just list the relevant passage you kindly label for me.

Besides, I have to mention two points as follows:

1. That at the time of the "interregnum" Ratnowsky together with Epstein represented theoretical physics in the form of teaching assignments

2. That he is a titular professor; nevertheless, I guess that there does not seem to be the slightest reason for the local faculty to exclude him from lecturing. That is one of the reasons why I believe in an intrigue.

Of course, both these circumstances are not decisive for your judgment—but all this is decisive, at least for me only. . . .

[**Our immediate comment:** It is important to explain here the sense of the term "titular professor." In German-speaking countries, this means the bearer "is entitled solely to bear the title of professor without associating the latter with a corresponding position." In particular, in Switzerland, this means that "titular professors may be lecturers from inside or outside the university. Lecturers from within the university would get their salaries following their actual employment agreements. For the lecturers from outside, the teaching obligations are, in principle, offered free of charge—if necessary, just a lump sum for travel expenses, etc., might be paid. Such an "amortization" should be done here solely if the said person bears the professor title in the respective professional environment." Ratnowsky was the latter case; he mostly remained outside the academic milieu (just a school teacher).]

Meanwhile, Sommerfeld had in a timely fashion fulfilled Schrödinger's request. Here is our English translation of his response (München, DM, Archive NL 89, 025):

Sommerfeld to Schrödinger

Munich, November 12, 1924

Dear Schrödinger,

Due to Kleiner's trust in me,[b] I repeatedly have a connection to your faculty affairs. The first time I could say a strong word in favor of Einstein, and that was, of course, the most unambiguous story. Later, with good conscience and suasion, I stood out for Epstein against Ratnowsky. I recall that at that time, a piece of work by Ratnowsky was being discussed, in which he wanted to make the quanta dispensable. In my report, I wrote that I would be ready to discuss this work in private correspondence with Ratnowsky. Nevertheless, this did not happen.

If a teaching assignment is in question now, not the professorship, then the situation is, of course, different. Who has not been a

[b]{Alfred Kleiner (1849–1916)}.

sinner among all of us in the various stages of the quantum theory development?

I certainly do not feel free of guilt and do not want to pick up the stone against R. I think I remember that Laue, who knows R. better than me, judges him much better than I did. Therefore, if it were desired, for personal or educational reasons, to give Mr. R. a teaching assignment, it would be a pity if my previous remarks in other contexts were still in the way.

What I am saying here is absolutely unimportant. I leave it to you to make use of that which seems right to you. . . .

Then, Sommerfeld also got a due response from Schrödinger:

Schrödinger to Sommerfeld

Zürich, November 19, 1924

Dear Esteemed Professor!

Thank you so much for your immediate and kind response to my request. I subsequently regretted having pushed you this way. The soil but appears to be so unfavorable that I have decided to wait until the winter semester, when I could better represent the teaching needs.

I would like Ratnowsky to be reading the mechanics course interchangeably with me so that I could come to a coherent course on theoretical physics. There is a commission set up for this problem, which will start working soon. Hence, your letter, along with a similar one from Planck, will be of immense value to me, for they will undoubtedly complement the letters of those days, or even without them, containing but serious objections to Ratnowsky based on them. If the result would nonetheless come out unwell—well, I can't beat it, but I could be in danger of being killed by an uncontrollable gossip with the references to you and Planck—so, the whole story could be that bad for me myself as well. . . .

We see that Ratnowsky's life was not easy as a result. No wonder that in total he could spend solely 61 years among us. Remarkable is also an obituary on Ratnowsky's behalf, which was posted by Meyer in the official yearly chronicle at the University of Zürich, for 1944–1945. Here we publish our English translation of that obituary.

Professor Simon Ratnowsky: September 8, 1884, to February 6, 1945

Simon Ratnowsky was born on September 8, 1884, in the Russian Empire, city of Rostov-on-Don. Until his eighteenth year, he enjoyed private lessons in his parents' house, and then he came to Switzerland, where he passed his Matura exam in Zürich. In 1903 he went to Bern to undertake philosophical studies. In 1907 he enrolled at the Faculty of Sciences in Nancy, but in the same year, he returned to Switzerland and finished his university studies in Geneva in 1910 by obtaining there his diploma of Docteur des Sciences Physiques.

In the Geneva laboratory, Professor Charles-Eugène Guye had inspired Ratnowsky to study one of the most fundamental questions of physics, namely to experimentally test the dependence of the mass of a body on its velocity, as suggested by A. Einstein. Ratnowsky carried out the experiments with cathode rays by determining the specific charge of the cathode ray particles, that is, the ratio of their charge to their mass, at different velocities. The result of the experiments could confirm Einstein's theory. It is interesting that in the same year (1910) in America, a work by C. A. Proctor on the same topic appeared but came to an opposite (false) result. The work of Ratnowsky, which at that time still experienced truly great experimental difficulties, was long considered to be the best and most accurate investigation in this field—until the later much higher experimental effort using newer methods and apparatus to exceed the accuracy of the measurements. Nevertheless, Ratnowsky's result has remained an undoubtedly successful step.

In autumn 1910, Ratnowsky came to Zürich, where he worked in the laboratory of the ETH, in addition to his further studies in theoretical physics.

In 1912 (this year, he also acquired Swiss citizenship), he became an assistant to Professor A. Kleiner at the Physical Institute of the University of Zurich. Here he began an experimental work "To prove the existence of preformed electric dipoles in liquid dielectrics," and with this work, he completed his habilitation in 1913 as a lecturer in physics at the University of Zürich. The supreme authorities of the Faculty of Arts and Humanities 11 have approved the habilitation in their report characterizing that work as "throughout independent and of great scientific value."

This concludes Ratnowsky's experimental work. Although there are only two studies, they both are well above the middle level; they both have made their undoubted contribution to the development of physics.

How did it happen that Ratnowsky, who had started so successfully, decided to leave the field of experimental physics? The reason was that our world is inadequate, that in addition to the mind, the body also needs food. The merciless barrier of materiality diverted Ratnowsky's development. He was forced to look for more rewarding earnings. Moreover, here, his great pedagogical talent came to his aid. In addition to his acting as Privatdozent, he went to secondary school teaching, while working in the field of theoretical physics on the side.

Already in 1912, he had published the paper entitled "The equation of state of a monatomic solid and the quantum theory," followed in 1913 by the paper entitled "The theory of the solid-state." 1914 onward, he was working only theoretically.

Ratnowsky's work dealt with a wide variety of problems: the melting processes, the Planck–Einstein energy formula, the entropy of solids and gases, and the theory of molecular and interatomic processes. Also, two dissertations were carried out at his suggestion and under his direction: one by Jan von Weyssenhoff (who later on became professor of theoretical physics in Krakow), entitled "The application of quantum theory to rotating structures and the theory of paramagnetism," and one by Mrs. S. Rotszajn, entitled "The application of Planck's extension of the quantum hypothesis to two-degree-of-freedom rotating devices in a directional field."

As this ought to be the broad overview of Ratnowsky's achievements in the field of science, then we must also mention his merits in the field of teaching at the University of Zürich.

As in the time between Max von Laue's departure and Erwin Schrödinger's arrival, the Chair of Theoretical Physics was vacant; Ratnowsky gladly accepted the task to represent theoretical physics at our university. In recognition of this activity and his scientific merits, he received in 1921 the appointment as a titular professor.

Truly versatile were the themes he lectured on.

To enumerate them would be going too far here. We must mention here only that Ratnowsky also lectured on the common area of physics and philosophy.

After he had previously taught at the Kanton School Zürich and the Institute of Tschulok, in 1926, the Regierungsrat elected him to be a professor of physics and mathematics at the Kanton School Winterthur. 1930 onward, he regularly held lectures on the methods and modalities of physical education at our university.

On a large scale, that was Ratnowsky's scientific career and field of activity. Here is not the best place to describe his vibrant personality, both his intellect and his heart-and-soul qualities. He was, what we may state after Goethe, a harmoniously developed man. His unique charm was that he stood at the height of Western European culture without having lost the vast and deeply emotional world of his Russian homeland. All who encountered him appreciated him; those who knew him were his friends.

He has left a good memory; many mourn deeply after him.

To sum up, the above story describes, in fact, an intentional, systematic effort to make use and then subsequently, gradually suffocate an undesired visitor whom one does not like but cannot merely throw away at once. As a result, Prof. Ratnowsky stopped publishing anything after 1918. His academic publications show that he was close to analyzing the statistical-mechanical picture of the EEC but could not pursue this research direction. His mentor, Guye, was trying to lend him at least a bit of support, by including him as a coauthor of the sequel to their earlier experimental work in connection with Einstein's theories. That took place in 1921 and was just a drop in the ocean. Still, Ratnowsky was looking into the future. He would have liked his apprentices to drive his trains of thought to further conceptual stations. Meanwhile, his efforts could not be 100% fruitful due to the following reasons:

- Weyssenhoff (1889–1972) could publish the text of his PhD thesis in a renowned journal [43]. The reaction to it was not quite inspiring, to the extent that he decided to switch to a wide variety of other themes having much more perspective and/or being much trendier (like relativity theory, for example).
- Rotszajn (1873–?) had most probably similar experiences after publishing materials of her PhD thesis [44], and she chose to quit the academic field in order to devote herself to her family, as a result.

There seems to be no more "driving force" to overcome "entropy"? Is this the ultimate end of the story?

Great posers, though they sound like rhetorical ones.

[**Our immediate comment:** Above is a story similar to the stories of Peter Boas Freuchen, in Denmark; Henrik Petrini, Nils Engelbrektsson, Karl Franzén, and David Enskog, in Sweden; Max

Bernhard Weinstein in Germany; George Augustus Linhart, Edgar Buckingham, and William Walker Strong in the United States; and many, many other (more or less) known colleagues all over the world (see Ref. [1] of Chapter 1). They were/are never "the research people in the accepted sense of the phrase." All of them were/are "more of the detective sort" according to the prophetic words of Prof. Gooch to Linhart's address (see Ref. [1] of Chapter 1). As long as such detectives continue to come to this world, scientific research across the world will never stop, although dwarfs may keep climbing onto the shoulders of master giants and rising to their dwarfish heights, in turn, shout to the world, "See how big I am!" Such are the performances that could lead to the ultimate stop of any scientific research and education.]

References

1. H. Barnum, J. Barrett, L. O. Clark, M. Leifer, R. Spekkens, N. Stepanik, A. Wilce, R. Wilke (2010). Entropy and information causality in general probabilistic theories, *New J. Phys.*, **12**, p. 033024.
2. J. S. Wicken (1981). Causal explanations in classical and statistical thermodynamics, *Philos. Sci.*, **48**, pp. 65–77.
3. H. Margenau (1953). Methodology of modern physics, *Philos. Sci.*, **2**, pp. 48–72.
4. H. Margenau (1953). Methodology of modern physics, *Philos. Sci.*, **2**, pp. 164–187.
5. D. C. Williams (1945). On the derivation of probabilities from frequencies, *Philos. Phenomenol. Res.*, **5**, pp. 449–484.
6. S. R. Bhatt (2019). *Quantum Reality and Theory of Śūnya*, Springer Nature Singapore.
7. D. Wallace (2012). *The Emergent Multiverse. Quantum Theory According to the Everett Interpretation*, Oxford University Press, Oxford, UK.
8. D. Wallace (2018). Lessons from realistic physics for the metaphysics of quantum theory, *Synthese*, doi: https://doi.org/10.1007/s11229-018-1706-y.
9. J. Horgan (2015). *End of Science. Facing the Limits of Knowledge in the Twilight of the Scientific Age*, Basic Books, Perseus Books Group, New York, USA.

10. M. Planck (1933). *Where Is Science Going*, George Allen & Unwin, London, UK.

11. M. Planck (1963). *The Philosophy of Physics*, W. W. Norton & Company, New York, USA.

12. S. Gimbel, A. Walz (2006). *Defending Einstein: Hans Reichenbach's Writings on Space, Time, and Motion*, Cambridge University Press, Cambridge, New York, Melbourne, Madrid, Cape Town, Singapore, São Paulo.

13. H. Reichenbach (1956). *The Direction of Time*, University of California Press, Los Angeles, USA.

14. W. C. Salmon (1980). Probabilistic causality, *Pac. Philos. Q.*, **61**, pp. 50–74.

15. F. S. Ellett, D. P. Ericson (1983). On Reichenbach's principle of the common cause, *Pac. Philos. Q.*, **64**, pp. 330–340.

16. H. Krips (1989). A propensity interpretation for quantum probabilities, *Philos. Q.*, **39**, pp. 308–333.

17. G. Hofer-Szabó, M. Rédei, L. E. Szabó (2013). *The Principle of Common Cause*, Cambridge University Press, Cambridge, New York, Melbourne, Madrid, Cape Town, Singapore, São Paulo.

18. J. Pearl (1988). *Probabilistic Reasoning in Intelligent Systems: Networks of Plausible Inference*, Morgan Kaufmann Publishers, San Mateo, California, USA.

19. L. E. Sucar, D. F. Gillies, D. A. Gillies (1993). Objective probabilities in expert systems, *Artif. Intell.*, **61**, pp. 187–208.

20. C.-K. Kwoh, D. F. Gillies (1995). Estimating the initial values of unobservable variables in probabilistic visual networks, in: *Computer Analysis of Images and Patterns (CAIP 1995), Lecture Notes in Computer Science*, V. Hlaváč, R. Šára (eds.), Springer-Verlag, Berlin Heidelberg, Germany, pp. 326–333.

21. C. Spearman (1904). The proof and measurement of association between two things, *Am. J. Psychol.*, **15**, pp. 72–101.

22. C. Spearman (1904). "General intelligence," objectively determined and measured, *Am. J. Psychol.*, **15**, pp. 201–292.

23. S. J. Press (2005). *Applied Multivariate Analysis: Using Bayesian and Frequentist Methods of Inference*, Dover Publications, Mineola, New York, USA.

24. R. A. Reyment, K. G. Jöreskog (1993). *Applied Factor Analysis in the Natural Sciences*, Cambridge University Press, Cambridge.

25. A. Amadei, A. B. M. Linssen, H. J. C. Berendsen (1993). Essential dynamics of proteins, *Proteins*, **17**, pp. 412–425.

26. F. Cordes, E. B. Starikov, W. Saenger (1995). Initial state of an enzymic reaction. Theoretical prediction of complex formation in the active site of RNase T1, *J. Am. Chem. Soc.*, **117**, pp. 10365–10372.

27. S. Hayward, N. Gõ (1995). Collective variable description of native protein dynamics, *Annu. Rev. Phys. Chem.*, **46**, pp. 223–250.

28. A. Kitao, N. Gõ (1999). Investigating protein dynamics in collective coordinate space, *Curr. Opin. Struct. Biol.*, **9**, pp. 164–169.

29. P. W. Pan, R. J. Dickson, H. L. Gordon, S. M. Rothstein, S. Tanaka (2005). Functionally relevant protein motions: extracting basin-specific collective coordinates from molecular dynamics trajectories, *J. Chem. Phys.*, **122**, p. 034904.

30. E. B. Starikov, M. A. Semenov, V. Ya. Maleev, A. I. Gasan (1991). Evidential study of correlated events in biochemistry: physicochemical mechanisms of nucleic acid hydration as revealed by factor analysis, *Biopolymers*, **31**, pp. 255–273.

31. E. B. Starikov, K. Braesicke, E.-W. Knapp, W. Saenger (2001). Negative solubility coefficient of methylated cyclodextrins in water: a theoretical study, *Chem. Phys. Lett.*, **336**, pp. 504–510.

32. G. T. Fechner (1864). Über die physikalische und philosophische Atomenlehre, *Zweite vermehrte Auflage*, Hermann Mendelssohn, Leipzig, Germany.

33. G. T. Fechner (1897). *Kollektivmaßlehre: Im Auftrage der Königlich sächsischen Gesellschaft der Wissenschaften herausgegeben von Gottlob Friedrich Lipps*, Verlag von Wilhelm Engelmann, Leipzig, Germany.

34. G. F. Lipps (1901). Theorie der Kollektivgegenstände, in: *Philosophische Studien, herausgegeben von Wilhelm Wundt, der Siebzehnte Band*, Seiten 78–184, 467–575.

35. G. F. Lipps (1902). *Die Theorie der Collectivgegenstände*, Verlag von Wilhelm Engelmann, Leipzig, Germany.

36. E. H. Bruns (1906). *Wahrscheinlichkeitsrechnung und Kollektivmasslehre*, Druck und Verlag von B. G. Teubner, Leipzig und Berlin, Germany.

37. E. Czuber (1907). *Die Kollektivmaßlehre, Mitteilungen über Gegenstände der Artillerie und Geniewesens*, zehntes Heft, Kaiserliche und königliche Hof-Buchdruckerei und Hof-Verlags-Buchhandlung Carl Fromme, Wien und Leipzig.

38. R. Greiner (1909). Über das Fehlersystem der Kollektivmaßlehre, *Z. Math. Phys.*, **57**, pp. 121–158, 225–260, 337–373.

39. E. Lysinski (1955). Berichtigung des Weber-Fechnerschen Gesetzes, *Psychologische Beiträge*, **2**, pp. 239–253.

40. C. A. Proctor (1910). The variation with velocity of e/m for cathode rays, *Phys. Rev. (Ser. I)*, **30**, pp. 53–62.

41. K. von Meyenn (2011). *Eine Entdeckung von ganz außerordentlicher Tragweite. Schrödingers Briefwechsel zur Wellenmechanik und zum Katzenparadoxon*, Springer, Heidelberg, Dordrecht, London, New York.

42. J. Herweg (1920). Die elektrischen Dipole in flüssigen Dielektricis, *Z. Phys.*, **3**, pp. 36–47.

43. J. von Weyssenhoff (1916). Die Anwendung der Quantentheorie auf rotierende Gebilde und die Theorie des Paramagnetismus, *Ann. Phys.*, **356**, pp. 285–326.

44. S. Rotszajn (1918). Die Anwendung der Planck'schen Erweiterung der Quantenhypothese auf rotierende Gebilde mit zwei Freiheitsgraden in einem Richtungsfelde, *Ann. Phys.*, **362**, pp. 81–123.

Chapter 2

Polynomial Exploratory Factor Analysis on Molecular Dynamics Trajectory of the Ras-GAP System: A Possible Theoretical Approach to Enzyme Engineering

E. B. Starikov,* Kohei Shimamura, Shota Matsunaga, and Shigenori Tanaka
Graduate School of System Informatics, Kobe University, 1-1 Rokkodai, Nada, Kobe 657-8501, Japan
starikow@port.kobe-u.ac.jp

A method has been proposed to interpret the molecular dynamical (MD) trajectories of macro- and supramolecular complexes widely relevant to nanoscience and nanotechnology. The approach is based upon the polynomial exploratory factor analysis of pairwise correlations between all the possible local and global

EBS: In memory of my beloved relatives who were victims of cancer:
My mom, Lilia Pavlovna Starikova (1939–1973)
My grandma Marina Naumovna Gorer (1918–1991)
My stepdaughter-in-law, Maria Alexandrovna Eremicheva (1993–2018)
My uncle Yurij Davidovich Volisson (1949–2020)

Entropy-Enthalpy Compensation: Finding a Methodological Common Denominator through Probability, Statistics, and Physics
Edited by Evgeni B. Starikov, Bengt Nordén, and Shigenori Tanaka

ISBN 978-981-4877-30-5 (Hardcover), 978-1-003-05625-6 (eBook)
www.jennystanford.com

dynamical variables of relevance to the functioning of the macro- or supramolecule under study. The microscopic modalities of the enzymatic mechanisms might readily be explored this way. The application of the method is illustrated by processing the MD trajectory of the hydrated electroneutral Ras-GAP-GDP-Pi complex.

2.1 Introduction

Recently, protein engineering—especially enzyme engineering—has become the field at the very forefront of modern bioinspired nanotechnology. Though mostly a valuable collection of sophisticated experimental approaches [1], it does acquire pertinent theoretical/computational branches little by little [2].

Of special interest in the latter sense is thermodynamics. Meanwhile, the trends of using this truly powerful instrument do not seem quite encouraging, for even the works to earn the master's degree at highly esteemed scientific centers (see, e.g., Ref. [3]) still demand the "theoretical foundations" in the form of purely mathematical games around the entropy formulae

$$S = k \cdot \ln(W) \text{ and/or } S(X) = \sum_{x \in \aleph} w_X(x) \cdot \ln(w_X(x)).$$

In fact, these formulae are known to be physically seminal, whereas the actual physical sense of both W (or w) and S in them seems to be widely unclear, with the former one being advertised as "Her Majesty Probability" (and only the Lord Almighty knows the probability *of what scilicet*) and the latter one as "Her Majesty Information" (and only the Lord Almighty knows the information *about what scilicet*). This mystic "information" may sometimes be advertised as "the entropy," and everybody must learn by rote—it is just this formula that correctly describes "the very thermodynamic entropy"; this looks but like a definite conceptual blind corner.

Nonetheless, the story is not that hopeless. Several colleagues were successfully working on clarifying the above-mentioned puzzles [4]. See Chapter 1 for a detailed methodological analysis of the situation. This encourages consistent rethinking and reshaping of the interpretation schemes of the relevant experimental and/or computer-aided-simulation results.

Here we suggest a method to systematically and thoroughly interpret trajectories of (bio)macromolecules, together with their full hydration-counterion sheaths, in the all-atomic representation. Such trajectories are the standard outcomes of the conventional computer-aided molecular dynamics (MD) simulations.

We consider discrete time points of the simulated trajectory as delivering the snapshots depicting the details of the relevant structural changes. Therefore, we might treat the collection of such snapshots as a statistical data set. As we deal here with the vast number of time-dependent variables, it is throughout possible to apply the multidimensional statistical approaches. This is just the topic of the work at hand. Bearing in mind the discussion in Chapter 1, we suggest that using the exploratory factor analysis (EFA) should help us reveal the modalities of the macromolecular functional dynamics hidden behind the chaotic thermal motion.

Finally, we illustrate the application of this method by processing a full-atomic MD trajectory of an enzymatic system. Specifically, here we consider the modalities of structure-function relationships for one of the GTPases in complex with its activation protein, which is of immense importance for cancer research.

We have taken the known structure of the Ras-GTPase-activating protein (GAP) complex with the inhibitor AlF_3 [5] as the initial point for our simulations. Figure 2.1 shows this structure. To model the final stage of the catalyzed reaction under study, we substitute the AlF_3 residue with the phosphate trianion (Pi) and construct this way the reaction product guanosine diphosphate (GDP)+Pi buried in the enzyme's active site.

Of interest is to try monitoring the dynamical fate of the reaction product (the result we expect ought to be the excision of the Pi from the enzyme's active site).

The small GTPases of the Ras type (H-, K-, and N-Ras) are the products of the *ras* proto-oncogenes. In the cells, they function as guanine nucleotide-dependent molecular switches regulating cell growth, development, and apoptosis via cyclic transitions between the guanosine triphosphate (GTP)-bound (active, Ras-GTP) and GDP-bound (inactive, Ras-GDP) forms [6–14]. The active Ras transmits the biological signals in the intracellular network to the downstream effectors. The Ras-type proteins are frequently activated much too much in a wide variety of human cancers, which ought to render them some of the most promising targets for anticancer drug development.

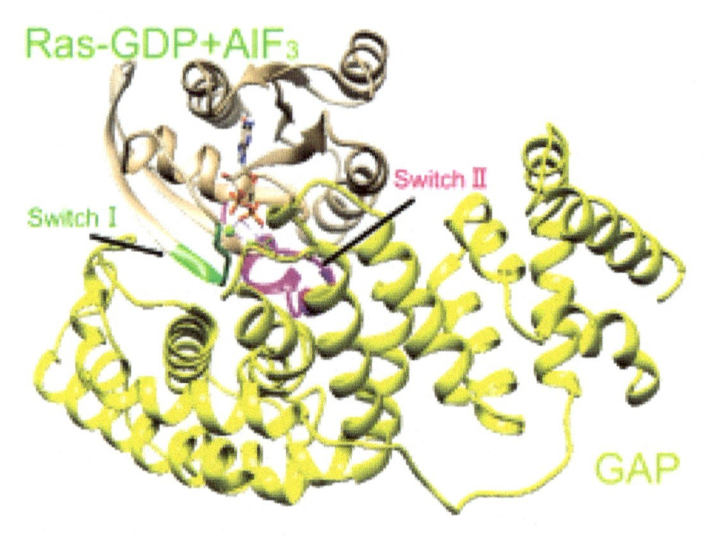

Figure 2.1 Crystal structure of Ras-GAP protein complex (PDB entry: 1WQ1 [5]). GAP is shown in yellow and the Switch I and Switch II regions of Ras in green and magenta, respectively. GDP and AlF_3 are shown by the stick model, whereas Mg^{2+} is shown by a green sphere.

There are two kinds of proteins that regulate the state transitions between GTP-bound and GDP-bound forms. GAP [5, 15] accelerates the intrinsic GTP-hydrolyzing activity of Ras, thereby making Ras inactive. On the other hand, guanine nucleotide exchange factor catalyzes the exchange of GDP for GTP. However, it is not well understood how Ras-GTP transforms to Ras-GDP after GAP-mediated hydrolysis of GTP. To develop molecular-targeted drugs, it is important to analyze the state transition mechanism of Ras protein. Therefore, we performed MD simulations on the structures of the Ras-GTP/GAP complex and the Ras-GDP+Pi/GAP complex to analyze structural changes of Ras-GAP before and after the hydrolysis. The work is still underway, and we shall report the results in due course.

Of immense importance for any rational design activity is to try revealing the detailed picture of the allosteric effects during the enzymatic reactions under study. We could show that our approach is throughout capable of depriving the ubiquitous chaotic thermal motion of its top secrets and delivering the desired picture. To go in for the allosteric effects, we must compare functional dynamics for the starting and final stages of the reaction involved.

This clearly demonstrates the practical value of our approach. Of considerable interest would also be to apply it to tackle tricky posers in the fields of pharmacy and materials science.

2.2 Results and Discussion

2.2.1 Linear Exploratory Factor Analysis Results

Linear factor analysis can reveal up to seven latent factors explaining the pairwise correlations between all the dynamical variables. To assign some physical sense to the latter latent factors, it is wise to present the factor loadings (the correlation coefficients between the variables and the factors) in a kind of 2D "phase diagrams," where the *x* and *y* axes correspond to the factors of interest.

Figure 2.2 presents such a "phase diagram"—a graph depicting the positions of all the 3000 variables under study with respect to Factors 1 and 2. Interestingly, the graphs corresponding to the analogous relationships between any other couple among the latent Factors 1–7 are identical to that in Fig. 2.2.

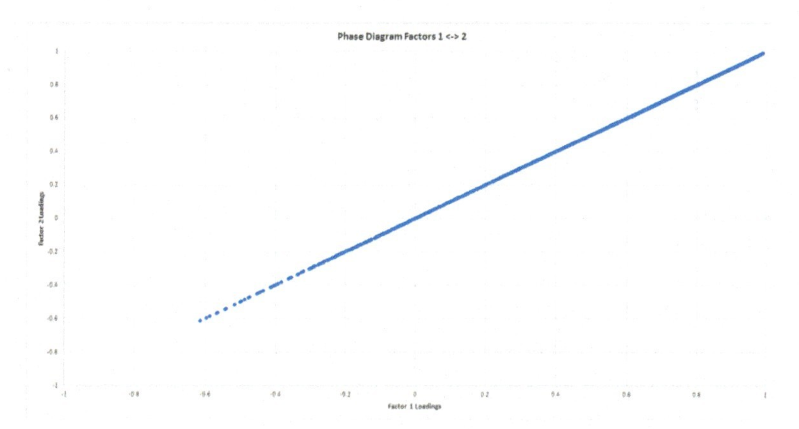

Figure 2.2 The disposition of the 3000 dynamical variables in the "factor phase diagram": The *x* axis contains the Factor 1 loadings and the *y* axis the Factor 2 loadings.

Usually, diagrams like that in Fig. 2.2 are used to interpret findings, in that the physical sense might be assigned to factors by collecting all the loadings near the "phase axes." This means that we are looking for collections of points with coordinates $(\sim 0; y)$ and $(x; \sim 0)$. The former may help assign physical sense to the factor corresponding to the y axis, whereas the latter helps in assigning physical sense to the factor corresponding to the x axis.

The standard factor analysis procedure performs orthogonal rotation of the eigenvectors of the pairwise correlation matrix with the aim to *maximize* the total number of "phase" points with the coordinates $(\sim 0; y)$ and $(x; \sim 0)$.

Such a result has been long well known in factor analysis as the "simple structure of the factor loading matrix," and it is arriving at the latter that represents the actual prerequisite of the meaningful interpretation (for more details on this very important topic see, for example, Refs. [16–19]).

Meanwhile, oblique rotation of factor axes might still be necessary to successfully complete the latter *maximization* process. In such cases, the final angle between the factor axes ought to be related to the coefficient of correlation between the relevant factors themselves. If the resulting factor coordinate system becomes non-orthogonal, it means that it is possible to discuss the correlations between/among the factors themselves (for more details on this theme, see, for example, Ref. [19]).

In our case, however, there is absolutely no chance to classify the factors revealed, because they look tightly correlated among themselves.

Physically, the MD trajectory depicts the thermal motion, but the system under study here is an enzymatic complex with its educt. Therefore, we expect some dynamics beyond the conventional thermal noise because we know that after completing the chemical reaction (in our case it is the reaction of cleaving GTP into GDP and Pi), the educt should be swiftly thrown "into garbage."

Of interest would be to separate such processes from the conventional thermal noise, to analyze which residues of the enzymatic complex are participating in this, how they are participating, and what the role is of the enzymes' counterion-hydration sheath modeling the enzymes' natural surroundings.

Meanwhile, it is clear that the linear factor analysis model is not enough to successfully investigate the above posers. So we have undertaken nonlinear EFA by employing the correspondingly modified algorithm [20].

2.2.2 Nonlinear Exploratory Factor Analysis Results

We could successfully perform a polynomial EFA with the same data set. We have even introduced the number of factors obtained from the linear EFA study described above by reducing the possible number of factors to 5.

Our decision to reduce the number of factors has been dictated by the idea that it is not important to retain the number of latent factors obtained but to try and find the approach that might help reveal and describe some dynamical factor (factors) that would go beyond the conventional thermal noise.

Once we find such a method, refining the results—or, at the very least, possible ways to do so—could become possible. The result is presented in Fig. 2.3.

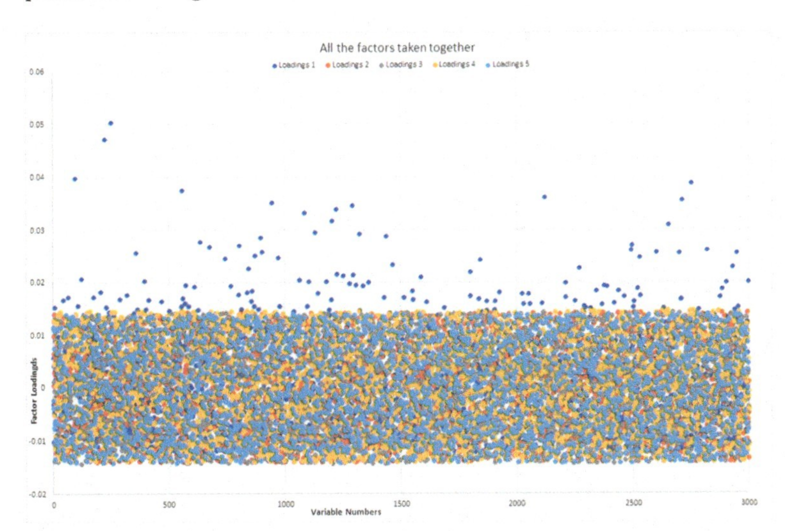

Figure 2.3 The disposition of the 3000 dynamical variables among all the factors studied: The *x* axis contains the variable numbers according to Table 2.1 (for the complete table, refer to the supplementary material) while the *y* axis measures all the relevant factor loadings. The results for loadings 1–5 are depicted by different colors.

At first glance, strictly mathematically speaking, we may immediately envisage the success of applying a somewhat more sophisticated form of polynomial factor analysis because linear factor analysis is, in effect, a method based upon polynomials as well and the highest degree of the relevant polynomial is unity.

The routine [20] allows us to choose the number and degrees of fitting polynomials. We have gone along using 20 polynomials with randomly chosen degrees from 1 to 5. Remarkably, the approach [20, 21] allows for a flexible choice of the fitting polynomial, whereas the number of factors involved should be restricted from the outset.

To provide a plausible first approximation of the desired estimate for this crucial parameter in nonlinear factor analysis, it is advisable to perform the conventional linear factor analysis beforehand.

We can immediately see that Factor 1 exhibits clearly different factor loading distribution as compared to Factors 2–5. Thus, the first idea would be to consider the latter four as the conventional thermal noise.

Factor 1 has positive factor loadings from several variables, which exceed the border of [−0.015; +0.015]. Remarkably, the factor loadings from all the factors do sum up to approximately zero (up to the fifth position after the decimal sign).

This means that some dynamical variables promote Factor 1, for their loadings are clearly positive. Some of them are quite neutral, for their loadings are near zero, whereas some of them hinder the factor, in that their loadings are clearly negative.

Hence, adopting the "energetic" representation of thermodynamics (for a detailed discussion on thermodynamics in the form of energetics, see Ref. [4] and Chapter 1), it is possible to speak about the perfect entropy-enthalpy compensation (EEC).

Indeed, if we assume that the latent factor corresponds to some dynamical process taking part in the supramolecular complex and going beyond the conventional thermal noise, there ought to be some key players that define the process's agenda, so to speak; some helpers; some indifferent spectators; and last but not the least, the mischievous monsters, who interfere with the process or are striving to halt or veto the process.

Therefore, Fig. 2.3 tells us the following story. The variable loadings for Factors 2–5 hold on the borders of [−0.015; +0.015], whereas the total of loadings for every factor is about zero. We might then assign these factors to conventional thermal noise.

Instead, in Factor 1, there are many variables whose positive loadings go beyond the borders. These variables could be assigned to the key players; they ought to define the clear physical direction of the corresponding process. Interestingly, none of the negative loadings go beyond the border, so the key players and the helpers can equilibrate/overcome the hindrances due to the mischievous monsters and successfully bring their desired process to the ultimate end.

By consequently considering the loadings of Factor 1, it is possible to draw some conclusions about the simulated process. Table 2.2 (for the complete table, please refer to the supplementary material) presents a list of the key players. Here, we confine ourselves to the most telling part of this table.

Table 2.2

Variable no.	Factor 1 load	Residue	Residue no.	Variable
2930	0.022772	PI	489	CenterM-Y
2949	0.025523	NA^+	496	CenterM-Z
2953	0.015137	NA^+	498	CenterM-X
2999	0.019929	Total	511	T_M_ZY

Table 2.2 highlights the "brightest" key players in red. This means that the process described by Factor 1 is "throwing away the loose Pi residue," that is one of the educts of the catalyzed reaction GTP $\longrightarrow$ GDP + Pi. The Na^+ cations from the surrounding take part in this process. A key player is one of the coordinates of one of the total inertia principal axes. This means that the supramolecular complex in question changes its overall geometric form somewhat. Apparently, such a change is necessary to facilitate the excision of the Pi to the maximum extent. Which amino acid residues play the role in the latter change, how important might this role be, and what is the physical sense of its role? All these posers could be answered by checking the height of the pertinent factor loadings and the physical sense of these variables.

Here, we would just like to show what the mischievous monsters are with respect to Factor 1. Table 2.3 (for the complete table, please refer to the supplementary material) presents the list of the variables having noticeably negative loadings to Factor 1. Here, we confine ourselves to the most telling part of this table.

Table 2.3

Variable no.	Factor 1 load	Residue	Residue no.	Variable
2917	−0.010146	MG^{++}	487	CenterM-X
2936	−0.011742	NA^{+}	491	CenterM-Y
2976	−0.01207	NA^{+}	505	CenterM-Z
2983	−0.01061	Total	507	Total_CM
2992	−0.010325	Total	509	T_M_XZ
2998	−0.01082	Total	511	T_M_ZX
3000	−0.010459	Total	511	T_M_ZZ

In Table 2.3, we have presented the factor loadings on the borders of [−0.01; −0.015], just to reveal the actual mischievous monsters, by leaving the indifferent spectators aside. With red, we have highlighted the most telling ones.

Interestingly, the Mg^{2+} cation from inside and a couple of Na^{+} cations from outside the complex interfere with the process of Pi excision. Moreover, there are some "negative" aspects of the change in the overall geometrical form of the enzymatic complex that oppose the excision as well. It is possible to investigate which amino acid residues contribute to the latter, to what extent, and how.

An attentive reader might exclaim here that the above deliberations sound like an apparent stretch, for the actual relationship between thermodynamics and the factor analytical approach ought to be far-fetched. Is the latter approach the right one to investigate EEC?

When looking through the literature on applied factor analysis, we may find some useful indications for conclusively answering this poser that might indeed seem truly unexpected for (bio)physical chemists.

In political science, the work of one of the outstanding specialists in the field, Richard Newton Rosecrance, entitled *Action and Reaction in World Politics* [22] is well known and thoroughly discussed [23, 24].

Of immediate methodological interest for our present discussion ought to be the call by Rosecrance to consider the conflicts among and inside nations from the standpoint of Newtonian mechanics.

Remarkably, Rosecrance arrived at some seminal conclusions that could be analyzed by the factor analysis of correlations. For example, Zinnes [23] says:

> While we may be unhappy with the analysis, we are still forced to recognize the fact that Rosecrance has provided us with a number of significant and intriguing insights which furnish meat for future hypotheses. We can now take his conclusions, restate them as hypotheses, and continue this research. It is hoped that additional studies will be undertaken by Rosecrance and others to further develop and explore these ideas.
>
> To some extent this has already been done by R. J. Rummel.[a] He found, contrary to Rosecrance's proposition, no correlation between various indices of internal conflict and external conflict, i.e., wars (Rummel, 1963[b]). Furthermore, in recent analyses he has discovered no correlation between any internal domestic variable (e.g., length of elite tenure, type or stability of government) and external conflict (unpublished results of ongoing NSF-supported research on the Dimensions of Nations).

Importantly, Rummel's main working instrument was *factor analysis of correlations*. In his publications, he did his best to duly present and explain to his colleagues the modalities of this analytic tool [25–30]. His works are highly recommended to active and proactive research workers in any scientific field.

As EEC in thermodynamics might well be viewed as a kind of "conflict" as well (see Ref. [4] and Chapter 1 for further details), it is clear that application of factor analysis in (bio)physical chemistry should anyway be considered seriously.

2.3 Detailed Description of the Method

The method consists in applying the EFA of correlations [16]. In other words, we look for pairwise correlations between all the values of the dynamical variables set that characterize the macromolecular system under study.

---

[a]Rudolph Joseph Rummel (1932–2014).

[b]R. J. Rummel (1963). *Dimensions of Conflict Behavior within and between Nations*, General Systems Yearbook, L. von Bertalanffy, A. Rapoport (eds.), Vol. 8, pp. 1–50.

To get these values, we take the MD trajectory into a statistically significant set of discrete snapshots. The snapshots are in fact the full sets of all the atomic coordinates at some time point along the MD trajectory simulated. We use these coordinate values to compute the values of their proper combinations at all the time points chosen. For example, we might be interested in the trajectories of the amino acids and the eventual heteroatomic groups. We compute the centers of mass and the principal axes of the inertia/gyration tensors for every structural fragment thus chosen. As a result, we arrive at the set of the molecular dynamical variables, that is, the time-dependent variables having direct molecular structural relevance.

Alongside, at every time point of the trajectory, we might consider evaluating the center of mass and the principal components of the inertia/gyration tensor for the macromolecular complex, plus all the pertinent supplementary kinematic information, like the Euler angles and the principal axes' coordinates for the total inertia/gyration tensor. By including the latter information into our dynamic variables set, we hope to find the contributions of the local variables into the global dynamics of the complex under study. We must look for the factors common for the local and global dynamics. In other words, we evaluate the factor loadings for all the variables chosen (i.e., the correlation coefficients between the variables and the factors, in other words, the contributions of the variables to the factors) and decide which loadings are the highest ones for some factor.

Meanwhile, there is a true diversity of factor analytical approaches. Which one to choose depends on the problem we would like to tackle. Hence, first it is necessary to discuss the differences between the various approaches at hand.

2.3.1 Difference between Confirmatory and Exploratory Factor Analysis

With all the above in mind, it is important to stress the basic difference between the two known forms of the method [17]:

- **Confirmatory factor analysis** is a handy method for data sets reduction. We look for pairwise correlations throughout the set of the observed cases—MD trajectory time points in our case—to properly classify the latter. We assume that we know the classification rule (conventionally, with the resulting factors being linear combinations of the contributions from the observed cases). Every observed case might be characterized by a considerable number of measured variables. We form the matrix of pairwise correlations between all the cases along the set of variables. The resulting matrix is subject to eigenvalue-eigenvector analysis. The number of the *highest* eigenvalues (highest according to some criterion) is the number of the factors (i.e., of the observed cases' groups), with the corresponding eigenvector being the weights of the cases in the groups. This way, we confirm the linear classification rule assumed. This method is also known as the principal components analysis (PCA). In 1993, there was a widespread call for using the confirmatory factor analysis (CFA)/PCA in the natural sciences [18]. This same year, the school of Herman Berendsen also called for widely employing PCA in processing the MD simulation results for biopolymers [31]. The work in Ref. [32] was one of the first examples of using CFA/PCA to study the MD trajectory of an enzyme, the RNase T1, to gain insights into the mechanisms of its activity. At the same time, the school of Nobuhiro Gõ presented the early general introduction to and early comprehensive review of the topic [33, 34]. Since then, applications of CFA/PCA in processing the results of the MD simulations could gradually become truly widespread (see, e.g., Ref. [35] and the references therein).

- Instead, the only conventional assumption of EFA is the linearity of the classification rule. Though being mathematically identical to CFA/PCA, EFA looks for pairwise correlations between all the observables/variables estimated along the set of observed cases (i.e., between the dynamical variables in our case). In this way, we look for the so-called latent factors to explain the correlations observed among the variables. We assume that the correlation observed between, say, some phenomena A and B is not just due to the trivial cause-and-effect relationships of the form "A is the cause of B, so that B is the effect of A," or vice versa. We assume that there is some realistic phenomenon C that is the intrinsic common cause of both A and B (or even their intrinsic common effect).

This factor may not be observed or explicitly simulated because of the natural constraints of the experimental method or of the computer-aided simulation approach. As a result, unlike CFA/PCA, EFA might be considered the classification of the variables but not the classification of the observed cases. The number of the *highest* eigenvalues (highest according to some criterion) is the number of the factors underlying the observed correlations between the variables, with the corresponding eigenvectors being the carrier of pairwise correlation coefficients between all the variables and the pertinent latent factors. This is just the basic difference between EFA and CFA/PCA. Remarkably, applications of EFA in the field of (bio)physical chemistry (to explore the experimental data or computer-aided simulation results) are rather rare [36, 37].

Of primary interest for our present discussion would be answering this poser: Might any form of factor analysis be helpful in investigating the mechanisms of the biophysically relevant phenomena?

Here, we mean factor analysis as a systematic tool in such fundamental fields of physical chemistry as thermodynamics and statistical and quantum mechanics. Their actual, as well as potential, application directions are well known to be truly diverse or easily recognizable to be unlimited [38–42].

Meanwhile, widely unknown (or, perhaps, simply forgotten?) contributions to the fields just mentioned have surfaced most recently [4], which do require careful and attentive reconsideration of the interpretational tools at hand.

Among other interesting topics, of true importance ought to be the phenomenon of EEC, which is still not well understood because of the vague understanding of the actual physical sense of the entropy notion. Nonetheless, EEC ought to be of direct and basic relevance to the actual mechanisms of (bio)physical-chemical phenomena.

Indeed, the widespread concept of entropy as a mystic "measure of disorder"—and all the related vivid speculations around this slogan—is based upon the long-standing ignorance of the fact that the W (*Wahrscheinlichkeit,* the probability) in the good old tried-and-true formula $S = k \cdot \ln(W)$ has a very clear physical sense. Regarding this, the actual physical notion of the thermodynamic entropy *S* described by the given formula ought to be reconsidered as a measure of the ubiquitous hindrances, obstacles, and resistances in the way of any kind of progress, be it of physical, chemical, biological, or any other nature (see Ref. [4], the references therein, and Chapter 1 here for more details).

Thus, interesting for taking account of entropic effects would be any kind of approach that can reveal the "difficulties on the way to progress." Interestingly, factor analysis of correlations has been long well known to be such a method.

2.3.2 Difficulty Factors in Factor Analysis

The research activity in an effort to reveal the "difficulty factors" using factor analysis of correlation was triggered in the field of psychology by a private opinion expressed by Ferguson in 1941 [43]. He pioneered discussing "the influence of test difficulty on the correlation between test items and between tests. The greater the difference in difficulty between two test items or between two tests the smaller the maximum correlation between them. In general, the greater the number of degrees of difficulty among the items in a test or among the tests in a battery, the higher the rank of the matrix of intercorrelations; that is, differences in difficulty are represented in the factorial configuration as additional factors." His seminal suggestion was that "if all tests included in a battery are roughly homogeneous with respect to difficulty, existing hierarchies will be more clearly defined and meaningful psychological interpretation of factors more readily attained."

To our sincere regret, from 1939 till 1945, the entire world had other more serious problems to solve than just to theoretically tackle the poser of the difficulty factors. Only in 1945, Carroll could have published a thorough study about the success of revealing the difficulty factors along the lines of CFA/PCA and/or EFA. Carroll wrote [44]:

> The correlation coefficient has frequently been used as an indication of the extent to which two items or two tests measure the same ability. It is the purpose of this paper to show that correlations

between items and between tests are affected by the difficulties of the items involved, and that to the degree that the items or tests are dissimilar in difficulty, conventional correlational statistics tend not to give a correct indication of the true overlap of ability. This demonstration is made through the analysis of a theoretical limiting case, that is, one in which a set of items measure a single ability, but it can also be shown to apply whenever items have any factorial overlap.

Assume that we have a set of (n) perfectly reliable items of varying difficulty, which all measure a single ability and only a single ability. Difficulty is measured by the proportion (k) of individuals failing each item. It is assumed that to pass an item the individual must have true mastery of the task involved; that is, the probability of chance success, **c**, is assumed to be zero. It is further assumed that each item has been presented to all individuals under constant conditions (*). The ability measured is of such a nature that success at any level of difficulty implies success at all lower levels of difficulty, and failure at any level implies failure at all higher levels of difficulty (**). Thus, it will be found that all individuals who pass an item at any given level of difficulty will pass all items of less difficulty, and that all individuals who fail an item at any given level of difficulty will fail all items of greater difficulty (***).

(*) This assumption precludes the analysis of a set of items in a time-limit test where the subjects are exposed to varying numbers of items.

(**) Most factors of ability, but not necessarily all, are probably of this nature.

(***) If one moves out of the context of these assumptions, however, the fact that any given pair of items is characterized by such a relation does not guarantee that the items are factorially homogeneous.

Accordingly, Carroll considered in detail the problem of "the effect of difficulty and chance success on (pairwise) correlations between items or between tests." In modern terms, we can ask: Which approach to factor analysis ought to be the most skillful one in estimating the degree of difficulties involved in the phenomenon/process under study: CFA/PCA (when considering the correlations among all the items tested) or EFA (when considering the correlations among all the tests themselves)?

Carroll's conclusion on the theme is as follows:

> A study is made of the extent to which correlations between items and between tests are affected by the difficulties of the items involved and by chance success through guessing. The Pearsonian product-moment coefficient does not necessarily give a correct indication of the relation between items or sets of items, since it tends to decrease as the items or tests become less similar in difficulty. It is suggested that the tetrachoric correlation coefficient can properly be used for estimating the correlation between the continua underlying items or sets of items even though they differ in difficulty, and a method for correcting a 2 X 2 table for the effect of chance is proposed.

The Pearsonian product-moment coefficient, r_{XY}, due to Pearson, is nowadays the standard representation of two data sets, say X and Y, ($X = x_i$ and $Y = y_i$, where $i = 1, \ldots n$), where

$$r_{XY} = \frac{Cov(X,Y)}{\sigma_X \cdot \sigma_Y} = \frac{\sum_{i=1}^{n}\left(x_i - \bar{x}\right)\cdot\left(y_i - \bar{y}\right)}{\sum_{i=1}^{n}\left(x_i - \bar{x}\right)^2 \cdot \sum_{i=1}^{n}\left(y_i - \bar{y}\right)^2}; \bar{x} = \frac{1}{n}\cdot\sum_{i=1}^{n} x_i; \bar{y} = \frac{1}{n}\cdot\sum_{i=1}^{n} y_i,$$

with Cov called covariance and σ called variance.

Tetrachoric correlation is used to measure the agreement for binary data. A binary data set contains data with two possible answers, usually right or wrong. The tetrachoric correlation estimates what the correlation would be if measured on a continuous scale. It is used for a variety of reasons, for example, for the analysis of scores in item response theory and converting comorbidity statistics to correlation coefficients. This type of correlation has the advantage that the actual numbers of rating levels, or the marginal proportions for rating levels, do not affect the result.

The term "comorbidity" comes from experimental/statistical medicine and means the presence of one or more additional conditions co-occurring with (i.e., concomitant or concurrent with) the primary condition. In the "tetrachoric" case, we study some further changing parameter—taking the binary values (say, Yes/No)—and introducing thus some additional dichotomy into the primary case. In the first line, this should be connected with the planning of experiments.

Moreover, the term "tetrachoric correlation" comes from the so-called tetrachoric series, a numerical method used before the advent of computers. While nowadays it is common to estimate correlations with methods like maximum likelihood estimation, there is a basic tetrachoric formula we may still use (for more details on this theme, see Ref. [45]).

Meanwhile, we are interested in interpreting the results of computer-aided simulations. So, the tetrachoric story does not seem to be of primary relevance here.

Important for our discussion would be to learn about the plausible mathematical approaches to interpret the conventional Pearsonian correlation coefficients. It is known that factor analysis might be based upon linear and nonlinear models. This is just what we would like to consider in more detail.

2.3.3 Difference between Linear and Nonlinear Factor Analysis

We might summarize the consideration in the previous paragraph as follows: The basic notion of orthogonal factor analysis is that a set of item scores (CFA/PCA) or test scores (EFA), or, to put it in general terms, manifest variates, can be expressed as functions of a set of mutually independent factor scores (CFA/PCA) or latent variates (EFA), together with components unique to each manifest variate. Finding the functions in question parallels the regression analysis.

In mathematical practice these regression functions have generally been assumed to be linear. However, cases where nonlinear relations might be suspected are well recognized as well. Systematic work to extend the scope of factor analysis started in the late 1950s to the early 1960s and lasted till the 1980s [46–57].

The works by McDonald and his school [48–57] are of special relevance in this regard, for he proposed tackling the problem of the difficulty factors by means of nonlinear factor analysis (see his work in Ref. [49] and a more detailed account in Ref. [57]).

Of interest here are also the works by Carroll's school—Carroll was not dealing with the difficulty factors in the first line but considered in detail the practical application modalities of nonlinear factor analysis.

Parallel to McDonald's efforts, he introduced the idea of polynomial factor analysis as follows [58]:

> Polynomial factor analysis (PFA) extends factor analysis to the case in which the relation between underlying factors and observed variables is mediated by polynomial (rather than by merely linear) functions. The particular polynomial model is specified by a matrix of exponents, which define a set of polynomial components (with the factors as arguments) to be included in the model. The method can easily be generalized to include possible non-polynomial functions, such as harmonic functions. PFA is described, it is contrasted with other methods of nonlinear factor analysis, and it is illustrated by an application to artificial data.

Carroll's school has also published a FORTRAN code embodying these ideas and freely available on the Internet [20]. A detailed representation of the algorithm has been published elsewhere [21] and has been discussed in the later review work by McDonald's school [57].

In effect, the original FORTRAN code [20] by Chang and Carroll embodies the nonlinear CFA/PCA. We have re-engineered their code to tackle EFA problems (the result is freely available from the authors upon request).

Here we apply both linear (standard FORTRAN subroutines) and nonlinear (modified Chang–Carroll routine) factor analysis to analyze the snapshots of the Ras-GAP-GDP-Pi complex. We describe this system in detail next.

2.3.4 The System under Study: Choosing the Proper Variables to Analyze the Macromolecular Dynamics

On the basis of known X-ray structure of the Ras-GDP+AlF_3/GAP complex (PDB entry: 1WQ1) [5], the structures of two types of Ras-GAP complexes, a Ras-GTP/GAP complex and a Ras-GDP+Pi/GAP complex, could have been created. AlF_3 was replaced by $P_\gamma O_3$, and then combining γ-phosphate with β-phosphate of GDP produced the Ras-GTP/GAP complex; replacing AlF_3 with $H_2PO_4^-$ created the Ras-GDP+Pi/GAP complex. MD simulations were performed on each of the structures of these Ras-GAP complexes.

The residue numbers of Ras are 1–166 and those of GAP are 167–486. There are two switch regions (Switch I: 32–38, Switch II: 60–75) in Ras, and a Mg^{2+} ion and GTP (or GDP+Pi) exist in between. We added 15,575 water molecules in the Ras-GTP/GAP system and 15,574 water molecules, excluding one water molecule used in the hydrolysis, in the Ras-GDP+Pi/GAP system. The resulting hydrated complex was placed into a box of size 86 × 72 × 86 Å^3, which served as the unit cell throughout the 20 ns NPT MD simulation. Finally, to keep the charge neutrality inside the simulation cell under periodic boundary condition, 16 Na^+ ions were added to both protein systems.

The technical details of the MD simulation itself will be reported in the next section, but first we would like to analyze the kinematics of the enzymatic complex under study in order to choose the proper, the most telling set of the dynamic variables that ought to represent the detailed framework for the functional activity of the enzyme under study.

In the following analysis, we mainly employ the MD results for the Ras-GDP+Pi/GAP complex. Table 2.1 (for the complete table, refer to the supplementary material) presents a comprehensive list of all the relevant dynamical variables. We have chosen the 3000 variables listed according to the following idea.

It is of interest to characterize every residue of the complex, that is, its amino acid residues and the relevant heteroatom (HETATM) groups. In our case, to the latter belong the GDP residue, Pi (the loose phosphate anion), water molecules, as well as Mg^{2+} and Na^+ cations. For polyatomic residues, like amino acids, GDP, and Pi, it is of interest to consider the dynamics of their centers of mass as well as that of the principal components of their gyration/inertia tensors. Thus, their dynamics in relation to each other as well as the dynamics of their geometrical shapes should be considered properly. For monoatomic cations, it is enough to consider their dynamics in the 3D space. As to the water molecules, we take it that although themselves they are three-atomic molecules, they form the hydration sheath. It is of definite interest to consider the overall dynamics of the latter so that we choose its common center of mass and the principal components of its inertia tensor (rows 2977–2982 in Table 2.1 [supplementary material]). Of interest would be the overall dynamics of the complex taken as a whole, that is, including itself and its counterion-hydration sheath (rows 2983–3000 in Table 2.1 [supplementary material]).

2.3.5 Technical Details of the MD Simulation and Data Processing

Recently (see also Ref. [59]), we successfully checked whether our MD simulation procedure might be applicable to studying the enzyme system of interest here.

2.3.5.1 The system setup

We employed an ff14SB force field [60] for the protein part and that prepared in the Amber parameter database for GTP, GDP, and Mg^{2+} [61, 62]. The force field of $H_2PO_4^-$ was prepared using the force field creation module antechamber implemented in AmberTools. The TIP3P water model [63, 64] and JC ion parameters adjusted for the TIP3P model [65, 66] were employed for water molecules and sodium ions, respectively.

All simulations were performed under the periodic boundary condition. Electrostatic interaction was treated by the particle mesh Ewald method, where the real space cutoff was set to 9 Å. The vibrational motions associated with hydrogen atoms were ignored. The translational center-of-mass motion of the entire system was removed by every 500 steps to keep the entire system around the origin, avoiding the overflow of coordinate information from the MD trajectory format. Each of the molecular mechanics (MM) and MD simulations was performed using the Amber 16 [67] GPU-version PMEMD module based on the SPFP algorithm [68] with NVIDIA GeForce GTX1080Ti.

2.3.5.2 MD simulation procedure

First, we energetically minimized each of the Ras-GAP systems by the following three-step MM simulations: the first MM simulation was for Mg^{2+}, GTP, and inorganic phosphate, the second one was for all the hydrogen atoms, and the third one was for all the atoms in the system. Each MM simulation consists of 1500 steps of the steepest descent method followed by 48,500 steps of the conjugate gradient method.

To energetically relax the position of ions and water molecules, the following five MD simulations were performed for each system:

NVT (0.001 to 1 [K], 0.1 [ps]) → NVT (1 [K], 0.1 [ps]) → NVT (1 to 300 [K], 20 [ps]) → NVT (300 [K], 20 [ps]) → NPT (300 [K], 300 [ps], 1 [bar]). In each MD simulation, the atomic coordinates except for water and Na^+ molecules were restrained by harmonic potential around the initial atomic coordinates of the simulation with a force constant of 100 kcal/mol/Å^2. The first two NVT simulations and the remaining ones were performed using the time steps of 0.01 fs and 2 fs for the integration, respectively. The first two NVT simulations and the following two were performed using the Berendsen thermostat [69], with a coupling constant equal to 0.001 ps, and the Langevin thermostat, with a collision coefficient equal to 1 ps^{-1}. In the first, second, and third NVT simulations, the reference temperature was linearly increased along the time course. The NPT simulation was performed using the Berendsen thermostat [69], with a coupling constant equal to 2 ps.

Using the atomic coordinates obtained from the NPT simulation, the following nine-step MD simulations were performed for each system: NVT (0.001 to 1 [K], 0.1 [ps], 100 [kcal/mol/Å^2]) → NVT (1 to 300 [K], 0.1 [ps], 100 [kcal/mol/Å^2]) → NVT (300 [K], 0.1 [ps], 100 [kcal/mol/Å^2]) → NVT (300 [K], 40 [ps], 10 [kcal/mol/Å^2]) → NVT (300 [K], 40 [ps], 5 [kcal/mol/Å^2]) → NVT (300 [K], 40 [ps], 1 [kcal/mol/Å^2]) → NVT (300 [K], 40 [ps]) → NPT (300 [K], 1 [bar], 20 [ns]). The last 20 ns NPT simulation was used for the following analyses. The first three NVT simulations and the remaining ones were performed using the time steps of 0.01 fs and 2 fs for the integration, respectively. In the first two NVT simulations, the reference temperature was linearly increased along the time course. Except for the last two steps, Ras, GAP, GTP, GDP, Pi, and Mg^{2+} were restrained by harmonic potential around the initial atomic coordinates. In each NVT simulation, the temperature was regulated using the Langevin thermostat, with a collision coefficient equal to 1 ps^{-1}. In the last 20 ns NPT simulation, temperature and pressure were regulated by the Berendsen thermostat [69], with a coupling constant equal to 2 ps and 5 ps, respectively. The initial atomic velocities were randomly assigned from a Maxwellian distribution at 0.001 K. MD trajectories were recorded at every 10 ps interval for the following analyses. For each of Ras-GAP systems, this procedure was repeated 50 times with a separate set of atomic velocities.

2.3.5.3 Analyses of MD trajectories

We calculated the root mean square deviations by using the ptraj module in the AmberTools 1.5 package [67].

After generating the entire MD trajectory, we divided it into 20 fragments corresponding to 1 ns stretches during the total simulation time, each of the resulting stretches containing a consecutive series of the structural snapshots for the enzymatic complex under study, representing the structural states of the latter at some particular time points.

To perform factor analysis, we chose 100 particular snapshots from different stretches after generating 100 pseudorandom numbers in the range [1; 100]: {*i*1, *i*2, . . . *i*100}. Then, from the trajectory stretch number 1, we chose the snapshot number *i*1 and so on. This way we avoided finding a numerical error in the simulation as a separate systematic factor. The risk of getting the latter result could be much higher if we would instead take 100 neighboring snapshots from a particular MD trajectory stretch.

2.4 Conclusion

To sum up, we are proposing herewith a method to interpret the simulated MD trajectories of macro- and supramolecular complexes of the widest relevance to nanoscience and nanotechnology.

Concerning the protein and enzyme engineering, a systematic application of our approach ought to help us get keen insights into the finest details of enzyme functioning and other protein functions.

We would not like to terminate our methodological studies at the stage described above. Using polynomial factor analysis, we could go beyond the capabilities of the linear factor model. Still, most of the factors studied here do land into the conventional thermal noise.

To our mind, it is throughout possible to enforce our method by considering nonlinear interactions among the factors [53, 54]. We thus hope to reveal more factors beyond the conventional thermal noise.

Another plausible possibility to refine our factor model would be allowing the degrees of the fitting polynomials to adopt even non-integer values.

Here we have applied our approach to the MD trajectory snapshots for all the atomic coordinates. Thus, we investigate the factor modalities for potential energy, which is dependent on the atomic coordinates.

It is also of interest to analyze the MD trajectory snapshots for all the atomic velocities. Thus, we might investigate the factor modalities for kinetic energy, which is dependent on the atomic velocities. This way, we might get a better insight into the microscopic modalities of the EEC, because kinetic energy is the actual physical basis of the driving force when speaking of some directed MD beyond the conventional thermal noise. In the first line, this driving force is used to overcome the effects of the mischievous monsters (i.e., the entropic effects, with the driving force representing, accordingly, the enthalpic effects). Hence, the factors revealed in the analysis of the atomic velocity snapshots should confirm, and, hopefully, even refine, the actual distribution of the dynamical variables among the key players, helpers, indifferent spectators, and mischievous monsters we get from the analysis of the "potential energy" snapshots.

Acknowledgments

We would like to acknowledge the grants-in-aid for scientific research (No. 17H06353 and 18K03825) from the Ministry of Education, Culture, Sports, Science and Technology, Japan (MEXT).

References

1. J. C. Samuelson (2013). *Enzyme Engineering. Methods and Protocols*, Springer Science+Business Media, New York, USA.
2. I. V. Korendovych (2018). Rational and semirational protein design, in: *Protein Engineering. Methods and Protocols*, U. T. Bornscheuer, M. Höhne (eds.), Springer, New York, USA, pp. 15–23.
3. J. W. Weis (2017). Artificial intelligence and protein engineering: information theoretical approaches to modeling enzymatic catalysis, MIT, Master's degree thesis, https://dspace.mit.edu/handle/1721.1/108969.
4. E. B. Starikov (2019). *A Different Thermodynamics and Its True Heroes*, Pan Stanford, Singapore.

5. K. Scheffzek, M. R. Ahmadian, W. Kabsch, L. Wiesmuller, A. Lautwein, F. Schmitz, A. Wittinghofer (1997). The Ras-RasGAP complex: structural basis for GTPase activation and its loss in oncogenic Ras mutants, *Science*, **277**, pp. 333–338.
6. M. Geyer, T. Schweins, C. Herrmann, T. Prisner, A. Wittinghofer, H. R. Kalbitzer (1996). Conformational transitions in p21ras and in its complexes with the effector protein Raf-RBD and the GTPase activating protein GAP, *Biochemistry*, **35**, pp. 10308–10320.
7. P. A. Boriack-Sjodin, S. M. Margarit, D. Bar-Sagi, J. Kuriyan (1998). The structural basis of the activation of Ras by Sos, *Nature*, **394**, pp. 337–343.
8. M. Malumbres, M. Barbacid (2003). RAS oncogenes: the first 30 years, *Nat. Rev. Cancer*, **3**, pp. 459–465.
9. A. E. Karnoub, R. A. Weinberg (2008). Ras oncogenes: split personalities, *Nat. Rev. Mol. Cell Biol.*, **9**, pp. 517–531.
10. F. Shima, Y. Ijiri, S. Muraoka, J. Liao, M. Ye, M. Araki, K. Matsumoto, N. Yamamoto, T. Sugimoto, Y. Yoshikawa, T. Kumasaka, M. Yamamoto, A. Tamura, T. Kataoka (2010). Structural basis for conformational dynamics of GTP-bound Ras protein, *J. Biol. Chem.*, **285**, pp. 22696–22705.
11. M. Araki, F. Shima, Y. Yoshikawa, S. Muraoka, Y. Ijiri, Y. Nagahara, T. Shirono, T. Kataoka, A. Tamura (2011). Solution structure of the state 1 conformer of GTP-bound H-Ras protein and distinct dynamic properties between the state 1 and state 2 conformers, *J. Biol. Chem.*, **286**, pp. 39644–39653.
12. F. Shima, Y. Yoshikawa, M. Ye, M. Araki, S. Matsumoto, J. Liao, L. Hu, T. Sugimoto, Y. Ijiri, Y. Takeda, Y. Nishiyama, C. Sato, S. Muraoka, K. Tamura, T. Osoda, K. Tsuda, T. Miyakawa, H. Fukunishi, J. Shimada, T. Kumasaka, M. Yamamoto, T. Kataoka (2013). In silico discovery of small-molecule Ras inhibitors that display antitumor activity by blocking the Ras-effector interaction, *Proc. Natl. Acad. Sci. U.S.A.*, **110**, pp. 8182–8187.
13. R. Knihtila, G. Holzapfel, K. Weiss, F. Meilleur, C. Mattos (2015). Neutron crystal structure of RAS GTPase puts in question the protonation state of the GTP γ-phosphate, *J. Biol. Chem.*, **290**, pp. 31025–31036.
14. S. Matsumoto, N. Miyano, S. Baba, J. Liao, T. Kawamura, C. Tsuda, A. Takeda, M. Yamamoto, T. Kumasaka, T. Kataoka, F. Shima (2016). Molecular mechanism for conformational dynamics of Ras·GTP elucidated from in-situ structural transition in crystal, *Sci. Rep.*, **6**, pp. 1–12.

15. M. G. Khrenova, B. L. Grigorenko, A. B. Kolomeisky, A. V. Nemukhin (2015). Hydrolysis of guanosine triphosphate (GTP) by the Ras·GAP protein complex: reaction mechanism and kinetic scheme, *J. Phys. Chem.*, **119**, pp. 12838–12845.
16. L. R. Fabrigar, D. T. Wegener (2012). *Exploratory Factor Analysis*, Oxford University Press, New York, Auckland, Cape Town, Dar es Salaam, Hong Kong, Karachi, Kuala Lumpur, Madrid, Melbourne, Mexico City, Nairobi, New Delhi, Shanghai, Taipei, Toronto.
17. B. Thompson (2004). *Exploratory and Confirmatory Factor Analysis: Understanding Concepts and Applications*, American Psychological Association, Washington, DC, USA.
18. R. A. Reyment, K. G. Jöreskog (1993). *Applied Factor Analysis in the Natural Sciences*, Cambridge University Press, Cambridge.
19. K. Überla (1968). *Faktorenanalyse* (in German), Springer Verlag, Berlin, Germany.
20. http://www.netlib.org/mds/polyfac.f.
21. J. D. Carroll, J.-J. Chang (1970). Analysis of individual differences in multidimensional scaling via an n-way generalization of "Eckart-Young" decomposition, *Psychometrika*, **35**, pp. 283–319.
22. R. N. Rosecrance (1963). *Action and Reaction in World Politics*, Little, Brown, Boston, USA.
23. D. A. Zinnes (1964). The requisites for international stability: a review. Richard N. Rosecrance, action and reaction in world politics, *J. Conflict Resolut.*, **8**, pp. 301–305.
24. A. L. Burns (1964–1965). Review: action and reaction in world politics: international systems in perspective, by Richard N. Rosecrance, *Int. J.* (Canadian International Council), **20**, p. 119.
25. R. J. Rummel (1966). Dimensions of conflict behavior within nations, 1946–59, *J. Conflict Resolut.*, **10**, pp. 65–73.
26. R. J. Rummel (1967). Understanding factor analysis, *J. Conflict Resolut.*, **11**, pp. 444–480.
27. R. J. Rummel (1970). *Applied Factor Analysis*, Northwestern University Press, Evanston, Ill., USA.
28. R. J. Rummel (1971). Some dimensions in the foreign behavior of nations, in: *Comparative Foreign Policy: Theoretical Essays*, W. F. Hanrieder (ed.), David McKay Company, Inc., New York, USA, pp. 132–166.
29. R. J. Rummel (1972). *Dimensions of Nations*, Sage, Beverly Hills, USA.

30. R. J. Rummel (1976). *Understanding Correlation*, http://www.mega.nu:8080/ampp/rummel/uc.htm.

31. A. Amadei, A. B. M. Linssen, H. J. C. Berendsen (1993). Essential dynamics of proteins, *Proteins*, **17**, pp. 412–425.

32. F. Cordes, E. B. Starikov, W. Saenger (1995). Initial state of an enzymic reaction. theoretical prediction of complex formation in the active site of RNase T1, *J. Am. Chem. Soc.*, **117**, pp. 10365–10372.

33. S. Hayward, N. Gõ (1995). Collective variable description of native protein dynamics, *Annu. Rev. Phys. Chem.*, **46**, pp. 223–250.

34. A. Kitao, N. Gõ (1999). Investigating protein dynamics in collective coordinate space, *Curr. Opin. Struct. Biol.*, **9**, pp. 164–169.

35. P. W. Pan, R. J. Dickson, H. L. Gordon, S. M. Rothstein, S. Tanaka (2005). Functionally relevant protein motions: extracting basin-specific collective coordinates from molecular dynamics trajectories, *J. Chem. Phys.*, **122**, p. 034904.

36. E. B. Starikov, M. A. Semenov, V. Ya. Maleev, A. I. Gasan (1991). evidential study of correlated events in biochemistry: physicochemical mechanisms of nucleic acid hydration as revealed by factor analysis, *Biopolymers*, **31**, pp. 255–273.

37. E. B. Starikov, K. Braesicke, E.-W. Knapp, W. Saenger (2001). Negative solubility coefficient of methylated cyclodextrins in water: a theoretical study, *Chem. Phys. Lett.*, **336**, pp. 504–510.

38. E. B. Starikov, J. P. Lewis, S. Tanaka (2006). *Modern Methods for Theoretical Physical Chemistry of Biopolymers*, Elsevier Science, Amsterdam, the Netherlands.

39. J.-L. Rivail, M. Ruiz-Lopez, X. Assfeld (2015). *Quantum Modeling of Complex Molecular Systems*, Springer International, Cham, Heidelberg, New York, Dordrecht, London.

40. R. Q. Snurr, C. S. Adjiman, D. A. Kofke (2016). *Foundations of Molecular Modeling and Simulation: Selected Papers from FOMMS 2015*, Springer Science+Business Media, Singapore.

41. P. Norman, K. Ruud, T. Saue (2018). *Principles and Practices of Molecular Properties: Theory, Modeling, and Simulations*, Wiley & Sons Ltd., Hoboken, Chichester, Oxford.

42. J. L. Ram, P. Michael Conn (2018). *Conn's Handbook of Models for Human Aging*, Academic Press, London, San Diego, Cambridge, Oxford.

43. G. A. Ferguson (1941). The factorial interpretation of test difficulty, *Psychometrika*, **6**, pp. 323–329.

44. J. B. Carroll (1945). The effect of difficulty and chance success on correlations between items or between tests, *Psychometrika*, **10**, pp. 1–19.

45. S. Deviant (2010). *The Practically Cheating Statistics Handbook*, CreateSpace, North Charleston, South Carolina, USA.

46. W. A. Gibson (1959). Three multivariate models. factor analysis, latent structure analysis, and latent profile analysis, *Psychometrika*, **24**, pp. 229–252.

47. W. A. Gibson (1960). Nonlinear factors in two dimensions, *Psychometrika*, **25**, pp. 381–392.

48. R. P. McDonald (1962). A general approach to nonlinear factor analysis, *Psychometrika*, **27**, pp. 397–415.

49. R. P. McDonald (1965). Difficulty factors and non-linear factor analysis, *Br. J. Math. Stat. Psychol.*, **18**, pp. 11–23.

50. R. P. McDonald (1965). Some IBM 7090-94 programs for nonlinear factor analysis, ETS Research Memorandum, RM-65-11.

51. R. P. McDonald (1965). Numerical methods for polynomial models in nonlinear factor analysis, ETS Research Bulletin Series, RB-67-46.

52. R. P. McDonald (1967). Numerical methods for polynomial models in nonlinear factor analysis, *Psychometrika*, **32**, pp. 77–112.

53. R. P. McDonald (1967). Factor interaction in nonlinear factor analysis, ETS Research Bulletin Series, RB-67-46.

54. R. P. McDonald (1967). Factor interaction in nonlinear factor analysis, *Br. J. Math. Stat. Psychol.*, **20**, pp. 205–215.

55. R. P. McDonald (1967). *Nonlinear Factor Analysis*, The William Byrd Press, Richmond, VA, USA.

56. R. P. McDonald (1979). The structural analysis of multivariate data: a sketch of a general theory, *Multivar. Behav. Res.*, **14**, pp. 21–38.

57. J. Etezadi-Amoli, R. P. McDonald (1983). A second-generation nonlinear factor analysis, *Psychometrika*, **48**, pp. 315–342.

58. J. D. Carroll (1969). Polynomial factor analysis. *Proceedings of the 77th Annual Convention of the American Psychological Association*, **3**(Pt. 1), pp. 103–104.

59. S. Matsunaga, Y. Hano, Y. Saito, K. J. Fujimoto, T. Kumasaka, S. Matsumoto, T. Kataoka, F. Shima, S. Tanaka (2017). Structural transition of solvated H-Ras/GTP revealed by molecular dynamics simulation and local network entropy, *J. Mol. Graph. Model.*, **77**, pp. 51–63.

60. J. A. Maier, C. Martinez, K. Kasavajhala, L. Wickstrom, K. E. Hauser, C. Simmerling (2015). ff14SB: improving the accuracy of protein side chain and backbone parameters from ff99SB, *J. Chem. Theory Comput.*, **11**, pp. 3696–3713.

61. K. L. Meagher, L. T. Redman, H. A. Carlson (2003). Development of polyphosphate parameters for use with the AMBER force field, *J. Comput. Chem.*, **24**(9), pp. 1016–1025.

62. G. M. Bradbrook, T. Gleichmann, S. J. Harrop, J. Habash, J. Raftery, J. Kalb, J. Yariv, J. I. H. Hillier, J. R. Helliwell (1998). X-ray and molecular dynamics studies of concanavalin-a glucoside and mannoside complexes - relating structure to thermodynamics of binding, *J. Chem. Soc. Faraday Trans.*, **94**, pp. 1603–1611.

63. W. L. Jorgensen, J. Chandrasekhar, J. D. Madura, R. W. Impey, M. L. Klein (1983). Comparison of simple potential functions for simulating liquid water, *J. Chem. Phys.*, **79**, pp.926–935.

64. P. G. Kusalik, I. M. Svishchev (1994). The spatial structure in liquid water, *Science*, **265**, pp.1219–1221.

65. I. S. Joung, T. E. Cheatham (2008). Determination of alkali and halide monovalent ion parameters for use in explicitly solvated biomolecular simulations, *J. Phys. Chem. B*, **112**, pp. 9020–9041.

66. I. S. Joung, T. E. Cheatham III (2009). Molecular dynamics simulations of the dynamic and energetic properties of alkali and halide ions using water-model-specific ion parameters, *J. Phys. Chem. B*, **113**, pp. 13279–13290.

67. D. A. Case, R. M. Betz, D.S. Cerutti, T. E. Cheatham, III, T. A. Darden, R. E. Duke, T. J. Giese, H. Gohlke, A. W. Goetz, N. Homeyer, S. Izadi, P. Janowski, J. Kaus, A. Kovalenko, T. S. Lee, S. Le Grand, P. Li, C. Lin, T. Luchko, R. Luo, B. Madej, D. Mermelstein, K. M. Merz, G. Monard, H. Nguyen, H. T. Nguyen, I. Omelyan, A. Onufriev, D. R. Roe, A. Roitberg, C. Sagui, C. L. Simmerling, W. M. Botello-Smith, J. Swails, R. C. Walker, J. Wang, R. M. Wolf, X. Wu, L. Xiao, P. A. Kollman (2016). AMBER 2016, University of California, San Francisco.

68. S. Le Grand, A. W. Gotz, R. C. Walker (2013). SPFP: speed without compromise - a mixed precision model for GPU accelerated molecular dynamics simulations, *Comput. Phys. Commun.*, **184**, pp. 374–380.

69. H. J. C. Berendsen, J. P. M. Postma, W. F. Vangunsteren, A. Dinola, J. R. Haak (1984). Molecular-dynamics with coupling to an external bath, *J. Chem. Phys.*, **81**, pp. 3684–3690.

Supplementary Material to Chapter 2

This part of our report contains the tables with results of our computations. In particular, Table 2.1 presents the assignments and physical descriptions of the kinematic parameters for all the amino acid residues, all the HETATM groups, the hydration shell as a whole, Mg^{2+} and Na^{+} counterions, and last but not least, the kinematic parameters for the enzymatic complex under study as a whole (i.e., including itself plus its hydration-counterionic sheath). This is relevant to Section 2.2. of the main text.

Table 2.1 The list of kinematic variables in the GAP-GDP-Pi complex, their ordinal numbers (Column 1), their physical sense (Column 4), and their assignment to the proper residue or the part thereof in the complex (Columns 2 and 3, numbers and names of the residues/parts, respectively)

Variable no.	Residue no.	Residue	Variable
1	1	MET	InertiaM-X
2	1	MET	InertiaM-Y
3	1	MET	InertiaM-Z
4	1	MET	CenterM-X
5	1	MET	CenterM-Y
6	1	MET	CenterM-Z
7	2	THR	InertiaM-X
8	2	THR	InertiaM-Y
9	2	THR	InertiaM-Z
10	2	THR	CenterM-X
11	2	THR	CenterM-Y
12	2	THR	CenterM-Z
13	3	GLU	InertiaM-X
14	3	GLU	InertiaM-Y
15	3	GLU	InertiaM-Z
16	3	GLU	CenterM-X
17	3	GLU	CenterM-Y
18	3	GLU	CenterM-Z
19	4	TYR	InertiaM-X

Variable no.	Residue no.	Residue	Variable
20	4	TYR	InertiaM-Y
21	4	TYR	InertiaM-Z
22	4	TYR	CenterM-X
23	4	TYR	CenterM-Y
24	4	TYR	CenterM-Z
25	5	LYS	InertiaM-X
26	5	LYS	InertiaM-Y
27	5	LYS	InertiaM-Z
28	5	LYS	CenterM-X
29	5	LYS	CenterM-Y
30	5	LYS	CenterM-Z
31	6	LEU	InertiaM-X
32	6	LEU	InertiaM-Y
33	6	LEU	InertiaM-Z
34	6	LEU	CenterM-X
35	6	LEU	CenterM-Y
36	6	LEU	CenterM-Z
37	7	VAL	InertiaM-X
38	7	VAL	InertiaM-Y
39	7	VAL	InertiaM-Z
40	7	VAL	CenterM-X
41	7	VAL	CenterM-Y
42	7	VAL	CenterM-Z
43	8	VAL	InertiaM-X
44	8	VAL	InertiaM-Y
45	8	VAL	InertiaM-Z
46	8	VAL	CenterM-X
47	8	VAL	CenterM-Y
48	8	VAL	CenterM-Z
49	9	VAL	InertiaM-X
50	9	VAL	InertiaM-Y
51	9	VAL	InertiaM-Z

(*Continued*)

Table 2.1 (*Continued*)

Variable no.	Residue no.	Residue	Variable
52	9	VAL	CenterM-X
53	9	VAL	CenterM-Y
54	9	VAL	CenterM-Z
55	10	GLY	InertiaM-X
56	10	GLY	InertiaM-Y
57	10	GLY	InertiaM-Z
58	10	GLY	CenterM-X
59	10	GLY	CenterM-Y
60	10	GLY	CenterM-Z
61	11	ALA	InertiaM-X
62	11	ALA	InertiaM-Y
63	11	ALA	InertiaM-Z
64	11	ALA	CenterM-X
65	11	ALA	CenterM-Y
66	11	ALA	CenterM-Z
67	12	GLY	InertiaM-X
68	12	GLY	InertiaM-Y
69	12	GLY	InertiaM-Z
70	12	GLY	CenterM-X
71	12	GLY	CenterM-Y
72	12	GLY	CenterM-Z
73	13	GLY	InertiaM-X
74	13	GLY	InertiaM-Y
75	13	GLY	InertiaM-Z
76	13	GLY	CenterM-X
77	13	GLY	CenterM-Y
78	13	GLY	CenterM-Z
79	14	VAL	InertiaM-X
80	14	VAL	InertiaM-Y
81	14	VAL	InertiaM-Z
82	14	VAL	CenterM-X

Variable no.	Residue no.	Residue	Variable
83	14	VAL	CenterM-Y
84	14	VAL	CenterM-Z
85	15	GLY	InertiaM-X
86	15	GLY	InertiaM-Y
87	15	GLY	InertiaM-Z
88	15	GLY	CenterM-X
89	15	GLY	CenterM-Y
90	16	GLY	CenterM-Z
91	16	LYS	InertiaM-X
92	16	LYS	InertiaM-Y
93	16	LYS	InertiaM-Z
94	16	LYS	CenterM-X
95	16	LYS	CenterM-Y
96	17	LYS	CenterM-Z
97	17	SER	InertiaM-X
98	17	SER	InertiaM-Y
99	17	SER	InertiaM-Z
100	17	SER	CenterM-X
101	17	SER	CenterM-Y
102	18	SER	CenterM-Z
103	18	ALA	InertiaM-X
104	18	ALA	InertiaM-Y
105	18	ALA	InertiaM-Z
106	18	ALA	CenterM-X
107	18	ALA	CenterM-Y
108	19	ALA	CenterM-Z
109	19	LEU	InertiaM-X
110	19	LEU	InertiaM-Y
111	19	LEU	InertiaM-Z
112	19	LEU	CenterM-X
113	19	LEU	CenterM-Y
114	20	LEU	CenterM-Z

(*Continued*)

Table 2.1 (*Continued*)

Variable no.	Residue no.	Residue	Variable
115	20	THR	InertiaM-X
116	20	THR	InertiaM-Y
117	20	THR	InertiaM-Z
118	20	THR	CenterM-X
119	20	THR	CenterM-Y
120	20	THR	CenterM-Z
121	21	ILE	InertiaM-X
122	21	ILE	InertiaM-Y
123	21	ILE	InertiaM-Z
124	21	ILE	CenterM-X
125	21	ILE	CenterM-Y
126	21	ILE	CenterM-Z
127	22	GLN	InertiaM-X
128	22	GLN	InertiaM-Y
129	22	GLN	InertiaM-Z
130	22	GLN	CenterM-X
131	22	GLN	CenterM-Y
132	22	GLN	CenterM-Z
133	23	LEU	InertiaM-X
134	23	LEU	InertiaM-Y
135	23	LEU	InertiaM-Z
136	23	LEU	CenterM-X
137	23	LEU	CenterM-Y
138	23	LEU	CenterM-Z
139	24	ILE	InertiaM-X
140	24	ILE	InertiaM-Y
141	24	ILE	InertiaM-Z
142	24	ILE	CenterM-X
143	24	ILE	CenterM-Y
144	24	ILE	CenterM-Z
145	25	GLN	InertiaM-X

Variable no.	Residue no.	Residue	Variable
146	25	GLN	InertiaM-Y
147	25	GLN	InertiaM-Z
148	25	GLN	CenterM-X
149	25	GLN	CenterM-Y
150	25	GLN	CenterM-Z
151	26	ASN	InertiaM-X
152	26	ASN	InertiaM-Y
153	26	ASN	InertiaM-Z
154	26	ASN	CenterM-X
155	26	ASN	CenterM-Y
156	26	ASN	CenterM-Z
157	27	HIE	InertiaM-X
158	27	HIE	InertiaM-Y
159	27	HIE	InertiaM-Z
160	27	HIE	CenterM-X
161	27	HIE	CenterM-Y
162	27	HIE	CenterM-Z
163	28	PHE	InertiaM-X
164	28	PHE	InertiaM-Y
165	28	PHE	InertiaM-Z
166	28	PHE	CenterM-X
167	28	PHE	CenterM-Y
168	28	PHE	CenterM-Z
169	29	VAL	InertiaM-X
170	29	VAL	InertiaM-Y
171	29	VAL	InertiaM-Z
172	29	VAL	CenterM-X
173	29	VAL	CenterM-Y
174	29	VAL	CenterM-Z
175	30	ASP	InertiaM-X
176	30	ASP	InertiaM-Y
177	30	ASP	InertiaM-Z

(*Continued*)

Table 2.1 (*Continued*)

Variable no.	Residue no.	Residue	Variable
178	30	ASP	CenterM-X
179	30	ASP	CenterM-Y
180	30	ASP	CenterM-Z
181	31	GLU	InertiaM-X
182	31	GLU	InertiaM-Y
183	31	GLU	InertiaM-Z
184	31	GLU	CenterM-X
185	31	GLU	CenterM-Y
186	31	GLU	CenterM-Z
187	32	TYR	InertiaM-X
188	32	TYR	InertiaM-Y
189	32	TYR	InertiaM-Z
190	32	TYR	CenterM-X
191	32	TYR	CenterM-Y
192	32	TYR	CenterM-Z
193	33	ASP	InertiaM-X
194	33	ASP	InertiaM-Y
195	33	ASP	InertiaM-Z
196	33	ASP	CenterM-X
197	33	ASP	CenterM-Y
198	33	ASP	CenterM-Z
199	34	PRO	InertiaM-X
200	34	PRO	InertiaM-Y
201	34	PRO	InertiaM-Z
202	34	PRO	CenterM-X
203	34	PRO	CenterM-Y
204	34	PRO	CenterM-Z
205	35	THR	InertiaM-X
206	35	THR	InertiaM-Y
207	35	THR	InertiaM-Z
208	35	THR	CenterM-X

Variable no.	Residue no.	Residue	Variable
209	35	THR	CenterM-Y
210	35	THR	CenterM-Z
211	36	ILE	InertiaM-X
212	36	ILE	InertiaM-Y
213	36	ILE	InertiaM-Z
214	36	ILE	CenterM-X
215	36	ILE	CenterM-Y
216	36	ILE	CenterM-Z
217	37	GLU	InertiaM-X
218	37	GLU	InertiaM-Y
219	37	GLU	InertiaM-Z
220	37	GLU	CenterM-X
221	37	GLU	CenterM-Y
222	37	GLU	CenterM-Z
223	38	ASP	InertiaM-X
224	38	ASP	InertiaM-Y
225	38	ASP	InertiaM-Z
226	38	ASP	CenterM-X
227	38	ASP	CenterM-Y
228	38	ASP	CenterM-Z
229	39	SER	InertiaM-X
230	39	SER	InertiaM-Y
231	39	SER	InertiaM-Z
232	39	SER	CenterM-X
233	39	SER	CenterM-Y
234	39	SER	CenterM-Z
235	40	TYR	InertiaM-X
236	40	TYR	InertiaM-Y
237	40	TYR	InertiaM-Z
238	40	TYR	CenterM-X
239	40	TYR	CenterM-Y
240	40	TYR	CenterM-Z

(*Continued*)

Table 2.1 (*Continued*)

Variable no.	Residue no.	Residue	Variable
241	41	ARG	InertiaM-X
242	41	ARG	InertiaM-Y
243	41	ARG	InertiaM-Z
244	41	ARG	CenterM-X
245	41	ARG	CenterM-Y
246	41	ARG	CenterM-Z
247	42	LYS	InertiaM-X
248	42	LYS	InertiaM-Y
249	42	LYS	InertiaM-Z
250	42	LYS	CenterM-X
251	42	LYS	CenterM-Y
252	42	LYS	CenterM-Z
253	43	GLN	InertiaM-X
254	43	GLN	InertiaM-Y
255	43	GLN	InertiaM-Z
256	43	GLN	CenterM-X
257	43	GLN	CenterM-Y
258	43	GLN	CenterM-Z
259	44	VAL	InertiaM-X
260	44	VAL	InertiaM-Y
261	44	VAL	InertiaM-Z
262	44	VAL	CenterM-X
263	44	VAL	CenterM-Y
264	44	VAL	CenterM-Z
265	45	VAL	InertiaM-X
266	45	VAL	InertiaM-Y
267	45	VAL	InertiaM-Z
268	45	VAL	CenterM-X
269	45	VAL	CenterM-Y
270	45	VAL	CenterM-Z
271	46	ILE	InertiaM-X

Variable no.	Residue no.	Residue	Variable
272	46	ILE	InertiaM-Y
273	46	ILE	InertiaM-Z
274	46	ILE	CenterM-X
275	46	ILE	CenterM-Y
276	46	ILE	CenterM-Z
277	47	ASP	InertiaM-X
278	47	ASP	InertiaM-Y
279	47	ASP	InertiaM-Z
280	47	ASP	CenterM-X
281	47	ASP	CenterM-Y
282	47	ASP	CenterM-Z
283	48	GLY	InertiaM-X
284	48	GLY	InertiaM-Y
285	48	GLY	InertiaM-Z
286	48	GLY	CenterM-X
287	48	GLY	CenterM-Y
288	48	GLY	CenterM-Z
289	49	GLU	InertiaM-X
290	49	GLU	InertiaM-Y
291	49	GLU	InertiaM-Z
292	49	GLU	CenterM-X
293	49	GLU	CenterM-Y
294	49	GLU	CenterM-Z
295	50	THR	InertiaM-X
296	50	THR	InertiaM-Y
297	50	THR	InertiaM-Z
298	50	THR	CenterM-X
299	50	THR	CenterM-Y
300	50	THR	CenterM-Z
301	51	CYS	InertiaM-X
302	51	CYS	InertiaM-Y
303	51	CYS	InertiaM-Z

(*Continued*)

Table 2.1 (*Continued*)

Variable no.	Residue no.	Residue	Variable
304	51	CYS	CenterM-X
305	51	CYS	CenterM-Y
306	51	CYS	CenterM-Z
307	52	LEU	InertiaM-X
308	52	LEU	InertiaM-Y
309	52	LEU	InertiaM-Z
310	52	LEU	CenterM-X
311	52	LEU	CenterM-Y
312	52	LEU	CenterM-Z
313	53	LEU	InertiaM-X
314	53	LEU	InertiaM-Y
315	53	LEU	InertiaM-Z
316	53	LEU	CenterM-X
317	53	LEU	CenterM-Y
318	53	LEU	CenterM-Z
319	54	ASP	InertiaM-X
320	54	ASP	InertiaM-Y
321	54	ASP	InertiaM-Z
322	54	ASP	CenterM-X
323	54	ASP	CenterM-Y
324	54	ASP	CenterM-Z
325	55	ILE	InertiaM-X
326	55	ILE	InertiaM-Y
327	55	ILE	InertiaM-Z
328	55	ILE	CenterM-X
329	55	ILE	CenterM-Y
330	55	ILE	CenterM-Z
331	56	LEU	InertiaM-X
332	56	LEU	InertiaM-Y
333	56	LEU	InertiaM-Z
334	56	LEU	CenterM-X

Variable no.	Residue no.	Residue	Variable
335	56	LEU	CenterM-Y
336	56	LEU	CenterM-Z
337	57	ASP	InertiaM-X
338	57	ASP	InertiaM-Y
339	57	ASP	InertiaM-Z
340	57	ASP	CenterM-X
341	57	ASP	CenterM-Y
342	57	ASP	CenterM-Z
343	58	THR	InertiaM-X
344	58	THR	InertiaM-Y
345	58	THR	InertiaM-Z
346	58	THR	CenterM-X
347	58	THR	CenterM-Y
348	58	THR	CenterM-Z
349	59	ALA	InertiaM-X
350	59	ALA	InertiaM-Y
351	59	ALA	InertiaM-Z
352	59	ALA	CenterM-X
353	59	ALA	CenterM-Y
354	59	ALA	CenterM-Z
355	60	GLY	InertiaM-X
356	60	GLY	InertiaM-Y
357	60	GLY	InertiaM-Z
358	60	GLY	CenterM-X
359	60	GLY	CenterM-Y
360	60	GLY	CenterM-Z
361	61	GLN	InertiaM-X
362	61	GLN	InertiaM-Y
363	61	GLN	InertiaM-Z
364	61	GLN	CenterM-X
365	61	GLN	CenterM-Y
366	61	GLN	CenterM-Z

(*Continued*)

Table 2.1 (*Continued*)

Variable no.	Residue no.	Residue	Variable
367	62	GLU	InertiaM-X
368	62	GLU	InertiaM-Y
369	62	GLU	InertiaM-Z
370	62	GLU	CenterM-X
371	62	GLU	CenterM-Y
372	62	GLU	CenterM-Z
373	63	GLU	InertiaM-X
374	63	GLU	InertiaM-Y
375	63	GLU	InertiaM-Z
376	63	GLU	CenterM-X
377	63	GLU	CenterM-Y
378	63	GLU	CenterM-Z
379	64	TYR	InertiaM-X
380	64	TYR	InertiaM-Y
381	64	TYR	InertiaM-Z
382	64	TYR	CenterM-X
383	64	TYR	CenterM-Y
384	64	TYR	CenterM-Z
385	65	SER	InertiaM-X
386	65	SER	InertiaM-Y
387	65	SER	InertiaM-Z
388	65	SER	CenterM-X
389	65	SER	CenterM-Y
390	65	SER	CenterM-Z
391	66	ALA	InertiaM-X
392	66	ALA	InertiaM-Y
393	66	ALA	InertiaM-Z
394	66	ALA	CenterM-X
395	66	ALA	CenterM-Y
396	66	ALA	CenterM-Z
397	67	MET	InertiaM-X

Variable no.	Residue no.	Residue	Variable
398	67	MET	InertiaM-Y
399	67	MET	InertiaM-Z
400	67	MET	CenterM-X
401	67	MET	CenterM-Y
402	67	MET	CenterM-Z
403	68	ARG	InertiaM-X
404	68	ARG	InertiaM-Y
405	68	ARG	InertiaM-Z
406	68	ARG	CenterM-X
407	68	ARG	CenterM-Y
408	68	ARG	CenterM-Z
409	69	ASP	InertiaM-X
410	69	ASP	InertiaM-Y
411	69	ASP	InertiaM-Z
412	69	ASP	CenterM-X
413	69	ASP	CenterM-Y
414	69	ASP	CenterM-Z
415	70	GLN	InertiaM-X
416	70	GLN	InertiaM-Y
417	70	GLN	InertiaM-Z
418	70	GLN	CenterM-X
419	70	GLN	CenterM-Y
420	70	GLN	CenterM-Z
421	71	TYR	InertiaM-X
422	71	TYR	InertiaM-Y
423	71	TYR	InertiaM-Z
424	71	TYR	CenterM-X
425	71	TYR	CenterM-Y
426	71	TYR	CenterM-Z
427	72	MET	InertiaM-X
428	72	MET	InertiaM-Y
429	72	MET	InertiaM-Z

(*Continued*)

Table 2.1 (*Continued*)

Variable no.	Residue no.	Residue	Variable
430	72	MET	CenterM-X
431	72	MET	CenterM-Y
432	72	MET	CenterM-Z
433	73	ARG	InertiaM-X
434	73	ARG	InertiaM-Y
435	73	ARG	InertiaM-Z
436	73	ARG	CenterM-X
437	73	ARG	CenterM-Y
438	73	ARG	CenterM-Z
439	74	THR	InertiaM-X
440	74	THR	InertiaM-Y
441	74	THR	InertiaM-Z
442	74	THR	CenterM-X
443	74	THR	CenterM-Y
444	74	THR	CenterM-Z
445	75	GLY	InertiaM-X
446	75	GLY	InertiaM-Y
447	75	GLY	InertiaM-Z
448	75	GLY	CenterM-X
449	75	GLY	CenterM-Y
450	75	GLY	CenterM-Z
451	76	GLU	InertiaM-X
452	76	GLU	InertiaM-Y
453	76	GLU	InertiaM-Z
454	76	GLU	CenterM-X
455	76	GLU	CenterM-Y
456	76	GLU	CenterM-Z
457	77	GLY	InertiaM-X
458	77	GLY	InertiaM-Y
459	77	GLY	InertiaM-Z
460	77	GLY	CenterM-X

Variable no.	Residue no.	Residue	Variable
461	77	GLY	CenterM-Y
462	77	GLY	CenterM-Z
463	78	PHE	InertiaM-X
464	78	PHE	InertiaM-Y
465	78	PHE	InertiaM-Z
466	78	PHE	CenterM-X
467	78	PHE	CenterM-Y
468	78	PHE	CenterM-Z
469	79	LEU	InertiaM-X
470	79	LEU	InertiaM-Y
471	79	LEU	InertiaM-Z
472	79	LEU	CenterM-X
473	79	LEU	CenterM-Y
474	79	LEU	CenterM-Z
475	80	CYS	InertiaM-X
476	80	CYS	InertiaM-Y
477	80	CYS	InertiaM-Z
478	80	CYS	CenterM-X
479	80	CYS	CenterM-Y
480	80	CYS	CenterM-Z
481	81	VAL	InertiaM-X
482	81	VAL	InertiaM-Y
483	81	VAL	InertiaM-Z
484	81	VAL	CenterM-X
485	81	VAL	CenterM-Y
486	81	VAL	CenterM-Z
487	82	PHE	InertiaM-X
488	82	PHE	InertiaM-Y
489	82	PHE	InertiaM-Z
490	82	PHE	CenterM-X
491	82	PHE	CenterM-Y
492	82	PHE	CenterM-Z

(*Continued*)

Table 2.1 (*Continued*)

Variable no.	Residue no.	Residue	Variable
493	83	ALA	InertiaM-X
494	83	ALA	InertiaM-Y
495	83	ALA	InertiaM-Z
496	83	ALA	CenterM-X
497	83	ALA	CenterM-Y
498	83	ALA	CenterM-Z
499	84	ILE	InertiaM-X
500	84	ILE	InertiaM-Y
501	84	ILE	InertiaM-Z
502	84	ILE	CenterM-X
503	84	ILE	CenterM-Y
504	84	ILE	CenterM-Z
505	85	ASN	InertiaM-X
506	85	ASN	InertiaM-Y
507	85	ASN	InertiaM-Z
508	85	ASN	CenterM-X
509	85	ASN	CenterM-Y
510	85	ASN	CenterM-Z
511	86	ASN	InertiaM-X
512	86	ASN	InertiaM-Y
513	86	ASN	InertiaM-Z
514	86	ASN	CenterM-X
515	86	ASN	CenterM-Y
516	86	ASN	CenterM-Z
517	87	THR	InertiaM-X
518	87	THR	InertiaM-Y
519	87	THR	InertiaM-Z
520	87	THR	CenterM-X
521	87	THR	CenterM-Y
522	87	THR	CenterM-Z
523	88	LYS	InertiaM-X

Variable no.	Residue no.	Residue	Variable
524	88	LYS	InertiaM-Y
525	88	LYS	InertiaM-Z
526	88	LYS	CenterM-X
527	88	LYS	CenterM-Y
528	88	LYS	CenterM-Z
529	89	SER	InertiaM-X
530	89	SER	InertiaM-Y
531	89	SER	InertiaM-Z
532	89	SER	CenterM-X
533	89	SER	CenterM-Y
534	89	SER	CenterM-Z
535	90	PHE	InertiaM-X
536	90	PHE	InertiaM-Y
537	90	PHE	InertiaM-Z
538	90	PHE	CenterM-X
539	90	PHE	CenterM-Y
540	90	PHE	CenterM-Z
541	91	GLU	InertiaM-X
542	91	GLU	InertiaM-Y
543	91	GLU	InertiaM-Z
544	91	GLU	CenterM-X
545	91	GLU	CenterM-Y
546	91	GLU	CenterM-Z
547	92	ASP	InertiaM-X
548	92	ASP	InertiaM-Y
549	92	ASP	InertiaM-Z
550	92	ASP	CenterM-X
551	92	ASP	CenterM-Y
552	92	ASP	CenterM-Z
553	93	ILE	InertiaM-X
554	93	ILE	InertiaM-Y
555	93	ILE	InertiaM-Z

(*Continued*)

Table 2.1 (*Continued*)

Variable no.	Residue no.	Residue	Variable
556	93	ILE	CenterM-X
557	93	ILE	CenterM-Y
558	93	ILE	CenterM-Z
559	94	HIE	InertiaM-X
560	94	HIE	InertiaM-Y
561	94	HIE	InertiaM-Z
562	94	HIE	CenterM-X
563	94	HIE	CenterM-Y
564	94	HIE	CenterM-Z
565	95	GLN	InertiaM-X
566	95	GLN	InertiaM-Y
567	95	GLN	InertiaM-Z
568	95	GLN	CenterM-X
569	95	GLN	CenterM-Y
570	95	GLN	CenterM-Z
571	96	TYR	InertiaM-X
572	96	TYR	InertiaM-Y
573	96	TYR	InertiaM-Z
574	96	TYR	CenterM-X
575	96	TYR	CenterM-Y
576	96	TYR	CenterM-Z
577	97	ARG	InertiaM-X
578	97	ARG	InertiaM-Y
579	97	ARG	InertiaM-Z
580	97	ARG	CenterM-X
581	97	ARG	CenterM-Y
582	97	ARG	CenterM-Z
583	98	GLU	InertiaM-X
584	98	GLU	InertiaM-Y
585	98	GLU	InertiaM-Z
586	98	GLU	CenterM-X

Variable no.	Residue no.	Residue	Variable
587	98	GLU	CenterM-Y
588	98	GLU	CenterM-Z
589	99	GLN	InertiaM-X
590	99	GLN	InertiaM-Y
591	99	GLN	InertiaM-Z
592	99	GLN	CenterM-X
593	99	GLN	CenterM-Y
594	99	GLN	CenterM-Z
595	100	ILE	InertiaM-X
596	100	ILE	InertiaM-Y
597	100	ILE	InertiaM-Z
598	100	ILE	CenterM-X
599	100	ILE	CenterM-Y
600	100	ILE	CenterM-Z
601	101	LYS	InertiaM-X
602	101	LYS	InertiaM-Y
603	101	LYS	InertiaM-Z
604	101	LYS	CenterM-X
605	101	LYS	CenterM-Y
606	101	LYS	CenterM-Z
607	102	ARG	InertiaM-X
608	102	ARG	InertiaM-Y
609	102	ARG	InertiaM-Z
610	102	ARG	CenterM-X
611	102	ARG	CenterM-Y
612	102	ARG	CenterM-Z
613	103	VAL	InertiaM-X
614	103	VAL	InertiaM-Y
615	103	VAL	InertiaM-Z
616	103	VAL	CenterM-X
617	103	VAL	CenterM-Y
618	103	VAL	CenterM-Z

(*Continued*)

Table 2.1 (*Continued*)

Variable no.	Residue no.	Residue	Variable
619	104	LYS	InertiaM-X
620	104	LYS	InertiaM-Y
621	104	LYS	InertiaM-Z
622	104	LYS	CenterM-X
623	104	LYS	CenterM-Y
624	104	LYS	CenterM-Z
625	105	ASP	InertiaM-X
626	105	ASP	InertiaM-Y
627	105	ASP	InertiaM-Z
628	105	ASP	CenterM-X
629	105	ASP	CenterM-Y
630	105	ASP	CenterM-Z
631	106	SER	InertiaM-X
632	106	SER	InertiaM-Y
633	106	SER	InertiaM-Z
634	106	SER	CenterM-X
635	106	SER	CenterM-Y
636	106	SER	CenterM-Z
637	107	ASP	InertiaM-X
638	107	ASP	InertiaM-Y
639	107	ASP	InertiaM-Z
640	107	ASP	CenterM-X
641	107	ASP	CenterM-Y
642	107	ASP	CenterM-Z
643	108	ASP	InertiaM-X
644	108	ASP	InertiaM-Y
645	108	ASP	InertiaM-Z
646	108	ASP	CenterM-X
647	108	ASP	CenterM-Y
648	108	ASP	CenterM-Z
649	109	VAL	InertiaM-X

Variable no.	Residue no.	Residue	Variable
650	109	VAL	InertiaM-Y
651	109	VAL	InertiaM-Z
652	109	VAL	CenterM-X
653	109	VAL	CenterM-Y
654	109	VAL	CenterM-Z
655	110	PRO	InertiaM-X
656	110	PRO	InertiaM-Y
657	110	PRO	InertiaM-Z
658	110	PRO	CenterM-X
659	110	PRO	CenterM-Y
660	110	PRO	CenterM-Z
661	111	MET	InertiaM-X
662	111	MET	InertiaM-Y
663	111	MET	InertiaM-Z
664	111	MET	CenterM-X
665	111	MET	CenterM-Y
666	111	MET	CenterM-Z
667	112	VAL	InertiaM-X
668	112	VAL	InertiaM-Y
669	112	VAL	InertiaM-Z
670	112	VAL	CenterM-X
671	112	VAL	CenterM-Y
672	112	VAL	CenterM-Z
673	113	LEU	InertiaM-X
674	113	LEU	InertiaM-Y
675	113	LEU	InertiaM-Z
676	113	LEU	CenterM-X
677	113	LEU	CenterM-Y
678	113	LEU	CenterM-Z
679	114	VAL	InertiaM-X
680	114	VAL	InertiaM-Y
681	114	VAL	InertiaM-Z

(*Continued*)

Table 2.1 (*Continued*)

Variable no.	Residue no.	Residue	Variable
682	114	VAL	CenterM-X
683	114	VAL	CenterM-Y
684	114	VAL	CenterM-Z
685	115	GLY	InertiaM-X
686	115	GLY	InertiaM-Y
687	115	GLY	InertiaM-Z
688	115	GLY	CenterM-X
689	115	GLY	CenterM-Y
690	115	GLY	CenterM-Z
691	116	ASN	InertiaM-X
692	116	ASN	InertiaM-Y
693	116	ASN	InertiaM-Z
694	116	ASN	CenterM-X
695	116	ASN	CenterM-Y
696	116	ASN	CenterM-Z
697	117	LYS	InertiaM-X
698	117	LYS	InertiaM-Y
699	117	LYS	InertiaM-Z
700	117	LYS	CenterM-X
701	117	LYS	CenterM-Y
702	117	LYS	CenterM-Z
703	118	CYS	InertiaM-X
704	118	CYS	InertiaM-Y
705	118	CYS	InertiaM-Z
706	118	CYS	CenterM-X
707	118	CYS	CenterM-Y
708	118	CYS	CenterM-Z
709	119	ASP	InertiaM-X
710	119	ASP	InertiaM-Y
711	119	ASP	InertiaM-Z
712	119	ASP	CenterM-X

Variable no.	Residue no.	Residue	Variable
713	119	ASP	CenterM-Y
714	119	ASP	CenterM-Z
715	120	LEU	InertiaM-X
716	120	LEU	InertiaM-Y
717	120	LEU	InertiaM-Z
718	120	LEU	CenterM-X
719	120	LEU	CenterM-Y
720	120	LEU	CenterM-Z
721	121	ALA	InertiaM-X
722	121	ALA	InertiaM-Y
723	121	ALA	InertiaM-Z
724	121	ALA	CenterM-X
725	121	ALA	CenterM-Y
726	121	ALA	CenterM-Z
727	122	ALA	InertiaM-X
728	122	ALA	InertiaM-Y
729	122	ALA	InertiaM-Z
730	122	ALA	CenterM-X
731	122	ALA	CenterM-Y
732	122	ALA	CenterM-Z
733	123	ARG	InertiaM-X
734	123	ARG	InertiaM-Y
735	123	ARG	InertiaM-Z
736	123	ARG	CenterM-X
737	123	ARG	CenterM-Y
738	123	ARG	CenterM-Z
739	124	THR	InertiaM-X
740	124	THR	InertiaM-Y
741	124	THR	InertiaM-Z
742	124	THR	CenterM-X
743	124	THR	CenterM-Y
744	124	THR	CenterM-Z

(*Continued*)

Table 2.1 (*Continued*)

Variable no.	Residue no.	Residue	Variable
745	125	VAL	InertiaM-X
746	125	VAL	InertiaM-Y
747	125	VAL	InertiaM-Z
748	125	VAL	CenterM-X
749	125	VAL	CenterM-Y
750	125	VAL	CenterM-Z
751	126	GLU	InertiaM-X
752	126	GLU	InertiaM-Y
753	126	GLU	InertiaM-Z
754	126	GLU	CenterM-X
755	126	GLU	CenterM-Y
756	126	GLU	CenterM-Z
757	127	SER	InertiaM-X
758	127	SER	InertiaM-Y
759	127	SER	InertiaM-Z
760	127	SER	CenterM-X
761	127	SER	CenterM-Y
762	127	SER	CenterM-Z
763	128	ARG	InertiaM-X
764	128	ARG	InertiaM-Y
765	128	ARG	InertiaM-Z
766	128	ARG	CenterM-X
767	128	ARG	CenterM-Y
768	128	ARG	CenterM-Z
769	129	GLN	InertiaM-X
770	129	GLN	InertiaM-Y
771	129	GLN	InertiaM-Z
772	129	GLN	CenterM-X
773	129	GLN	CenterM-Y
774	129	GLN	CenterM-Z
775	130	ALA	InertiaM-X

Variable no.	Residue no.	Residue	Variable
776	130	ALA	InertiaM-Y
777	130	ALA	InertiaM-Z
778	130	ALA	CenterM-X
779	130	ALA	CenterM-Y
780	130	ALA	CenterM-Z
781	131	GLN	InertiaM-X
782	131	GLN	InertiaM-Y
783	131	GLN	InertiaM-Z
784	131	GLN	CenterM-X
785	131	GLN	CenterM-Y
786	131	GLN	CenterM-Z
787	132	ASP	InertiaM-X
788	132	ASP	InertiaM-Y
789	132	ASP	InertiaM-Z
790	132	ASP	CenterM-X
791	132	ASP	CenterM-Y
792	132	ASP	CenterM-Z
793	133	LEU	InertiaM-X
794	133	LEU	InertiaM-Y
795	133	LEU	InertiaM-Z
796	133	LEU	CenterM-X
797	133	LEU	CenterM-Y
798	133	LEU	CenterM-Z
799	134	ALA	InertiaM-X
800	134	ALA	InertiaM-Y
801	134	ALA	InertiaM-Z
802	134	ALA	CenterM-X
803	134	ALA	CenterM-Y
804	134	ALA	CenterM-Z
805	135	ARG	InertiaM-X
806	135	ARG	InertiaM-Y
807	135	ARG	InertiaM-Z

(Continued)

Table 2.1 (*Continued*)

Variable no.	Residue no.	Residue	Variable
808	135	ARG	CenterM-X
809	135	ARG	CenterM-Y
810	135	ARG	CenterM-Z
811	136	SER	InertiaM-X
812	136	SER	InertiaM-Y
813	136	SER	InertiaM-Z
814	136	SER	CenterM-X
815	136	SER	CenterM-Y
816	136	SER	CenterM-Z
817	137	TYR	InertiaM-X
818	137	TYR	InertiaM-Y
819	137	TYR	InertiaM-Z
820	137	TYR	CenterM-X
821	137	TYR	CenterM-Y
822	137	TYR	CenterM-Z
823	138	GLY	InertiaM-X
824	138	GLY	InertiaM-Y
825	138	GLY	InertiaM-Z
826	138	GLY	CenterM-X
827	138	GLY	CenterM-Y
828	138	GLY	CenterM-Z
829	139	ILE	InertiaM-X
830	139	ILE	InertiaM-Y
831	139	ILE	InertiaM-Z
832	139	ILE	CenterM-X
833	139	ILE	CenterM-Y
834	139	ILE	CenterM-Z
835	140	PRO	InertiaM-X
836	140	PRO	InertiaM-Y
837	140	PRO	InertiaM-Z
838	140	PRO	CenterM-X

Variable no.	Residue no.	Residue	Variable
839	140	PRO	CenterM-Y
840	140	PRO	CenterM-Z
841	141	TYR	InertiaM-X
842	141	TYR	InertiaM-Y
843	141	TYR	InertiaM-Z
844	141	TYR	CenterM-X
845	141	TYR	CenterM-Y
846	141	TYR	CenterM-Z
847	142	ILE	InertiaM-X
848	142	ILE	InertiaM-Y
849	142	ILE	InertiaM-Z
850	142	ILE	CenterM-X
851	142	ILE	CenterM-Y
852	142	ILE	CenterM-Z
853	143	GLU	InertiaM-X
854	143	GLU	InertiaM-Y
855	143	GLU	InertiaM-Z
856	143	GLU	CenterM-X
857	143	GLU	CenterM-Y
858	143	GLU	CenterM-Z
859	144	THR	InertiaM-X
860	144	THR	InertiaM-Y
861	144	THR	InertiaM-Z
862	144	THR	CenterM-X
863	144	THR	CenterM-Y
864	144	THR	CenterM-Z
865	145	SER	InertiaM-X
866	145	SER	InertiaM-Y
867	145	SER	InertiaM-Z
868	145	SER	CenterM-X
869	145	SER	CenterM-Y
870	145	SER	CenterM-Z

(*Continued*)

Table 2.1 (*Continued*)

Variable no.	Residue no.	Residue	Variable
871	146	ALA	InertiaM-X
872	146	ALA	InertiaM-Y
873	146	ALA	InertiaM-Z
874	146	ALA	CenterM-X
875	146	ALA	CenterM-Y
876	146	ALA	CenterM-Z
877	147	LYS	InertiaM-X
878	147	LYS	InertiaM-Y
879	147	LYS	InertiaM-Z
880	147	LYS	CenterM-X
881	147	LYS	CenterM-Y
882	147	LYS	CenterM-Z
883	148	THR	InertiaM-X
884	148	THR	InertiaM-Y
885	148	THR	InertiaM-Z
886	148	THR	CenterM-X
887	148	THR	CenterM-Y
888	148	THR	CenterM-Z
889	149	ARG	InertiaM-X
890	149	ARG	InertiaM-Y
891	149	ARG	InertiaM-Z
892	149	ARG	CenterM-X
893	149	ARG	CenterM-Y
894	149	ARG	CenterM-Z
895	150	GLN	InertiaM-X
896	150	GLN	InertiaM-Y
897	150	GLN	InertiaM-Z
898	150	GLN	CenterM-X
899	150	GLN	CenterM-Y
900	150	GLN	CenterM-Z
901	151	GLY	InertiaM-X

Variable no.	Residue no.	Residue	Variable
902	151	GLY	InertiaM-Y
903	151	GLY	InertiaM-Z
904	151	GLY	CenterM-X
905	151	GLY	CenterM-Y
906	151	GLY	CenterM-Z
907	152	VAL	InertiaM-X
908	152	VAL	InertiaM-Y
909	152	VAL	InertiaM-Z
910	152	VAL	CenterM-X
911	152	VAL	CenterM-Y
912	152	VAL	CenterM-Z
913	153	GLU	InertiaM-X
914	153	GLU	InertiaM-Y
915	153	GLU	InertiaM-Z
916	153	GLU	CenterM-X
917	153	GLU	CenterM-Y
918	153	GLU	CenterM-Z
919	154	ASP	InertiaM-X
920	154	ASP	InertiaM-Y
921	154	ASP	InertiaM-Z
922	154	ASP	CenterM-X
923	154	ASP	CenterM-Y
924	154	ASP	CenterM-Z
925	155	ALA	InertiaM-X
926	155	ALA	InertiaM-Y
927	155	ALA	InertiaM-Z
928	155	ALA	CenterM-X
929	155	ALA	CenterM-Y
930	155	ALA	CenterM-Z
931	156	PHE	InertiaM-X
932	156	PHE	InertiaM-Y
933	156	PHE	InertiaM-Z

(*Continued*)

Table 2.1 (*Continued*)

Variable no.	Residue no.	Residue	Variable
934	156	PHE	CenterM-X
935	156	PHE	CenterM-Y
936	156	PHE	CenterM-Z
937	157	TYR	InertiaM-X
938	157	TYR	InertiaM-Y
939	157	TYR	InertiaM-Z
940	157	TYR	CenterM-X
941	157	TYR	CenterM-Y
942	157	TYR	CenterM-Z
943	158	THR	InertiaM-X
944	158	THR	InertiaM-Y
945	158	THR	InertiaM-Z
946	158	THR	CenterM-X
947	158	THR	CenterM-Y
948	158	THR	CenterM-Z
949	159	LEU	InertiaM-X
950	159	LEU	InertiaM-Y
951	159	LEU	InertiaM-Z
952	159	LEU	CenterM-X
953	159	LEU	CenterM-Y
954	159	LEU	CenterM-Z
955	160	VAL	InertiaM-X
956	160	VAL	InertiaM-Y
957	160	VAL	InertiaM-Z
958	160	VAL	CenterM-X
959	160	VAL	CenterM-Y
960	160	VAL	CenterM-Z
961	161	ARG	InertiaM-X
962	161	ARG	InertiaM-Y
963	161	ARG	InertiaM-Z
964	161	ARG	CenterM-X

Variable no.	Residue no.	Residue	Variable
965	161	ARG	CenterM-Y
966	161	ARG	CenterM-Z
967	162	GLU	InertiaM-X
968	162	GLU	InertiaM-Y
969	162	GLU	InertiaM-Z
970	162	GLU	CenterM-X
971	162	GLU	CenterM-Y
972	162	GLU	CenterM-Z
973	163	ILE	InertiaM-X
974	163	ILE	InertiaM-Y
975	163	ILE	InertiaM-Z
976	163	ILE	CenterM-X
977	163	ILE	CenterM-Y
978	163	ILE	CenterM-Z
979	164	ARG	InertiaM-X
980	164	ARG	InertiaM-Y
981	164	ARG	InertiaM-Z
982	164	ARG	CenterM-X
983	164	ARG	CenterM-Y
984	164	ARG	CenterM-Z
985	165	GLN	InertiaM-X
986	165	GLN	InertiaM-Y
987	165	GLN	InertiaM-Z
988	165	GLN	CenterM-X
989	165	GLN	CenterM-Y
990	165	GLN	CenterM-Z
991	166	HIE	InertiaM-X
992	166	HIE	InertiaM-Y
993	166	HIE	InertiaM-Z
994	166	HIE	CenterM-X
995	166	HIE	CenterM-Y
996	166	HIE	CenterM-Z

(*Continued*)

Table 2.1 (*Continued*)

Variable no.	Residue no.	Residue	Variable
997	167	MET	InertiaM-X
998	167	MET	InertiaM-Y
999	167	MET	InertiaM-Z
1000	167	MET	CenterM-X
1001	167	MET	CenterM-Y
1002	167	MET	CenterM-Z
1003	168	PRO	InertiaM-X
1004	168	PRO	InertiaM-Y
1005	168	PRO	InertiaM-Z
1006	168	PRO	CenterM-X
1007	168	PRO	CenterM-Y
1008	168	PRO	CenterM-Z
1009	169	GLU	InertiaM-X
1010	169	GLU	InertiaM-Y
1011	169	GLU	InertiaM-Z
1012	169	GLU	CenterM-X
1013	169	GLU	CenterM-Y
1014	169	GLU	CenterM-Z
1015	170	GLU	InertiaM-X
1016	170	GLU	InertiaM-Y
1017	170	GLU	InertiaM-Z
1018	170	GLU	CenterM-X
1019	170	GLU	CenterM-Y
1020	170	GLU	CenterM-Z
1021	171	GLU	InertiaM-X
1022	171	GLU	InertiaM-Y
1023	171	GLU	InertiaM-Z
1024	171	GLU	CenterM-X
1025	171	GLU	CenterM-Y
1026	171	GLU	CenterM-Z
1027	172	TYR	InertiaM-X

Variable no.	Residue no.	Residue	Variable
1028	172	TYR	InertiaM-Y
1029	172	TYR	InertiaM-Z
1030	172	TYR	CenterM-X
1031	172	TYR	CenterM-Y
1032	172	TYR	CenterM-Z
1033	173	SER	InertiaM-X
1034	173	SER	InertiaM-Y
1035	173	SER	InertiaM-Z
1036	173	SER	CenterM-X
1037	173	SER	CenterM-Y
1038	173	SER	CenterM-Z
1039	174	GLU	InertiaM-X
1040	174	GLU	InertiaM-Y
1041	174	GLU	InertiaM-Z
1042	174	GLU	CenterM-X
1043	174	GLU	CenterM-Y
1044	174	GLU	CenterM-Z
1045	175	PHE	InertiaM-X
1046	175	PHE	InertiaM-Y
1047	175	PHE	InertiaM-Z
1048	175	PHE	CenterM-X
1049	175	PHE	CenterM-Y
1050	175	PHE	CenterM-Z
1051	176	LYS	InertiaM-X
1052	176	LYS	InertiaM-Y
1053	176	LYS	InertiaM-Z
1054	176	LYS	CenterM-X
1055	176	LYS	CenterM-Y
1056	176	LYS	CenterM-Z
1057	177	GLU	InertiaM-X
1058	177	GLU	InertiaM-Y
1059	177	GLU	InertiaM-Z

(*Continued*)

Table 2.1 (*Continued*)

Variable no.	Residue no.	Residue	Variable
1060	177	GLU	CenterM-X
1061	177	GLU	CenterM-Y
1062	177	GLU	CenterM-Z
1063	178	LEU	InertiaM-X
1064	178	LEU	InertiaM-Y
1065	178	LEU	InertiaM-Z
1066	178	LEU	CenterM-X
1067	178	LEU	CenterM-Y
1068	178	LEU	CenterM-Z
1069	179	ILE	InertiaM-X
1070	179	ILE	InertiaM-Y
1071	179	ILE	InertiaM-Z
1072	179	ILE	CenterM-X
1073	179	ILE	CenterM-Y
1074	179	ILE	CenterM-Z
1075	180	LEU	InertiaM-X
1076	180	LEU	InertiaM-Y
1077	180	LEU	InertiaM-Z
1078	180	LEU	CenterM-X
1079	180	LEU	CenterM-Y
1080	180	LEU	CenterM-Z
1081	181	GLN	InertiaM-X
1082	181	GLN	InertiaM-Y
1083	181	GLN	InertiaM-Z
1084	181	GLN	CenterM-X
1085	181	GLN	CenterM-Y
1086	181	GLN	CenterM-Z
1087	182	LYS	InertiaM-X
1088	182	LYS	InertiaM-Y
1089	182	LYS	InertiaM-Z
1090	182	LYS	CenterM-X

Variable no.	Residue no.	Residue	Variable
1091	182	LYS	CenterM-Y
1092	182	LYS	CenterM-Z
1093	183	GLU	InertiaM-X
1094	183	GLU	InertiaM-Y
1095	183	GLU	InertiaM-Z
1096	183	GLU	CenterM-X
1097	183	GLU	CenterM-Y
1098	183	GLU	CenterM-Z
1099	184	LEU	InertiaM-X
1100	184	LEU	InertiaM-Y
1101	184	LEU	InertiaM-Z
1102	184	LEU	CenterM-X
1103	184	LEU	CenterM-Y
1104	184	LEU	CenterM-Z
1105	185	HIE	InertiaM-X
1106	185	HIE	InertiaM-Y
1107	185	HIE	InertiaM-Z
1108	185	HIE	CenterM-X
1109	185	HIE	CenterM-Y
1110	185	HIE	CenterM-Z
1111	186	VAL	InertiaM-X
1112	186	VAL	InertiaM-Y
1113	186	VAL	InertiaM-Z
1114	186	VAL	CenterM-X
1115	186	VAL	CenterM-Y
1116	186	VAL	CenterM-Z
1117	187	VAL	InertiaM-X
1118	187	VAL	InertiaM-Y
1119	187	VAL	InertiaM-Z
1120	187	VAL	CenterM-X
1121	187	VAL	CenterM-Y
1122	187	VAL	CenterM-Z

(*Continued*)

Table 2.1 (*Continued*)

Variable no.	Residue no.	Residue	Variable
1123	188	TYR	InertiaM-X
1124	188	TYR	InertiaM-Y
1125	188	TYR	InertiaM-Z
1126	188	TYR	CenterM-X
1127	188	TYR	CenterM-Y
1128	188	TYR	CenterM-Z
1129	189	ALA	InertiaM-X
1130	189	ALA	InertiaM-Y
1131	189	ALA	InertiaM-Z
1132	189	ALA	CenterM-X
1133	189	ALA	CenterM-Y
1134	189	ALA	CenterM-Z
1135	190	LEU	InertiaM-X
1136	190	LEU	InertiaM-Y
1137	190	LEU	InertiaM-Z
1138	190	LEU	CenterM-X
1139	190	LEU	CenterM-Y
1140	190	LEU	CenterM-Z
1141	191	SER	InertiaM-X
1142	191	SER	InertiaM-Y
1143	191	SER	InertiaM-Z
1144	191	SER	CenterM-X
1145	191	SER	CenterM-Y
1146	191	SER	CenterM-Z
1147	192	HIE	InertiaM-X
1148	192	HIE	InertiaM-Y
1149	192	HIE	InertiaM-Z
1150	192	HIE	CenterM-X
1151	192	HIE	CenterM-Y
1152	192	HIE	CenterM-Z
1153	193	VAL	InertiaM-X

Variable no.	Residue no.	Residue	Variable
1154	193	VAL	InertiaM-Y
1155	193	VAL	InertiaM-Z
1156	193	VAL	CenterM-X
1157	193	VAL	CenterM-Y
1158	193	VAL	CenterM-Z
1159	194	CYS	InertiaM-X
1160	194	CYS	InertiaM-Y
1161	194	CYS	InertiaM-Z
1162	194	CYS	CenterM-X
1163	194	CYS	CenterM-Y
1164	194	CYS	CenterM-Z
1165	195	GLY	InertiaM-X
1166	195	GLY	InertiaM-Y
1167	195	GLY	InertiaM-Z
1168	195	GLY	CenterM-X
1169	195	GLY	CenterM-Y
1170	195	GLY	CenterM-Z
1171	196	GLN	InertiaM-X
1172	196	GLN	InertiaM-Y
1173	196	GLN	InertiaM-Z
1174	196	GLN	CenterM-X
1175	196	GLN	CenterM-Y
1176	196	GLN	CenterM-Z
1177	197	ASP	InertiaM-X
1178	197	ASP	InertiaM-Y
1179	197	ASP	InertiaM-Z
1180	197	ASP	CenterM-X
1181	197	ASP	CenterM-Y
1182	197	ASP	CenterM-Z
1183	198	ARG	InertiaM-X
1184	198	ARG	InertiaM-Y
1185	198	ARG	InertiaM-Z

(*Continued*)

Table 2.1 (*Continued*)

Variable no.	Residue no.	Residue	Variable
1186	198	ARG	CenterM-X
1187	198	ARG	CenterM-Y
1188	198	ARG	CenterM-Z
1189	199	THR	InertiaM-X
1190	199	THR	InertiaM-Y
1191	199	THR	InertiaM-Z
1192	199	THR	CenterM-X
1193	199	THR	CenterM-Y
1194	199	THR	CenterM-Z
1195	200	LEU	InertiaM-X
1196	200	LEU	InertiaM-Y
1197	200	LEU	InertiaM-Z
1198	200	LEU	CenterM-X
1199	200	LEU	CenterM-Y
1200	200	LEU	CenterM-Z
1201	201	LEU	InertiaM-X
1202	201	LEU	InertiaM-Y
1203	201	LEU	InertiaM-Z
1204	201	LEU	CenterM-X
1205	201	LEU	CenterM-Y
1206	201	LEU	CenterM-Z
1207	202	ALA	InertiaM-X
1208	202	ALA	InertiaM-Y
1209	202	ALA	InertiaM-Z
1210	202	ALA	CenterM-X
1211	202	ALA	CenterM-Y
1212	202	ALA	CenterM-Z
1213	203	SER	InertiaM-X
1214	203	SER	InertiaM-Y
1215	203	SER	InertiaM-Z
1216	203	SER	CenterM-X

Variable no.	Residue no.	Residue	Variable
1217	203	SER	CenterM-Y
1218	203	SER	CenterM-Z
1219	204	ILE	InertiaM-X
1220	204	ILE	InertiaM-Y
1221	204	ILE	InertiaM-Z
1222	204	ILE	CenterM-X
1223	204	ILE	CenterM-Y
1224	204	ILE	CenterM-Z
1225	205	LEU	InertiaM-X
1226	205	LEU	InertiaM-Y
1227	205	LEU	InertiaM-Z
1228	205	LEU	CenterM-X
1229	205	LEU	CenterM-Y
1230	205	LEU	CenterM-Z
1231	206	LEU	InertiaM-X
1232	206	LEU	InertiaM-Y
1233	206	LEU	InertiaM-Z
1234	206	LEU	CenterM-X
1235	206	LEU	CenterM-Y
1236	206	LEU	CenterM-Z
1237	207	ARG	InertiaM-X
1238	207	ARG	InertiaM-Y
1239	207	ARG	InertiaM-Z
1240	207	ARG	CenterM-X
1241	207	ARG	CenterM-Y
1242	207	ARG	CenterM-Z
1243	208	ILE	InertiaM-X
1244	208	ILE	InertiaM-Y
1245	208	ILE	InertiaM-Z
1246	208	ILE	CenterM-X
1247	208	ILE	CenterM-Y
1248	208	ILE	CenterM-Z

(*Continued*)

Table 2.1 (*Continued*)

Variable no.	Residue no.	Residue	Variable
1249	209	PHE	InertiaM-X
1250	209	PHE	InertiaM-Y
1251	209	PHE	InertiaM-Z
1252	209	PHE	CenterM-X
1253	209	PHE	CenterM-Y
1254	209	PHE	CenterM-Z
1255	210	LEU	InertiaM-X
1256	210	LEU	InertiaM-Y
1257	210	LEU	InertiaM-Z
1258	210	LEU	CenterM-X
1259	210	LEU	CenterM-Y
1260	210	LEU	CenterM-Z
1261	211	HIE	InertiaM-X
1262	211	HIE	InertiaM-Y
1263	211	HIE	InertiaM-Z
1264	211	HIE	CenterM-X
1265	211	HIE	CenterM-Y
1266	211	HIE	CenterM-Z
1267	212	GLU	InertiaM-X
1268	212	GLU	InertiaM-Y
1269	212	GLU	InertiaM-Z
1270	212	GLU	CenterM-X
1271	212	GLU	CenterM-Y
1272	212	GLU	CenterM-Z
1273	213	LYS	InertiaM-X
1274	213	LYS	InertiaM-Y
1275	213	LYS	InertiaM-Z
1276	213	LYS	CenterM-X
1277	213	LYS	CenterM-Y
1278	213	LYS	CenterM-Z
1279	214	LEU	InertiaM-X

Variable no.	Residue no.	Residue	Variable
1280	214	LEU	InertiaM-Y
1281	214	LEU	InertiaM-Z
1282	214	LEU	CenterM-X
1283	214	LEU	CenterM-Y
1284	214	LEU	CenterM-Z
1285	215	GLU	InertiaM-X
1286	215	GLU	InertiaM-Y
1287	215	GLU	InertiaM-Z
1288	215	GLU	CenterM-X
1289	215	GLU	CenterM-Y
1290	215	GLU	CenterM-Z
1291	216	SER	InertiaM-X
1292	216	SER	InertiaM-Y
1293	216	SER	InertiaM-Z
1294	216	SER	CenterM-X
1295	216	SER	CenterM-Y
1296	216	SER	CenterM-Z
1297	217	LEU	InertiaM-X
1298	217	LEU	InertiaM-Y
1299	217	LEU	InertiaM-Z
1300	217	LEU	CenterM-X
1301	217	LEU	CenterM-Y
1302	217	LEU	CenterM-Z
1303	218	LEU	InertiaM-X
1304	218	LEU	InertiaM-Y
1305	218	LEU	InertiaM-Z
1306	218	LEU	CenterM-X
1307	218	LEU	CenterM-Y
1308	218	LEU	CenterM-Z
1309	219	LEU	InertiaM-X
1310	219	LEU	InertiaM-Y
1311	219	LEU	InertiaM-Z

(*Continued*)

Table 2.1 (*Continued*)

Variable no.	Residue no.	Residue	Variable
1312	219	LEU	CenterM-X
1313	219	LEU	CenterM-Y
1314	219	LEU	CenterM-Z
1315	220	CYX	InertiaM-X
1316	220	CYX	InertiaM-Y
1317	220	CYX	InertiaM-Z
1318	220	CYX	CenterM-X
1319	220	CYX	CenterM-Y
1320	220	CYX	CenterM-Z
1321	221	THR	InertiaM-X
1322	221	THR	InertiaM-Y
1323	221	THR	InertiaM-Z
1324	221	THR	CenterM-X
1325	221	THR	CenterM-Y
1326	221	THR	CenterM-Z
1327	222	LEU	InertiaM-X
1328	222	LEU	InertiaM-Y
1329	222	LEU	InertiaM-Z
1330	222	LEU	CenterM-X
1331	222	LEU	CenterM-Y
1332	222	LEU	CenterM-Z
1333	223	ASN	InertiaM-X
1334	223	ASN	InertiaM-Y
1335	223	ASN	InertiaM-Z
1336	223	ASN	CenterM-X
1337	223	ASN	CenterM-Y
1338	223	ASN	CenterM-Z
1339	224	ASP	InertiaM-X
1340	224	ASP	InertiaM-Y
1341	224	ASP	InertiaM-Z
1342	224	ASP	CenterM-X

Variable no.	Residue no.	Residue	Variable
1343	224	ASP	CenterM-Y
1344	224	ASP	CenterM-Z
1345	225	ARG	InertiaM-X
1346	225	ARG	InertiaM-Y
1347	225	ARG	InertiaM-Z
1348	225	ARG	CenterM-X
1349	225	ARG	CenterM-Y
1350	225	ARG	CenterM-Z
1351	226	GLU	InertiaM-X
1352	226	GLU	InertiaM-Y
1353	226	GLU	InertiaM-Z
1354	226	GLU	CenterM-X
1355	226	GLU	CenterM-Y
1356	226	GLU	CenterM-Z
1357	227	ILE	InertiaM-X
1358	227	ILE	InertiaM-Y
1359	227	ILE	InertiaM-Z
1360	227	ILE	CenterM-X
1361	227	ILE	CenterM-Y
1362	227	ILE	CenterM-Z
1363	228	SER	InertiaM-X
1364	228	SER	InertiaM-Y
1365	228	SER	InertiaM-Z
1366	228	SER	CenterM-X
1367	228	SER	CenterM-Y
1368	228	SER	CenterM-Z
1369	229	MET	InertiaM-X
1370	229	MET	InertiaM-Y
1371	229	MET	InertiaM-Z
1372	229	MET	CenterM-X
1373	229	MET	CenterM-Y
1374	229	MET	CenterM-Z

(*Continued*)

Table 2.1 (*Continued*)

Variable no.	Residue no.	Residue	Variable
1375	230	GLU	InertiaM-X
1376	230	GLU	InertiaM-Y
1377	230	GLU	InertiaM-Z
1378	230	GLU	CenterM-X
1379	230	GLU	CenterM-Y
1380	230	GLU	CenterM-Z
1381	230	ASP	InertiaM-X
1382	230	ASP	InertiaM-Y
1383	231	ASP	InertiaM-Z
1384	231	ASP	CenterM-X
1385	231	ASP	CenterM-Y
1386	231	ASP	CenterM-Z
1387	232	GLU	InertiaM-X
1388	232	GLU	InertiaM-Y
1389	232	GLU	InertiaM-Z
1390	232	GLU	CenterM-X
1391	232	GLU	CenterM-Y
1392	232	GLU	CenterM-Z
1393	233	ALA	InertiaM-X
1394	233	ALA	InertiaM-Y
1395	233	ALA	InertiaM-Z
1396	233	ALA	CenterM-X
1397	233	ALA	CenterM-Y
1398	233	ALA	CenterM-Z
1399	234	THR	InertiaM-X
1400	234	THR	InertiaM-Y
1401	234	THR	InertiaM-Z
1402	234	THR	CenterM-X
1403	234	THR	CenterM-Y
1404	234	THR	CenterM-Z
1405	234	THR	InertiaM-X

Variable no.	Residue no.	Residue	Variable
1406	234	THR	InertiaM-Y
1407	235	THR	InertiaM-Z
1408	235	THR	CenterM-X
1409	235	THR	CenterM-Y
1410	235	THR	CenterM-Z
1411	236	LEU	InertiaM-X
1412	236	LEU	InertiaM-Y
1413	236	LEU	InertiaM-Z
1414	236	LEU	CenterM-X
1415	236	LEU	CenterM-Y
1416	236	LEU	CenterM-Z
1417	237	PHE	InertiaM-X
1418	237	PHE	InertiaM-Y
1419	237	PHE	InertiaM-Z
1420	237	PHE	CenterM-X
1421	237	PHE	CenterM-Y
1422	237	PHE	CenterM-Z
1423	238	ARG	InertiaM-X
1424	238	ARG	InertiaM-Y
1425	238	ARG	InertiaM-Z
1426	238	ARG	CenterM-X
1427	238	ARG	CenterM-Y
1428	238	ARG	CenterM-Z
1429	239	ALA	InertiaM-X
1430	239	ALA	InertiaM-Y
1431	239	ALA	InertiaM-Z
1432	239	ALA	CenterM-X
1433	239	ALA	CenterM-Y
1434	239	ALA	CenterM-Z
1435	240	THR	InertiaM-X
1436	240	THR	InertiaM-Y
1437	240	THR	InertiaM-Z

(*Continued*)

Table 2.1 (*Continued*)

Variable no.	Residue no.	Residue	Variable
1438	240	THR	CenterM-X
1439	240	THR	CenterM-Y
1440	240	THR	CenterM-Z
1441	241	THR	InertiaM-X
1442	241	THR	InertiaM-Y
1443	241	THR	InertiaM-Z
1444	241	THR	CenterM-X
1445	241	THR	CenterM-Y
1446	241	THR	CenterM-Z
1447	242	LEU	InertiaM-X
1448	242	LEU	InertiaM-Y
1449	242	LEU	InertiaM-Z
1450	242	LEU	CenterM-X
1451	242	LEU	CenterM-Y
1452	242	LEU	CenterM-Z
1453	243	ALA	InertiaM-X
1454	243	ALA	InertiaM-Y
1455	243	ALA	InertiaM-Z
1456	243	ALA	CenterM-X
1457	243	ALA	CenterM-Y
1458	243	ALA	CenterM-Z
1459	244	SER	InertiaM-X
1460	244	SER	InertiaM-Y
1461	244	SER	InertiaM-Z
1462	244	SER	CenterM-X
1463	244	SER	CenterM-Y
1464	244	SER	CenterM-Z
1465	245	THR	InertiaM-X
1466	245	THR	InertiaM-Y
1467	245	THR	InertiaM-Z
1468	245	THR	CenterM-X

Variable no.	Residue no.	Residue	Variable
1469	245	THR	CenterM-Y
1470	245	THR	CenterM-Z
1471	246	LEU	InertiaM-X
1472	246	LEU	InertiaM-Y
1473	246	LEU	InertiaM-Z
1474	246	LEU	CenterM-X
1475	246	LEU	CenterM-Y
1476	246	LEU	CenterM-Z
1477	247	MET	InertiaM-X
1478	247	MET	InertiaM-Y
1479	247	MET	InertiaM-Z
1480	247	MET	CenterM-X
1481	247	MET	CenterM-Y
1482	247	MET	CenterM-Z
1483	248	GLU	InertiaM-X
1484	248	GLU	InertiaM-Y
1485	248	GLU	InertiaM-Z
1486	248	GLU	CenterM-X
1487	248	GLU	CenterM-Y
1488	248	GLU	CenterM-Z
1489	249	GLN	InertiaM-X
1490	249	GLN	InertiaM-Y
1491	249	GLN	InertiaM-Z
1492	249	GLN	CenterM-X
1493	249	GLN	CenterM-Y
1494	249	GLN	CenterM-Z
1495	250	TYR	InertiaM-X
1496	250	TYR	InertiaM-Y
1497	250	TYR	InertiaM-Z
1498	250	TYR	CenterM-X
1499	250	TYR	CenterM-Y
1500	250	TYR	CenterM-Z

(*Continued*)

Table 2.1 (*Continued*)

Variable no.	Residue no.	Residue	Variable
1501	251	MET	InertiaM-X
1502	251	MET	InertiaM-Y
1503	251	MET	InertiaM-Z
1504	251	MET	CenterM-X
1505	251	MET	CenterM-Y
1506	251	MET	CenterM-Z
1507	252	LYS	InertiaM-X
1508	252	LYS	InertiaM-Y
1509	252	LYS	InertiaM-Z
1510	252	LYS	CenterM-X
1511	252	LYS	CenterM-Y
1512	252	LYS	CenterM-Z
1513	253	ALA	InertiaM-X
1514	253	ALA	InertiaM-Y
1515	253	ALA	InertiaM-Z
1516	253	ALA	CenterM-X
1517	253	ALA	CenterM-Y
1518	253	ALA	CenterM-Z
1519	254	THR	InertiaM-X
1520	254	THR	InertiaM-Y
1521	254	THR	InertiaM-Z
1522	254	THR	CenterM-X
1523	254	THR	CenterM-Y
1524	254	THR	CenterM-Z
1525	254	ALA	InertiaM-X
1526	254	ALA	InertiaM-Y
1527	255	ALA	InertiaM-Z
1528	255	ALA	CenterM-X
1529	255	ALA	CenterM-Y
1530	255	ALA	CenterM-Z
1531	256	THR	InertiaM-X

Variable no.	Residue no.	Residue	Variable
1532	256	THR	InertiaM-Y
1533	256	THR	InertiaM-Z
1534	256	THR	CenterM-X
1535	256	THR	CenterM-Y
1536	256	THR	CenterM-Z
1537	257	GLN	InertiaM-X
1538	257	GLN	InertiaM-Y
1539	257	GLN	InertiaM-Z
1540	257	GLN	CenterM-X
1541	257	GLN	CenterM-Y
1542	257	GLN	CenterM-Z
1543	258	PHE	InertiaM-X
1544	258	PHE	InertiaM-Y
1545	258	PHE	InertiaM-Z
1546	258	PHE	CenterM-X
1547	258	PHE	CenterM-Y
1548	258	PHE	CenterM-Z
1549	259	VAL	InertiaM-X
1550	259	VAL	InertiaM-Y
1551	259	VAL	InertiaM-Z
1552	259	VAL	CenterM-X
1553	259	VAL	CenterM-Y
1554	259	VAL	CenterM-Z
1555	260	HIE	InertiaM-X
1556	260	HIE	InertiaM-Y
1557	260	HIE	InertiaM-Z
1558	260	HIE	CenterM-X
1559	260	HIE	CenterM-Y
1560	260	HIE	CenterM-Z
1561	260	HIE	InertiaM-X
1562	260	HIE	InertiaM-Y
1563	261	HIE	InertiaM-Z

(*Continued*)

Table 2.1 (*Continued*)

Variable no.	Residue no.	Residue	Variable
1564	261	HIE	CenterM-X
1565	261	HIE	CenterM-Y
1566	261	HIE	CenterM-Z
1567	262	ALA	InertiaM-X
1568	262	ALA	InertiaM-Y
1569	262	ALA	InertiaM-Z
1570	262	ALA	CenterM-X
1571	262	ALA	CenterM-Y
1572	262	ALA	CenterM-Z
1573	263	LEU	InertiaM-X
1574	263	LEU	InertiaM-Y
1575	263	LEU	InertiaM-Z
1576	263	LEU	CenterM-X
1577	263	LEU	CenterM-Y
1578	263	LEU	CenterM-Z
1579	264	LYS	InertiaM-X
1580	264	LYS	InertiaM-Y
1581	264	LYS	InertiaM-Z
1582	264	LYS	CenterM-X
1583	264	LYS	CenterM-Y
1584	264	LYS	CenterM-Z
1585	265	ASP	InertiaM-X
1586	265	ASP	InertiaM-Y
1587	265	ASP	InertiaM-Z
1588	265	ASP	CenterM-X
1589	265	ASP	CenterM-Y
1590	265	ASP	CenterM-Z
1591	266	SER	InertiaM-X
1592	266	SER	InertiaM-Y
1593	266	SER	InertiaM-Z
1594	266	SER	CenterM-X

Variable no.	Residue no.	Residue	Variable
1595	266	SER	CenterM-Y
1596	266	SER	CenterM-Z
1597	267	ILE	InertiaM-X
1598	267	ILE	InertiaM-Y
1599	267	ILE	InertiaM-Z
1600	267	ILE	CenterM-X
1601	267	ILE	CenterM-Y
1602	267	ILE	CenterM-Z
1603	268	LEU	InertiaM-X
1604	268	LEU	InertiaM-Y
1605	268	LEU	InertiaM-Z
1606	268	LEU	CenterM-X
1607	268	LEU	CenterM-Y
1608	268	LEU	CenterM-Z
1609	269	LYS	InertiaM-X
1610	269	LYS	InertiaM-Y
1611	269	LYS	InertiaM-Z
1612	269	LYS	CenterM-X
1613	269	LYS	CenterM-Y
1614	269	LYS	CenterM-Z
1615	270	ILE	InertiaM-X
1616	270	ILE	InertiaM-Y
1617	270	ILE	InertiaM-Z
1618	270	ILE	CenterM-X
1619	270	ILE	CenterM-Y
1620	270	ILE	CenterM-Z
1621	271	MET	InertiaM-X
1622	271	MET	InertiaM-Y
1623	271	MET	InertiaM-Z
1624	271	MET	CenterM-X
1625	271	MET	CenterM-Y
1626	271	MET	CenterM-Z

(*Continued*)

Table 2.1 (*Continued*)

Variable no.	Residue no.	Residue	Variable
1627	272	GLU	InertiaM-X
1628	272	GLU	InertiaM-Y
1629	272	GLU	InertiaM-Z
1630	272	GLU	CenterM-X
1631	272	GLU	CenterM-Y
1632	272	GLU	CenterM-Z
1633	273	SER	InertiaM-X
1634	273	SER	InertiaM-Y
1635	273	SER	InertiaM-Z
1636	273	SER	CenterM-X
1637	273	SER	CenterM-Y
1638	273	SER	CenterM-Z
1639	274	LYS	InertiaM-X
1640	274	LYS	InertiaM-Y
1641	274	LYS	InertiaM-Z
1642	274	LYS	CenterM-X
1643	274	LYS	CenterM-Y
1644	274	LYS	CenterM-Z
1645	275	GLN	InertiaM-X
1646	275	GLN	InertiaM-Y
1647	275	GLN	InertiaM-Z
1648	275	GLN	CenterM-X
1649	275	GLN	CenterM-Y
1650	275	GLN	CenterM-Z
1651	276	SER	InertiaM-X
1652	276	SER	InertiaM-Y
1653	276	SER	InertiaM-Z
1654	276	SER	CenterM-X
1655	276	SER	CenterM-Y
1656	276	SER	CenterM-Z
1657	277	CYS	InertiaM-X

Variable no.	Residue no.	Residue	Variable
1658	277	CYS	InertiaM-Y
1659	277	CYS	InertiaM-Z
1660	277	CYS	CenterM-X
1661	277	CYS	CenterM-Y
1662	277	CYS	CenterM-Z
1663	278	GLU	InertiaM-X
1664	278	GLU	InertiaM-Y
1665	278	GLU	InertiaM-Z
1666	278	GLU	CenterM-X
1667	278	GLU	CenterM-Y
1668	278	GLU	CenterM-Z
1669	279	LEU	InertiaM-X
1670	279	LEU	InertiaM-Y
1671	279	LEU	InertiaM-Z
1672	279	LEU	CenterM-X
1673	279	LEU	CenterM-Y
1674	279	LEU	CenterM-Z
1675	280	SER	InertiaM-X
1676	280	SER	InertiaM-Y
1677	280	SER	InertiaM-Z
1678	280	SER	CenterM-X
1679	280	SER	CenterM-Y
1680	280	SER	CenterM-Z
1681	281	PRO	InertiaM-X
1682	281	PRO	InertiaM-Y
1683	281	PRO	InertiaM-Z
1684	281	PRO	CenterM-X
1685	281	PRO	CenterM-Y
1686	281	PRO	CenterM-Z
1687	282	SER	InertiaM-X
1688	282	SER	InertiaM-Y
1689	282	SER	InertiaM-Z

(*Continued*)

Table 2.1 (*Continued*)

Variable no.	Residue no.	Residue	Variable
1690	282	SER	CenterM-X
1691	282	SER	CenterM-Y
1692	282	SER	CenterM-Z
1693	283	LYS	InertiaM-X
1694	283	LYS	InertiaM-Y
1695	283	LYS	InertiaM-Z
1696	283	LYS	CenterM-X
1697	283	LYS	CenterM-Y
1698	283	LYS	CenterM-Z
1699	284	LEU	InertiaM-X
1700	284	LEU	InertiaM-Y
1701	284	LEU	InertiaM-Z
1702	284	LEU	CenterM-X
1703	284	LEU	CenterM-Y
1704	284	LEU	CenterM-Z
1705	285	GLU	InertiaM-X
1706	285	GLU	InertiaM-Y
1707	285	GLU	InertiaM-Z
1708	285	GLU	CenterM-X
1709	285	GLU	CenterM-Y
1710	285	GLU	CenterM-Z
1711	286	LYS	InertiaM-X
1712	286	LYS	InertiaM-Y
1713	286	LYS	InertiaM-Z
1714	286	LYS	CenterM-X
1715	286	LYS	CenterM-Y
1716	286	LYS	CenterM-Z
1717	287	ASN	InertiaM-X
1718	287	ASN	InertiaM-Y
1719	287	ASN	InertiaM-Z
1720	287	ASN	CenterM-X

Variable no.	Residue no.	Residue	Variable
1721	287	ASN	CenterM-Y
1722	287	ASN	CenterM-Z
1723	288	GLU	InertiaM-X
1724	288	GLU	InertiaM-Y
1725	288	GLU	InertiaM-Z
1726	288	GLU	CenterM-X
1727	288	GLU	CenterM-Y
1728	288	GLU	CenterM-Z
1729	289	ASP	InertiaM-X
1730	289	ASP	InertiaM-Y
1731	289	ASP	InertiaM-Z
1732	289	ASP	CenterM-X
1733	289	ASP	CenterM-Y
1734	289	ASP	CenterM-Z
1735	290	VAL	InertiaM-X
1736	290	VAL	InertiaM-Y
1737	290	VAL	InertiaM-Z
1738	290	VAL	CenterM-X
1739	290	VAL	CenterM-Y
1740	290	VAL	CenterM-Z
1741	291	ASN	InertiaM-X
1742	291	ASN	InertiaM-Y
1743	291	ASN	InertiaM-Z
1744	291	ASN	CenterM-X
1745	291	ASN	CenterM-Y
1746	291	ASN	CenterM-Z
1747	291	THR	InertiaM-X
1748	292	THR	InertiaM-Y
1749	292	THR	InertiaM-Z
1750	292	THR	CenterM-X
1751	292	THR	CenterM-Y
1752	292	THR	CenterM-Z

(*Continued*)

Table 2.1 (*Continued*)

Variable no.	Residue no.	Residue	Variable
1753	293	ASN	InertiaM-X
1754	293	ASN	InertiaM-Y
1755	293	ASN	InertiaM-Z
1756	293	ASN	CenterM-X
1757	293	ASN	CenterM-Y
1758	293	ASN	CenterM-Z
1759	293	LEU	InertiaM-X
1760	293	LEU	InertiaM-Y
1761	294	LEU	InertiaM-Z
1762	294	LEU	CenterM-X
1763	294	LEU	CenterM-Y
1764	294	LEU	CenterM-Z
1765	295	THR	InertiaM-X
1766	295	THR	InertiaM-Y
1767	295	THR	InertiaM-Z
1768	295	THR	CenterM-X
1769	295	THR	CenterM-Y
1770	295	THR	CenterM-Z
1771	296	HIE	InertiaM-X
1772	296	HIE	InertiaM-Y
1773	296	HIE	InertiaM-Z
1774	296	HIE	CenterM-X
1775	296	HIE	CenterM-Y
1776	296	HIE	CenterM-Z
1777	297	LEU	InertiaM-X
1778	297	LEU	InertiaM-Y
1779	297	LEU	InertiaM-Z
1780	297	LEU	CenterM-X
1781	297	LEU	CenterM-Y
1782	297	LEU	CenterM-Z
1783	298	LEU	InertiaM-X

Variable no.	Residue no.	Residue	Variable
1784	298	LEU	InertiaM-Y
1785	298	LEU	InertiaM-Z
1786	298	LEU	CenterM-X
1787	298	LEU	CenterM-Y
1788	298	LEU	CenterM-Z
1789	299	ASN	InertiaM-X
1790	299	ASN	InertiaM-Y
1791	299	ASN	InertiaM-Z
1792	299	ASN	CenterM-X
1793	299	ASN	CenterM-Y
1794	299	ASN	CenterM-Z
1795	300	ILE	InertiaM-X
1796	300	ILE	InertiaM-Y
1797	300	ILE	InertiaM-Z
1798	300	ILE	CenterM-X
1799	300	ILE	CenterM-Y
1800	300	ILE	CenterM-Z
1801	301	LEU	InertiaM-X
1802	301	LEU	InertiaM-Y
1803	301	LEU	InertiaM-Z
1804	301	LEU	CenterM-X
1805	301	LEU	CenterM-Y
1806	301	LEU	CenterM-Z
1807	302	SER	InertiaM-X
1808	302	SER	InertiaM-Y
1809	302	SER	InertiaM-Z
1810	302	SER	CenterM-X
1811	302	SER	CenterM-Y
1812	302	SER	CenterM-Z
1813	303	GLU	InertiaM-X
1814	303	GLU	InertiaM-Y
1815	303	GLU	InertiaM-Z

(*Continued*)

Table 2.1 (*Continued*)

Variable no.	Residue no.	Residue	Variable
1816	303	GLU	CenterM-X
1817	303	GLU	CenterM-Y
1818	303	GLU	CenterM-Z
1819	304	LEU	InertiaM-X
1820	304	LEU	InertiaM-Y
1821	304	LEU	InertiaM-Z
1822	304	LEU	CenterM-X
1823	304	LEU	CenterM-Y
1824	304	LEU	CenterM-Z
1825	305	VAL	InertiaM-X
1826	305	VAL	InertiaM-Y
1827	305	VAL	InertiaM-Z
1828	305	VAL	CenterM-X
1829	305	VAL	CenterM-Y
1830	305	VAL	CenterM-Z
1831	306	GLU	InertiaM-X
1832	306	GLU	InertiaM-Y
1833	306	GLU	InertiaM-Z
1834	306	GLU	CenterM-X
1835	306	GLU	CenterM-Y
1836	306	GLU	CenterM-Z
1837	307	LYS	InertiaM-X
1838	307	LYS	InertiaM-Y
1839	307	LYS	InertiaM-Z
1840	307	LYS	CenterM-X
1841	307	LYS	CenterM-Y
1842	307	LYS	CenterM-Z
1843	308	ILE	InertiaM-X
1844	308	ILE	InertiaM-Y
1845	308	ILE	InertiaM-Z
1846	308	ILE	CenterM-X

Variable no.	Residue no.	Residue	Variable
1847	308	ILE	CenterM-Y
1848	308	ILE	CenterM-Z
1849	309	PHE	InertiaM-X
1850	309	PHE	InertiaM-Y
1851	309	PHE	InertiaM-Z
1852	309	PHE	CenterM-X
1853	309	PHE	CenterM-Y
1854	309	PHE	CenterM-Z
1855	310	MET	InertiaM-X
1856	310	MET	InertiaM-Y
1857	310	MET	InertiaM-Z
1858	310	MET	CenterM-X
1859	310	MET	CenterM-Y
1860	310	MET	CenterM-Z
1861	311	ALA	InertiaM-X
1862	311	ALA	InertiaM-Y
1863	311	ALA	InertiaM-Z
1864	311	ALA	CenterM-X
1865	311	ALA	CenterM-Y
1866	311	ALA	CenterM-Z
1867	312	SER	InertiaM-X
1868	312	SER	InertiaM-Y
1869	312	SER	InertiaM-Z
1870	312	SER	CenterM-X
1871	312	SER	CenterM-Y
1872	312	SER	CenterM-Z
1873	313	GLU	InertiaM-X
1874	313	GLU	InertiaM-Y
1875	313	GLU	InertiaM-Z
1876	313	GLU	CenterM-X
1877	313	GLU	CenterM-Y
1878	313	GLU	CenterM-Z

(*Continued*)

Table 2.1 (*Continued*)

Variable no.	Residue no.	Residue	Variable
1879	314	ILE	InertiaM-X
1880	314	ILE	InertiaM-Y
1881	314	ILE	InertiaM-Z
1882	314	ILE	CenterM-X
1883	314	ILE	CenterM-Y
1884	314	ILE	CenterM-Z
1885	315	LEU	InertiaM-X
1886	315	LEU	InertiaM-Y
1887	315	LEU	InertiaM-Z
1888	315	LEU	CenterM-X
1889	315	LEU	CenterM-Y
1890	315	LEU	CenterM-Z
1891	316	PRO	InertiaM-X
1892	316	PRO	InertiaM-Y
1893	316	PRO	InertiaM-Z
1894	316	PRO	CenterM-X
1895	316	PRO	CenterM-Y
1896	316	PRO	CenterM-Z
1897	317	PRO	InertiaM-X
1898	317	PRO	InertiaM-Y
1899	317	PRO	InertiaM-Z
1900	317	PRO	CenterM-X
1901	317	PRO	CenterM-Y
1902	317	PRO	CenterM-Z
1903	318	THR	InertiaM-X
1904	318	THR	InertiaM-Y
1905	318	THR	InertiaM-Z
1906	318	THR	CenterM-X
1907	318	THR	CenterM-Y
1908	318	THR	CenterM-Z
1909	319	LEU	InertiaM-X

Variable no.	Residue no.	Residue	Variable
1910	319	LEU	InertiaM-Y
1911	319	LEU	InertiaM-Z
1912	319	LEU	CenterM-X
1913	319	LEU	CenterM-Y
1914	319	LEU	CenterM-Z
1915	320	ARG	InertiaM-X
1916	320	ARG	InertiaM-Y
1917	320	ARG	InertiaM-Z
1918	320	ARG	CenterM-X
1919	320	ARG	CenterM-Y
1920	320	ARG	CenterM-Z
1921	321	TYR	InertiaM-X
1922	321	TYR	InertiaM-Y
1923	321	TYR	InertiaM-Z
1924	321	TYR	CenterM-X
1925	321	TYR	CenterM-Y
1926	321	TYR	CenterM-Z
1927	322	ILE	InertiaM-X
1928	322	ILE	InertiaM-Y
1929	322	ILE	InertiaM-Z
1930	322	ILE	CenterM-X
1931	322	ILE	CenterM-Y
1932	322	ILE	CenterM-Z
1933	323	TYR	InertiaM-X
1934	323	TYR	InertiaM-Y
1935	323	TYR	InertiaM-Z
1936	323	TYR	CenterM-X
1937	323	TYR	CenterM-Y
1938	323	TYR	CenterM-Z
1939	324	GLY	InertiaM-X
1940	324	GLY	InertiaM-Y
1941	324	GLY	InertiaM-Z

(*Continued*)

Table 2.1 (*Continued*)

Variable no.	Residue no.	Residue	Variable
1942	324	GLY	CenterM-X
1943	324	GLY	CenterM-Y
1944	324	GLY	CenterM-Z
1945	325	CYX	InertiaM-X
1946	325	CYX	InertiaM-Y
1947	325	CYX	InertiaM-Z
1948	325	CYX	CenterM-X
1949	325	CYX	CenterM-Y
1950	325	CYX	CenterM-Z
1951	326	LEU	InertiaM-X
1952	326	LEU	InertiaM-Y
1953	326	LEU	InertiaM-Z
1954	326	LEU	CenterM-X
1955	326	LEU	CenterM-Y
1956	326	LEU	CenterM-Z
1957	327	GLN	InertiaM-X
1958	327	GLN	InertiaM-Y
1959	327	GLN	InertiaM-Z
1960	327	GLN	CenterM-X
1961	327	GLN	CenterM-Y
1962	327	GLN	CenterM-Z
1963	328	LYS	InertiaM-X
1964	328	LYS	InertiaM-Y
1965	328	LYS	InertiaM-Z
1966	328	LYS	CenterM-X
1967	328	LYS	CenterM-Y
1968	328	LYS	CenterM-Z
1969	329	SER	InertiaM-X
1970	329	SER	InertiaM-Y
1971	329	SER	InertiaM-Z
1972	329	SER	CenterM-X

Variable no.	Residue no.	Residue	Variable
1973	329	SER	CenterM-Y
1974	329	SER	CenterM-Z
1975	330	VAL	InertiaM-X
1976	330	VAL	InertiaM-Y
1977	330	VAL	InertiaM-Z
1978	330	VAL	CenterM-X
1979	330	VAL	CenterM-Y
1980	330	VAL	CenterM-Z
1981	331	GLN	InertiaM-X
1982	331	GLN	InertiaM-Y
1983	331	GLN	InertiaM-Z
1984	331	GLN	CenterM-X
1985	331	GLN	CenterM-Y
1986	331	GLN	CenterM-Z
1987	332	HIE	InertiaM-X
1988	332	HIE	InertiaM-Y
1989	332	HIE	InertiaM-Z
1990	332	HIE	CenterM-X
1991	332	HIE	CenterM-Y
1992	332	HIE	CenterM-Z
1993	333	LYS	InertiaM-X
1994	333	LYS	InertiaM-Y
1995	333	LYS	InertiaM-Z
1996	333	LYS	CenterM-X
1997	333	LYS	CenterM-Y
1998	333	LYS	CenterM-Z
1999	334	TRP	InertiaM-X
2000	334	TRP	InertiaM-Y
2001	334	TRP	InertiaM-Z
2002	334	TRP	CenterM-X
2003	334	TRP	CenterM-Y
2004	334	TRP	CenterM-Z

(*Continued*)

Table 2.1 (*Continued*)

Variable no.	Residue no.	Residue	Variable
2005	335	PRO	InertiaM-X
2006	335	PRO	InertiaM-Y
2007	335	PRO	InertiaM-Z
2008	335	PRO	CenterM-X
2009	335	PRO	CenterM-Y
2010	335	PRO	CenterM-Z
2011	336	THR	InertiaM-X
2012	336	THR	InertiaM-Y
2013	336	THR	InertiaM-Z
2014	336	THR	CenterM-X
2015	336	THR	CenterM-Y
2016	336	THR	CenterM-Z
2017	337	ASN	InertiaM-X
2018	337	ASN	InertiaM-Y
2019	337	ASN	InertiaM-Z
2020	337	ASN	CenterM-X
2021	337	ASN	CenterM-Y
2022	337	ASN	CenterM-Z
2023	338	THR	InertiaM-X
2024	338	THR	InertiaM-Y
2025	338	THR	InertiaM-Z
2026	338	THR	CenterM-X
2027	338	THR	CenterM-Y
2028	338	THR	CenterM-Z
2029	339	THR	InertiaM-X
2030	339	THR	InertiaM-Y
2031	339	THR	InertiaM-Z
2032	339	THR	CenterM-X
2033	339	THR	CenterM-Y
2034	339	THR	CenterM-Z
2035	340	MET	InertiaM-X

Variable no.	Residue no.	Residue	Variable
2036	340	MET	InertiaM-Y
2037	340	MET	InertiaM-Z
2038	340	MET	CenterM-X
2039	340	MET	CenterM-Y
2040	340	MET	CenterM-Z
2041	341	ARG	InertiaM-X
2042	341	ARG	InertiaM-Y
2043	341	ARG	InertiaM-Z
2044	341	ARG	CenterM-X
2045	341	ARG	CenterM-Y
2046	341	ARG	CenterM-Z
2047	342	THR	InertiaM-X
2048	342	THR	InertiaM-Y
2049	342	THR	InertiaM-Z
2050	342	THR	CenterM-X
2051	342	THR	CenterM-Y
2052	342	THR	CenterM-Z
2053	343	ARG	InertiaM-X
2054	343	ARG	InertiaM-Y
2055	343	ARG	InertiaM-Z
2056	343	ARG	CenterM-X
2057	343	ARG	CenterM-Y
2058	343	ARG	CenterM-Z
2059	344	VAL	InertiaM-X
2060	344	VAL	InertiaM-Y
2061	344	VAL	InertiaM-Z
2062	344	VAL	CenterM-X
2063	344	VAL	CenterM-Y
2064	344	VAL	CenterM-Z
2065	345	VAL	InertiaM-X
2066	345	VAL	InertiaM-Y
2067	345	VAL	InertiaM-Z

(*Continued*)

Table 2.1 (*Continued*)

Variable no.	Residue no.	Residue	Variable
2068	345	VAL	CenterM-X
2069	345	VAL	CenterM-Y
2070	345	VAL	CenterM-Z
2071	346	SER	InertiaM-X
2072	346	SER	InertiaM-Y
2073	346	SER	InertiaM-Z
2074	346	SER	CenterM-X
2075	346	SER	CenterM-Y
2076	346	SER	CenterM-Z
2077	347	GLY	InertiaM-X
2078	347	GLY	InertiaM-Y
2079	347	GLY	InertiaM-Z
2080	347	GLY	CenterM-X
2081	347	GLY	CenterM-Y
2082	347	GLY	CenterM-Z
2083	348	PHE	InertiaM-X
2084	348	PHE	InertiaM-Y
2085	348	PHE	InertiaM-Z
2086	348	PHE	CenterM-X
2087	348	PHE	CenterM-Y
2088	348	PHE	CenterM-Z
2089	349	VAL	InertiaM-X
2090	349	VAL	InertiaM-Y
2091	349	VAL	InertiaM-Z
2092	349	VAL	CenterM-X
2093	349	VAL	CenterM-Y
2094	349	VAL	CenterM-Z
2095	350	PHE	InertiaM-X
2096	350	PHE	InertiaM-Y
2097	350	PHE	InertiaM-Z
2098	350	PHE	CenterM-X

Variable no.	Residue no.	Residue	Variable
2099	350	PHE	CenterM-Y
2100	350	PHE	CenterM-Z
2101	351	LEU	InertiaM-X
2102	351	LEU	InertiaM-Y
2103	351	LEU	InertiaM-Z
2104	351	LEU	CenterM-X
2105	351	LEU	CenterM-Y
2106	351	LEU	CenterM-Z
2107	352	ARG	InertiaM-X
2108	352	ARG	InertiaM-Y
2109	352	ARG	InertiaM-Z
2110	352	ARG	CenterM-X
2111	352	ARG	CenterM-Y
2112	352	ARG	CenterM-Z
2113	353	LEU	InertiaM-X
2114	353	LEU	InertiaM-Y
2115	353	LEU	InertiaM-Z
2116	353	LEU	CenterM-X
2117	353	LEU	CenterM-Y
2118	353	LEU	CenterM-Z
2119	354	ILE	InertiaM-X
2120	354	ILE	InertiaM-Y
2121	354	ILE	InertiaM-Z
2122	354	ILE	CenterM-X
2123	354	ILE	CenterM-Y
2124	354	ILE	CenterM-Z
2125	355	CYS	InertiaM-X
2126	355	CYS	InertiaM-Y
2127	355	CYS	InertiaM-Z
2128	355	CYS	CenterM-X
2129	355	CYS	CenterM-Y
2130	355	CYS	CenterM-Z

(*Continued*)

Table 2.1 (*Continued*)

Variable no.	Residue no.	Residue	Variable
2131	356	PRO	InertiaM-X
2132	356	PRO	InertiaM-Y
2133	356	PRO	InertiaM-Z
2134	356	PRO	CenterM-X
2135	356	PRO	CenterM-Y
2136	356	PRO	CenterM-Z
2137	357	ALA	InertiaM-X
2138	357	ALA	InertiaM-Y
2139	357	ALA	InertiaM-Z
2140	357	ALA	CenterM-X
2141	357	ALA	CenterM-Y
2142	357	ALA	CenterM-Z
2143	358	ILE	InertiaM-X
2144	358	ILE	InertiaM-Y
2145	358	ILE	InertiaM-Z
2146	358	ILE	CenterM-X
2147	358	ILE	CenterM-Y
2148	358	ILE	CenterM-Z
2149	359	LEU	InertiaM-X
2150	359	LEU	InertiaM-Y
2151	359	LEU	InertiaM-Z
2152	359	LEU	CenterM-X
2153	359	LEU	CenterM-Y
2154	359	LEU	CenterM-Z
2155	360	ASN	InertiaM-X
2156	360	ASN	InertiaM-Y
2157	360	ASN	InertiaM-Z
2158	360	ASN	CenterM-X
2159	360	ASN	CenterM-Y
2160	360	ASN	CenterM-Z
2161	361	PRO	InertiaM-X

Variable no.	Residue no.	Residue	Variable
2162	361	PRO	InertiaM-Y
2163	361	PRO	InertiaM-Z
2164	361	PRO	CenterM-X
2165	361	PRO	CenterM-Y
2166	361	PRO	CenterM-Z
2167	362	ARG	InertiaM-X
2168	362	ARG	InertiaM-Y
2169	362	ARG	InertiaM-Z
2170	362	ARG	CenterM-X
2171	362	ARG	CenterM-Y
2172	362	ARG	CenterM-Z
2173	363	MET	InertiaM-X
2174	363	MET	InertiaM-Y
2175	363	MET	InertiaM-Z
2176	363	MET	CenterM-X
2177	363	MET	CenterM-Y
2178	363	MET	CenterM-Z
2179	364	PHE	InertiaM-X
2180	364	PHE	InertiaM-Y
2181	364	PHE	InertiaM-Z
2182	364	PHE	CenterM-X
2183	364	PHE	CenterM-Y
2184	364	PHE	CenterM-Z
2185	365	ASN	InertiaM-X
2186	365	ASN	InertiaM-Y
2187	365	ASN	InertiaM-Z
2188	365	ASN	CenterM-X
2189	365	ASN	CenterM-Y
2190	365	ASN	CenterM-Z
2191	366	ILE	InertiaM-X
2192	366	ILE	InertiaM-Y
2193	366	ILE	InertiaM-Z

(*Continued*)

Table 2.1 (*Continued*)

Variable no.	Residue no.	Residue	Variable
2194	366	ILE	CenterM-X
2195	366	ILE	CenterM-Y
2196	366	ILE	CenterM-Z
2197	367	ILE	InertiaM-X
2198	367	ILE	InertiaM-Y
2199	367	ILE	InertiaM-Z
2200	367	ILE	CenterM-X
2201	367	ILE	CenterM-Y
2202	367	ILE	CenterM-Z
2203	368	SER	InertiaM-X
2204	368	SER	InertiaM-Y
2205	368	SER	InertiaM-Z
2206	368	SER	CenterM-X
2207	368	SER	CenterM-Y
2208	368	SER	CenterM-Z
2209	369	ASP	InertiaM-X
2210	369	ASP	InertiaM-Y
2211	369	ASP	InertiaM-Z
2212	369	ASP	CenterM-X
2213	369	ASP	CenterM-Y
2214	369	ASP	CenterM-Z
2215	370	SER	InertiaM-X
2216	370	SER	InertiaM-Y
2217	370	SER	InertiaM-Z
2218	370	SER	CenterM-X
2219	370	SER	CenterM-Y
2220	370	SER	CenterM-Z
2221	371	PRO	InertiaM-X
2222	371	PRO	InertiaM-Y
2223	371	PRO	InertiaM-Z
2224	371	PRO	CenterM-X

Variable no.	Residue no.	Residue	Variable
2225	371	PRO	CenterM-Y
2226	371	PRO	CenterM-Z
2227	372	SER	InertiaM-X
2228	372	SER	InertiaM-Y
2229	372	SER	InertiaM-Z
2230	372	SER	CenterM-X
2231	372	SER	CenterM-Y
2232	372	SER	CenterM-Z
2233	373	PRO	InertiaM-X
2234	373	PRO	InertiaM-Y
2235	373	PRO	InertiaM-Z
2236	373	PRO	CenterM-X
2237	373	PRO	CenterM-Y
2238	373	PRO	CenterM-Z
2239	374	ILE	InertiaM-X
2240	374	ILE	InertiaM-Y
2241	374	ILE	InertiaM-Z
2242	374	ILE	CenterM-X
2243	374	ILE	CenterM-Y
2244	374	ILE	CenterM-Z
2245	375	ALA	InertiaM-X
2246	375	ALA	InertiaM-Y
2247	375	ALA	InertiaM-Z
2248	375	ALA	CenterM-X
2249	375	ALA	CenterM-Y
2250	375	ALA	CenterM-Z
2251	376	ALA	InertiaM-X
2252	376	ALA	InertiaM-Y
2253	376	ALA	InertiaM-Z
2254	376	ALA	CenterM-X
2255	376	ALA	CenterM-Y
2256	376	ALA	CenterM-Z

(*Continued*)

Table 2.1 (*Continued*)

Variable no.	Residue no.	Residue	Variable
2257	377	ARG	InertiaM-X
2258	377	ARG	InertiaM-Y
2259	377	ARG	InertiaM-Z
2260	377	ARG	CenterM-X
2261	377	ARG	CenterM-Y
2262	377	ARG	CenterM-Z
2263	378	THR	InertiaM-X
2264	378	THR	InertiaM-Y
2265	378	THR	InertiaM-Z
2266	378	THR	CenterM-X
2267	378	THR	CenterM-Y
2268	378	THR	CenterM-Z
2269	379	LEU	InertiaM-X
2270	379	LEU	InertiaM-Y
2271	379	LEU	InertiaM-Z
2272	379	LEU	CenterM-X
2273	379	LEU	CenterM-Y
2274	379	LEU	CenterM-Z
2275	380	ILE	InertiaM-X
2276	380	ILE	InertiaM-Y
2277	380	ILE	InertiaM-Z
2278	380	ILE	CenterM-X
2279	380	ILE	CenterM-Y
2280	380	ILE	CenterM-Z
2281	381	LEU	InertiaM-X
2282	381	LEU	InertiaM-Y
2283	381	LEU	InertiaM-Z
2284	381	LEU	CenterM-X
2285	381	LEU	CenterM-Y
2286	381	LEU	CenterM-Z
2287	382	VAL	InertiaM-X

Variable no.	Residue no.	Residue	Variable
2288	382	VAL	InertiaM-Y
2289	382	VAL	InertiaM-Z
2290	382	VAL	CenterM-X
2291	382	VAL	CenterM-Y
2292	382	VAL	CenterM-Z
2293	383	ALA	InertiaM-X
2294	383	ALA	InertiaM-Y
2295	383	ALA	InertiaM-Z
2296	383	ALA	CenterM-X
2297	383	ALA	CenterM-Y
2298	383	ALA	CenterM-Z
2299	384	LYS	InertiaM-X
2300	384	LYS	InertiaM-Y
2301	384	LYS	InertiaM-Z
2302	384	LYS	CenterM-X
2303	384	LYS	CenterM-Y
2304	384	LYS	CenterM-Z
2305	385	SER	InertiaM-X
2306	385	SER	InertiaM-Y
2307	385	SER	InertiaM-Z
2308	385	SER	CenterM-X
2309	385	SER	CenterM-Y
2310	385	SER	CenterM-Z
2311	386	VAL	InertiaM-X
2312	386	VAL	InertiaM-Y
2313	386	VAL	InertiaM-Z
2314	386	VAL	CenterM-X
2315	386	VAL	CenterM-Y
2316	386	VAL	CenterM-Z
2317	387	GLN	InertiaM-X
2318	387	GLN	InertiaM-Y
2319	387	GLN	InertiaM-Z

(*Continued*)

Table 2.1 (*Continued*)

Variable no.	Residue no.	Residue	Variable
2320	387	GLN	CenterM-X
2321	387	GLN	CenterM-Y
2322	387	GLN	CenterM-Z
2323	388	ASN	InertiaM-X
2324	388	ASN	InertiaM-Y
2325	388	ASN	InertiaM-Z
2326	388	ASN	CenterM-X
2327	388	ASN	CenterM-Y
2328	388	ASN	CenterM-Z
2329	389	LEU	InertiaM-X
2330	389	LEU	InertiaM-Y
2331	389	LEU	InertiaM-Z
2332	389	LEU	CenterM-X
2333	389	LEU	CenterM-Y
2334	389	LEU	CenterM-Z
2335	390	ALA	InertiaM-X
2336	390	ALA	InertiaM-Y
2337	390	ALA	InertiaM-Z
2338	390	ALA	CenterM-X
2339	390	ALA	CenterM-Y
2340	390	ALA	CenterM-Z
2341	391	ASN	InertiaM-X
2342	391	ASN	InertiaM-Y
2343	391	ASN	InertiaM-Z
2344	391	ASN	CenterM-X
2345	391	ASN	CenterM-Y
2346	391	ASN	CenterM-Z
2347	392	LEU	InertiaM-X
2348	392	LEU	InertiaM-Y
2349	392	LEU	InertiaM-Z
2350	392	LEU	CenterM-X

Variable no.	Residue no.	Residue	Variable
2351	392	LEU	CenterM-Y
2352	392	LEU	CenterM-Z
2353	393	VAL	InertiaM-X
2354	393	VAL	InertiaM-Y
2355	393	VAL	InertiaM-Z
2356	393	VAL	CenterM-X
2357	393	VAL	CenterM-Y
2358	393	VAL	CenterM-Z
2359	394	GLU	InertiaM-X
2360	394	GLU	InertiaM-Y
2361	394	GLU	InertiaM-Z
2362	394	GLU	CenterM-X
2363	394	GLU	CenterM-Y
2364	394	GLU	CenterM-Z
2365	395	PHE	InertiaM-X
2366	395	PHE	InertiaM-Y
2367	395	PHE	InertiaM-Z
2368	395	PHE	CenterM-X
2369	395	PHE	CenterM-Y
2370	395	PHE	CenterM-Z
2371	396	GLY	InertiaM-X
2372	396	GLY	InertiaM-Y
2373	396	GLY	InertiaM-Z
2374	396	GLY	CenterM-X
2375	396	GLY	CenterM-Y
2376	396	GLY	CenterM-Z
2377	397	ALA	InertiaM-X
2378	397	ALA	InertiaM-Y
2379	397	ALA	InertiaM-Z
2380	397	ALA	CenterM-X
2381	397	ALA	CenterM-Y
2382	397	ALA	CenterM-Z

(*Continued*)

Table 2.1 (*Continued*)

Variable no.	Residue no.	Residue	Variable
2383	398	LYS	InertiaM-X
2384	398	LYS	InertiaM-Y
2385	398	LYS	InertiaM-Z
2386	398	LYS	CenterM-X
2387	398	LYS	CenterM-Y
2388	398	LYS	CenterM-Z
2389	399	GLU	InertiaM-X
2390	399	GLU	InertiaM-Y
2391	399	GLU	InertiaM-Z
2392	399	GLU	CenterM-X
2393	399	GLU	CenterM-Y
2394	399	GLU	CenterM-Z
2395	400	PRO	InertiaM-X
2396	400	PRO	InertiaM-Y
2397	400	PRO	InertiaM-Z
2398	400	PRO	CenterM-X
2399	400	PRO	CenterM-Y
2400	400	PRO	CenterM-Z
2401	401	TYR	InertiaM-X
2402	401	TYR	InertiaM-Y
2403	401	TYR	InertiaM-Z
2404	401	TYR	CenterM-X
2405	401	TYR	CenterM-Y
2406	401	TYR	CenterM-Z
2407	402	MET	InertiaM-X
2408	402	MET	InertiaM-Y
2409	402	MET	InertiaM-Z
2410	402	MET	CenterM-X
2411	402	MET	CenterM-Y
2412	402	MET	CenterM-Z
2413	403	GLU	InertiaM-X

Variable no.	Residue no.	Residue	Variable
2414	403	GLU	InertiaM-Y
2415	403	GLU	InertiaM-Z
2416	403	GLU	CenterM-X
2417	403	GLU	CenterM-Y
2418	403	GLU	CenterM-Z
2419	404	GLY	InertiaM-X
2420	404	GLY	InertiaM-Y
2421	404	GLY	InertiaM-Z
2422	404	GLY	CenterM-X
2423	404	GLY	CenterM-Y
2424	404	GLY	CenterM-Z
2425	405	VAL	InertiaM-X
2426	405	VAL	InertiaM-Y
2427	405	VAL	InertiaM-Z
2428	405	VAL	CenterM-X
2429	405	VAL	CenterM-Y
2430	405	VAL	CenterM-Z
2431	406	ASN	InertiaM-X
2432	406	ASN	InertiaM-Y
2433	406	ASN	InertiaM-Z
2434	406	ASN	CenterM-X
2435	406	ASN	CenterM-Y
2436	406	ASN	CenterM-Z
2437	407	PRO	InertiaM-X
2438	407	PRO	InertiaM-Y
2439	407	PRO	InertiaM-Z
2440	407	PRO	CenterM-X
2441	407	PRO	CenterM-Y
2442	407	PRO	CenterM-Z
2443	408	PHE	InertiaM-X
2444	408	PHE	InertiaM-Y
2445	408	PHE	InertiaM-Z

(*Continued*)

Table 2.1 (*Continued*)

Variable no.	Residue no.	Residue	Variable
2446	408	PHE	CenterM-X
2447	408	PHE	CenterM-Y
2448	408	PHE	CenterM-Z
2449	409	ILE	InertiaM-X
2450	409	ILE	InertiaM-Y
2451	409	ILE	InertiaM-Z
2452	409	ILE	CenterM-X
2453	409	ILE	CenterM-Y
2454	409	ILE	CenterM-Z
2455	410	LYS	InertiaM-X
2456	410	LYS	InertiaM-Y
2457	410	LYS	InertiaM-Z
2458	410	LYS	CenterM-X
2459	410	LYS	CenterM-Y
2460	410	LYS	CenterM-Z
2461	411	SER	InertiaM-X
2462	411	SER	InertiaM-Y
2463	411	SER	InertiaM-Z
2464	411	SER	CenterM-X
2465	411	SER	CenterM-Y
2466	411	SER	CenterM-Z
2467	412	ASN	InertiaM-X
2468	412	ASN	InertiaM-Y
2469	412	ASN	InertiaM-Z
2470	412	ASN	CenterM-X
2471	412	ASN	CenterM-Y
2472	412	ASN	CenterM-Z
2473	413	LYS	InertiaM-X
2474	413	LYS	InertiaM-Y
2475	413	LYS	InertiaM-Z
2476	413	LYS	CenterM-X

Variable no.	Residue no.	Residue	Variable
2477	413	LYS	CenterM-Y
2478	413	LYS	CenterM-Z
2479	414	HIE	InertiaM-X
2480	414	HIE	InertiaM-Y
2481	414	HIE	InertiaM-Z
2482	414	HIE	CenterM-X
2483	414	HIE	CenterM-Y
2484	414	HIE	CenterM-Z
2485	415	ARG	InertiaM-X
2486	415	ARG	InertiaM-Y
2487	415	ARG	InertiaM-Z
2488	415	ARG	CenterM-X
2489	415	ARG	CenterM-Y
2490	415	ARG	CenterM-Z
2491	416	MET	InertiaM-X
2492	416	MET	InertiaM-Y
2493	416	MET	InertiaM-Z
2494	416	MET	CenterM-X
2495	416	MET	CenterM-Y
2496	416	MET	CenterM-Z
2497	417	ILE	InertiaM-X
2498	417	ILE	InertiaM-Y
2499	417	ILE	InertiaM-Z
2500	417	ILE	CenterM-X
2501	417	ILE	CenterM-Y
2502	417	ILE	CenterM-Z
2503	418	MET	InertiaM-X
2504	418	MET	InertiaM-Y
2505	418	MET	InertiaM-Z
2506	418	MET	CenterM-X
2507	418	MET	CenterM-Y
2508	418	MET	CenterM-Z

(*Continued*)

Table 2.1 (*Continued*)

Variable no.	Residue no.	Residue	Variable
2509	419	PHE	InertiaM-X
2510	419	PHE	InertiaM-Y
2511	419	PHE	InertiaM-Z
2512	419	PHE	CenterM-X
2513	419	PHE	CenterM-Y
2514	419	PHE	CenterM-Z
2515	420	LEU	InertiaM-X
2516	420	LEU	InertiaM-Y
2517	420	LEU	InertiaM-Z
2518	420	LEU	CenterM-X
2519	420	LEU	CenterM-Y
2520	420	LEU	CenterM-Z
2521	421	ASP	InertiaM-X
2522	421	ASP	InertiaM-Y
2523	421	ASP	InertiaM-Z
2524	421	ASP	CenterM-X
2525	421	ASP	CenterM-Y
2526	421	ASP	CenterM-Z
2527	422	GLU	InertiaM-X
2528	422	GLU	InertiaM-Y
2529	422	GLU	InertiaM-Z
2530	422	GLU	CenterM-X
2531	422	GLU	CenterM-Y
2532	422	GLU	CenterM-Z
2533	423	LEU	InertiaM-X
2534	423	LEU	InertiaM-Y
2535	423	LEU	InertiaM-Z
2536	423	LEU	CenterM-X
2537	423	LEU	CenterM-Y
2538	423	LEU	CenterM-Z
2539	424	GLY	InertiaM-X

Variable no.	Residue no.	Residue	Variable
2540	424	GLY	InertiaM-Y
2541	424	GLY	InertiaM-Z
2542	424	GLY	CenterM-X
2543	424	GLY	CenterM-Y
2544	424	GLY	CenterM-Z
2545	425	ASN	InertiaM-X
2546	425	ASN	InertiaM-Y
2547	425	ASN	InertiaM-Z
2548	425	ASN	CenterM-X
2549	425	ASN	CenterM-Y
2550	425	ASN	CenterM-Z
2551	426	VAL	InertiaM-X
2552	426	VAL	InertiaM-Y
2553	426	VAL	InertiaM-Z
2554	426	VAL	CenterM-X
2555	426	VAL	CenterM-Y
2556	426	VAL	CenterM-Z
2557	427	PRO	InertiaM-X
2558	427	PRO	InertiaM-Y
2559	427	PRO	InertiaM-Z
2560	427	PRO	CenterM-X
2561	427	PRO	CenterM-Y
2562	427	PRO	CenterM-Z
2563	428	GLU	InertiaM-X
2564	428	GLU	InertiaM-Y
2565	428	GLU	InertiaM-Z
2566	428	GLU	CenterM-X
2567	428	GLU	CenterM-Y
2568	428	GLU	CenterM-Z
2569	429	LEU	InertiaM-X
2570	429	LEU	InertiaM-Y
2571	429	LEU	InertiaM-Z

(*Continued*)

Table 2.1 (*Continued*)

Variable no.	Residue no.	Residue	Variable
2572	429	LEU	CenterM-X
2573	429	LEU	CenterM-Y
2574	429	LEU	CenterM-Z
2575	430	PRO	InertiaM-X
2576	430	PRO	InertiaM-Y
2577	430	PRO	InertiaM-Z
2578	430	PRO	CenterM-X
2579	430	PRO	CenterM-Y
2580	430	PRO	CenterM-Z
2581	431	ASP	InertiaM-X
2582	431	ASP	InertiaM-Y
2583	431	ASP	InertiaM-Z
2584	431	ASP	CenterM-X
2585	431	ASP	CenterM-Y
2586	431	ASP	CenterM-Z
2587	432	THR	InertiaM-X
2588	432	THR	InertiaM-Y
2589	432	THR	InertiaM-Z
2590	432	THR	CenterM-X
2591	432	THR	CenterM-Y
2592	432	THR	CenterM-Z
2593	433	THR	InertiaM-X
2594	433	THR	InertiaM-Y
2595	433	THR	InertiaM-Z
2596	433	THR	CenterM-X
2597	433	THR	CenterM-Y
2598	433	THR	CenterM-Z
2599	434	GLU	InertiaM-X
2600	434	GLU	InertiaM-Y
2601	434	GLU	InertiaM-Z
2602	434	GLU	CenterM-X

Variable no.	Residue no.	Residue	Variable
2603	434	GLU	CenterM-Y
2604	434	GLU	CenterM-Z
2605	435	HIE	InertiaM-X
2606	435	HIE	InertiaM-Y
2607	435	HIE	InertiaM-Z
2608	435	HIE	CenterM-X
2609	435	HIE	CenterM-Y
2610	435	HIE	CenterM-Z
2611	436	SER	InertiaM-X
2612	436	SER	InertiaM-Y
2613	436	SER	InertiaM-Z
2614	436	SER	CenterM-X
2615	436	SER	CenterM-Y
2616	436	SER	CenterM-Z
2617	437	ARG	InertiaM-X
2618	437	ARG	InertiaM-Y
2619	437	ARG	InertiaM-Z
2620	437	ARG	CenterM-X
2621	437	ARG	CenterM-Y
2622	437	ARG	CenterM-Z
2623	438	THR	InertiaM-X
2624	438	THR	InertiaM-Y
2625	438	THR	InertiaM-Z
2626	438	THR	CenterM-X
2627	438	THR	CenterM-Y
2628	438	THR	CenterM-Z
2629	439	ASP	InertiaM-X
2630	439	ASP	InertiaM-Y
2631	439	ASP	InertiaM-Z
2632	439	ASP	CenterM-X
2633	439	ASP	CenterM-Y
2634	439	ASP	CenterM-Z

(*Continued*)

Table 2.1 (*Continued*)

Variable no.	Residue no.	Residue	Variable
2635	440	LEU	InertiaM-X
2636	440	LEU	InertiaM-Y
2637	440	LEU	InertiaM-Z
2638	440	LEU	CenterM-X
2639	440	LEU	CenterM-Y
2640	440	LEU	CenterM-Z
2641	441	SER	InertiaM-X
2642	441	SER	InertiaM-Y
2643	441	SER	InertiaM-Z
2644	441	SER	CenterM-X
2645	441	SER	CenterM-Y
2646	441	SER	CenterM-Z
2647	442	ARG	InertiaM-X
2648	442	ARG	InertiaM-Y
2649	442	ARG	InertiaM-Z
2650	442	ARG	CenterM-X
2651	442	ARG	CenterM-Y
2652	442	ARG	CenterM-Z
2653	443	ASP	InertiaM-X
2654	443	ASP	InertiaM-Y
2655	443	ASP	InertiaM-Z
2656	443	ASP	CenterM-X
2657	443	ASP	CenterM-Y
2658	443	ASP	CenterM-Z
2659	444	LEU	InertiaM-X
2660	444	LEU	InertiaM-Y
2661	444	LEU	InertiaM-Z
2662	444	LEU	CenterM-X
2663	444	LEU	CenterM-Y
2664	444	LEU	CenterM-Z
2665	445	ALA	InertiaM-X

Variable no.	Residue no.	Residue	Variable
2666	445	ALA	InertiaM-Y
2667	445	ALA	InertiaM-Z
2668	445	ALA	CenterM-X
2669	445	ALA	CenterM-Y
2670	445	ALA	CenterM-Z
2671	446	ALA	InertiaM-X
2672	446	ALA	InertiaM-Y
2673	446	ALA	InertiaM-Z
2674	446	ALA	CenterM-X
2675	446	ALA	CenterM-Y
2676	446	ALA	CenterM-Z
2677	447	LEU	InertiaM-X
2678	447	LEU	InertiaM-Y
2679	447	LEU	InertiaM-Z
2680	447	LEU	CenterM-X
2681	447	LEU	CenterM-Y
2682	447	LEU	CenterM-Z
2683	448	HIE	InertiaM-X
2684	448	HIE	InertiaM-Y
2685	448	HIE	InertiaM-Z
2686	448	HIE	CenterM-X
2687	448	HIE	CenterM-Y
2688	448	HIE	CenterM-Z
2689	449	GLU	InertiaM-X
2690	449	GLU	InertiaM-Y
2691	449	GLU	InertiaM-Z
2692	449	GLU	CenterM-X
2693	449	GLU	CenterM-Y
2694	449	GLU	CenterM-Z
2695	450	ILE	InertiaM-X
2696	450	ILE	InertiaM-Y
2697	450	ILE	InertiaM-Z

(*Continued*)

Table 2.1 (*Continued*)

Variable no.	Residue no.	Residue	Variable
2698	450	ILE	CenterM-X
2699	450	ILE	CenterM-Y
2700	450	ILE	CenterM-Z
2701	451	CYS	InertiaM-X
2702	451	CYS	InertiaM-Y
2703	451	CYS	InertiaM-Z
2704	451	CYS	CenterM-X
2705	451	CYS	CenterM-Y
2706	451	CYS	CenterM-Z
2707	452	VAL	InertiaM-X
2708	452	VAL	InertiaM-Y
2709	452	VAL	InertiaM-Z
2710	452	VAL	CenterM-X
2711	452	VAL	CenterM-Y
2712	452	VAL	CenterM-Z
2713	453	ALA	InertiaM-X
2714	453	ALA	InertiaM-Y
2715	453	ALA	InertiaM-Z
2716	453	ALA	CenterM-X
2717	453	ALA	CenterM-Y
2718	453	ALA	CenterM-Z
2719	454	HIE	InertiaM-X
2720	454	HIE	InertiaM-Y
2721	454	HIE	InertiaM-Z
2722	454	HIE	CenterM-X
2723	454	HIE	CenterM-Y
2724	454	HIE	CenterM-Z
2725	455	SER	InertiaM-X
2726	455	SER	InertiaM-Y
2727	455	SER	InertiaM-Z
2728	455	SER	CenterM-X

Variable no.	Residue no.	Residue	Variable
2729	455	SER	CenterM-Y
2730	455	SER	CenterM-Z
2731	456	ASP	InertiaM-X
2732	456	ASP	InertiaM-Y
2733	456	ASP	InertiaM-Z
2734	456	ASP	CenterM-X
2735	456	ASP	CenterM-Y
2736	456	ASP	CenterM-Z
2737	457	GLU	InertiaM-X
2738	457	GLU	InertiaM-Y
2739	457	GLU	InertiaM-Z
2740	457	GLU	CenterM-X
2741	457	GLU	CenterM-Y
2742	457	GLU	CenterM-Z
2743	458	LEU	InertiaM-X
2744	458	LEU	InertiaM-Y
2745	458	LEU	InertiaM-Z
2746	458	LEU	CenterM-X
2747	458	LEU	CenterM-Y
2748	458	LEU	CenterM-Z
2749	459	ARG	InertiaM-X
2750	459	ARG	InertiaM-Y
2751	459	ARG	InertiaM-Z
2752	459	ARG	CenterM-X
2753	459	ARG	CenterM-Y
2754	459	ARG	CenterM-Z
2755	460	THR	InertiaM-X
2756	460	THR	InertiaM-Y
2757	460	THR	InertiaM-Z
2758	460	THR	CenterM-X
2759	460	THR	CenterM-Y
2760	460	THR	CenterM-Z

(*Continued*)

Table 2.1 (*Continued*)

Variable no.	Residue no.	Residue	Variable
2761	461	LEU	InertiaM-X
2762	461	LEU	InertiaM-Y
2763	461	LEU	InertiaM-Z
2764	461	LEU	CenterM-X
2765	461	LEU	CenterM-Y
2766	461	LEU	CenterM-Z
2767	462	SER	InertiaM-X
2768	462	SER	InertiaM-Y
2769	462	SER	InertiaM-Z
2770	462	SER	CenterM-X
2771	462	SER	CenterM-Y
2772	462	SER	CenterM-Z
2773	463	ASN	InertiaM-X
2774	463	ASN	InertiaM-Y
2775	463	ASN	InertiaM-Z
2776	463	ASN	CenterM-X
2777	463	ASN	CenterM-Y
2778	463	ASN	CenterM-Z
2779	464	GLU	InertiaM-X
2780	464	GLU	InertiaM-Y
2781	464	GLU	InertiaM-Z
2782	464	GLU	CenterM-X
2783	464	GLU	CenterM-Y
2784	464	GLU	CenterM-Z
2785	465	ARG	InertiaM-X
2786	465	ARG	InertiaM-Y
2787	465	ARG	InertiaM-Z
2788	465	ARG	CenterM-X
2789	465	ARG	CenterM-Y
2790	465	ARG	CenterM-Z
2791	466	GLY	InertiaM-X

Variable no.	Residue no.	Residue	Variable
2792	466	GLY	InertiaM-Y
2793	466	GLY	InertiaM-Z
2794	466	GLY	CenterM-X
2795	466	GLY	CenterM-Y
2796	466	GLY	CenterM-Z
2797	467	ALA	InertiaM-X
2798	467	ALA	InertiaM-Y
2799	467	ALA	InertiaM-Z
2800	467	ALA	CenterM-X
2801	467	ALA	CenterM-Y
2802	467	ALA	CenterM-Z
2803	468	GLN	InertiaM-X
2804	468	GLN	InertiaM-Y
2805	468	GLN	InertiaM-Z
2806	468	GLN	CenterM-X
2807	468	GLN	CenterM-Y
2808	468	GLN	CenterM-Z
2809	469	GLN	InertiaM-X
2810	469	GLN	InertiaM-Y
2811	469	GLN	InertiaM-Z
2812	469	GLN	CenterM-X
2813	469	GLN	CenterM-Y
2814	469	GLN	CenterM-Z
2815	470	HIE	InertiaM-X
2816	470	HIE	InertiaM-Y
2817	470	HIE	InertiaM-Z
2818	470	HIE	CenterM-X
2819	470	HIE	CenterM-Y
2820	470	HIE	CenterM-Z
2821	471	VAL	InertiaM-X
2822	471	VAL	InertiaM-Y
2823	471	VAL	InertiaM-Z

(*Continued*)

Table 2.1 (*Continued*)

Variable no.	Residue no.	Residue	Variable
2824	471	VAL	CenterM-X
2825	471	VAL	CenterM-Y
2826	471	VAL	CenterM-Z
2827	472	LEU	InertiaM-X
2828	472	LEU	InertiaM-Y
2829	472	LEU	InertiaM-Z
2830	472	LEU	CenterM-X
2831	472	LEU	CenterM-Y
2832	472	LEU	CenterM-Z
2833	473	LYS	InertiaM-X
2834	473	LYS	InertiaM-Y
2835	473	LYS	InertiaM-Z
2836	473	LYS	CenterM-X
2837	473	LYS	CenterM-Y
2838	473	LYS	CenterM-Z
2839	474	LYS	InertiaM-X
2840	474	LYS	InertiaM-Y
2841	474	LYS	InertiaM-Z
2842	474	LYS	CenterM-X
2843	474	LYS	CenterM-Y
2844	474	LYS	CenterM-Z
2845	475	LEU	InertiaM-X
2846	475	LEU	InertiaM-Y
2847	475	LEU	InertiaM-Z
2848	475	LEU	CenterM-X
2849	475	LEU	CenterM-Y
2850	475	LEU	CenterM-Z
2851	476	LEU	InertiaM-X
2852	476	LEU	InertiaM-Y
2853	476	LEU	InertiaM-Z
2854	476	LEU	CenterM-X

Variable no.	Residue no.	Residue	Variable
2855	476	LEU	CenterM-Y
2856	476	LEU	CenterM-Z
2857	477	ALA	InertiaM-X
2858	477	ALA	InertiaM-Y
2859	477	ALA	InertiaM-Z
2860	477	ALA	CenterM-X
2861	477	ALA	CenterM-Y
2862	477	ALA	CenterM-Z
2863	478	ILE	InertiaM-X
2864	478	ILE	InertiaM-Y
2865	478	ILE	InertiaM-Z
2866	478	ILE	CenterM-X
2867	478	ILE	CenterM-Y
2868	478	ILE	CenterM-Z
2869	479	THR	InertiaM-X
2870	479	THR	InertiaM-Y
2871	479	THR	InertiaM-Z
2872	479	THR	CenterM-X
2873	479	THR	CenterM-Y
2874	479	THR	CenterM-Z
2875	480	GLU	InertiaM-X
2876	480	GLU	InertiaM-Y
2877	480	GLU	InertiaM-Z
2878	480	GLU	CenterM-X
2879	480	GLU	CenterM-Y
2880	480	GLU	CenterM-Z
2881	481	LEU	InertiaM-X
2882	481	LEU	InertiaM-Y
2883	481	LEU	InertiaM-Z
2884	481	LEU	CenterM-X
2885	481	LEU	CenterM-Y
2886	481	LEU	CenterM-Z

(*Continued*)

Table 2.1 (*Continued*)

Variable no.	Residue no.	Residue	Variable
2887	482	LEU	InertiaM-X
2888	482	LEU	InertiaM-Y
2889	482	LEU	InertiaM-Z
2890	482	LEU	CenterM-X
2891	482	LEU	CenterM-Y
2892	482	LEU	CenterM-Z
2893	483	GLN	InertiaM-X
2894	483	GLN	InertiaM-Y
2895	483	GLN	InertiaM-Z
2896	483	GLN	CenterM-X
2897	483	GLN	CenterM-Y
2898	483	GLN	CenterM-Z
2899	484	GLN	InertiaM-X
2900	484	GLN	InertiaM-Y
2901	484	GLN	InertiaM-Z
2902	484	GLN	CenterM-X
2903	484	GLN	CenterM-Y
2904	484	GLN	CenterM-Z
2905	485	LYS	InertiaM-X
2906	485	LYS	InertiaM-Y
2907	485	LYS	InertiaM-Z
2908	485	LYS	CenterM-X
2909	485	LYS	CenterM-Y
2910	485	LYS	CenterM-Z
2911	486	GLN	InertiaM-X
2912	486	GLN	InertiaM-Y
2913	486	GLN	InertiaM-Z
2914	486	GLN	CenterM-X
2915	486	GLN	CenterM-Y
2916	486	GLN	CenterM-Z
2917	487	MG^{++}	CenterM-X

Variable no.	Residue no.	Residue	Variable
2918	487	MG^{++}	CenterM-Y
2919	487	MG^{++}	CenterM-Z
2920	488	GDP	InertiaM-X
2921	488	GDP	InertiaM-Y
2922	488	GDP	InertiaM-Z
2923	488	GDP	CenterM-X
2924	488	GDP	CenterM-Y
2925	488	GDP	CenterM-Z
2926	488	PI	InertiaM-X
2927	489	PI	InertiaM-Y
2928	489	PI	InertiaM-Z
2929	489	PI	CenterM-X
2930	489	PI	CenterM-Y
2931	489	PI	CenterM-Z
2932	490	NA^{+}	CenterM-X
2933	490	NA^{+}	CenterM-Y
2934	490	NA^{+}	CenterM-Z
2935	491	NA^{+}	CenterM-X
2936	491	NA^{+}	CenterM-Y
2937	491	NA^{+}	CenterM-Z
2938	492	NA^{+}	CenterM-X
2939	492	NA^{+}	CenterM-Y
2940	492	NA^{+}	CenterM-Z
2941	493	NA^{+}	CenterM-X
2942	494	NA^{+}	CenterM-Y
2943	494	NA^{+}	CenterM-Z
2944	494	NA^{+}	CenterM-X
2945	495	NA^{+}	CenterM-Y
2946	495	NA^{+}	CenterM-Z
2947	496	NA^{+}	CenterM-X
2948	496	NA^{+}	CenterM-Y
2949	496	NA^{+}	CenterM-Z

(*Continued*)

Table 2.1 (*Continued*)

Variable no.	Residue no.	Residue	Variable
2950	497	NA^+	CenterM-X
2951	497	NA^+	CenterM-Y
2952	497	NA^+	CenterM-Z
2953	498	NA^+	CenterM-X
2954	498	NA^+	CenterM-Y
2955	498	NA^+	CenterM-Z
2956	499	NA^+	CenterM-X
2957	499	NA^+	CenterM-Y
2958	499	NA^+	CenterM-Z
2959	500	NA^+	CenterM-X
2960	500	NA^+	CenterM-Y
2961	500	NA^+	CenterM-Z
2962	501	NA^+	CenterM-X
2963	501	NA^+	CenterM-Y
2964	501	NA^+	CenterM-Z
2965	502	NA^+	CenterM-X
2966	502	NA^+	CenterM-Y
2967	502	NA^+	CenterM-Z
2968	503	NA^+	CenterM-X
2969	503	NA^+	CenterM-Y
2970	503	NA^+	CenterM-Z
2971	504	NA^+	CenterM-X
2972	504	NA^+	CenterM-Y
2973	504	NA^+	CenterM-Z
2974	505	NA^+	CenterM-X
2975	505	NA^+	CenterM-Y
2976	505	NA^+	CenterM-Z
2977	506	Hy_sh_H_2O	InertiaM-X
2978	506	Hy_sh_H_2O	InertiaM-Y
2979	506	Hy_sh_H_2O	InertiaM-Z
2980	506	Hy_sh_H_2O	CenterM-X

Variable no.	Residue no.	Residue	Variable
2981	506	Hy_sh_H_2O	CenterM-Y
2982	506	Hy_sh_H_2O	CenterM-Z
2983	507	Total	Total_CM
2984	507	Total	Total_CM
2985	507	Total	Total_CM
2986	508	Total	Total_Euler
2987	508	Total	Total_Euler
2988	508	Total	Total_Euler
2989	509	Total	Tot_Mom_X
2990	509	Total	T_M_XX
2991	509	Total	T_M_XY
2992	509	Total	T_M_XZ
2993	510	Total	Tot_Mom_Y
2994	510	Total	T_M_YX
2995	510	Total	T_M_YY
2996	510	Total	T_M_YZ
2997	511	Total	Tot_Mom_Z
2998	511	Total	T_M_ZX
2999	511	Total	T_M_ZY
3000	511	Total	T_M_ZZ

Note: InertiaM-X(Y,Z), principal components of the proper partial inertia tensor; CenterM-(X,Y,Z), coordinates of the proper partial center of mass; Total_CM, the center of mass of the whole hydrated complex; Total_Euler, Euler angles of the whole hydrated complex; Tot_Mom_(X,Y,Z), principal components of the total inertia tensor; T_M_(...), coordinates of the vectors corresponding to the latter components

Table 2.2 presents the results of the polynomial factor analysis and is relevant to Section 2.2 of the main text.

Table 2.2 A list of dynamical variables acting as the key players in Factor 1, for their positive loadings are noticeably greater than the limiting value of +0.015

Variable no.	Factor 1 load	Residue	Residue no.	Variable
8	0.01532	THR	2	InertiaM-Y
44	0.016687	VAL	8	InertiaM-Y
68	0.01717	GLY	12	InertiaM-Y
98	0.039789	SER	17	InertiaM-Y
109	0.015516	LEU	19	InertiaM-X
126	0.02069	ILE	21	CenterM-Z
178	0.017232	ASP	30	CenterM-X
206	0.018158	THR	35	InertiaM-Y
226	0.047183	ASP	38	CenterM-X
231	0.015235	SER	39	InertiaM-Z
252	0.05025	LYS	42	CenterM-Z
290	0.016868	GLU	49	InertiaM-Y
320	0.017674	ASP	54	InertiaM-Y
361	0.025687	GLN	61	InertiaM-X
396	0.020234	ALA	66	CenterM-Z
414	0.016726	ASP	69	CenterM-Z
470	0.016361	LEU	79	InertiaM-Y
555	0.015398	ILE	93	InertiaM-Z
558	0.037474	ILE	93	CenterM-Z
563	0.016986	HIE	94	CenterM-Y
570	0.016026	GLN	95	CenterM-Z
575	0.019454	TYR	96	CenterM-Y
587	0.015395	GLU	98	CenterM-Y
610	0.019183	ARG	102	CenterM-X
636	0.027682	SER	106	CenterM-Z
675	0.026691	LEU	113	InertiaM-Z
741	0.024546	THR	124	InertiaM-Z
765	0.019233	ARG	128	InertiaM-Z
800	0.017586	ALA	134	InertiaM-Y
803	0.026961	ALA	134	CenterM-Y

Variable no.	Factor 1 load	Residue	Residue no.	Variable
837	0.018051	PRO	140	InertiaM-Z
844	0.022637	TYR	141	CenterM-X
854	0.016328	GLU	143	InertiaM-Y
858	0.018272	GLU	143	CenterM-Z
867	0.01543	SER	145	InertiaM-Z
871	0.025038	ALA	146	InertiaM-X
895	0.028519	GLN	150	InertiaM-X
900	0.025721	GLN	150	CenterM-Z
943	0.035169	THR	158	InertiaM-X
971	0.024686	GLU	162	CenterM-Y
977	0.015466	ILE	163	CenterM-Y
1063	0.020379	LEU	178	InertiaM-X
1085	0.033198	GLN	181	CenterM-Y
1096	0.015456	GLU	183	CenterM-X
1103	0.020178	LEU	184	CenterM-Y
1129	0.029497	ALA	189	InertiaM-X
1143	0.017959	SER	191	InertiaM-Z
1179	0.020056	ASP	197	InertiaM-Z
1199	0.016682	LEU	200	CenterM-Y
1204	0.031658	LEU	201	CenterM-X
1222	0.033864	ILE	204	CenterM-X
1226	0.021481	LEU	205	InertiaM-Y
1251	0.021219	PHE	209	InertiaM-Z
1279	0.019715	LEU	214	InertiaM-X
1292	0.03453	SER	216	InertiaM-Y
1294	0.021383	SER	216	CenterM-X
1306	0.019477	LEU	218	CenterM-X
1321	0.029187	THR	221	InertiaM-X
1337	0.019257	ASN	223	CenterM-Y
1361	0.02	ILE	227	CenterM-Y
1428	0.017097	ARG	238	CenterM-Z
1439	0.02879	THR	240	CenterM-Y

(*Continued*)

Table 2.2 (*Continued*)

Variable no.	Factor 1 load	Residue	Residue no.	Variable
1465	0.023323	THR	245	InertiaM-X
1515	0.017095	ALA	253	InertiaM-Z
1552	0.018393	VAL	259	CenterM-X
1555	0.016608	HIE	260	InertiaM-X
1588	0.021003	ASP	265	CenterM-X
1614	0.016226	LYS	269	CenterM-Z
1690	0.015126	SER	282	CenterM-X
1801	0.021928	LEU	301	InertiaM-X
1803	0.017313	LEU	301	InertiaM-Z
1839	0.016828	LYS	307	InertiaM-Z
1845	0.024319	ILE	308	InertiaM-Z
1873	0.01646	GLU	313	InertiaM-X
1910	0.01641	LEU	319	InertiaM-Y
1928	0.017991	ILE	322	InertiaM-Y
2024	0.01772	THR	338	InertiaM-Y
2031	0.015905	THR	339	InertiaM-Z
2052	0.017709	THR	342	CenterM-Z
2060	0.0159	VAL	344	InertiaM-Y
2112	0.016023	ARG	352	CenterM-Z
2124	0.036074	ILE	354	CenterM-Z
2210	0.017028	ASP	369	InertiaM-Y
2212	0.019808	ASP	369	CenterM-X
2250	0.016413	ALA	375	CenterM-Z
2270	0.022655	LEU	379	InertiaM-Y
2292	0.015436	VAL	382	CenterM-Z
2297	0.015114	ALA	383	CenterM-Y
2319	0.015083	GLN	387	InertiaM-Z
2343	0.018389	ASN	391	InertiaM-Z
2368	0.015643	PHE	395	CenterM-X
2377	0.019289	ALA	397	InertiaM-X
2388	0.019097	LYS	398	CenterM-Z

Variable no.	Factor 1 load	Residue	Residue no.	Variable
2419	0.015956	GLY	404	InertiaM-X
2438	0.018383	PRO	407	InertiaM-Y
2463	0.017264	SER	411	InertiaM-Z
2493	0.025993	MET	416	InertiaM-Z
2496	0.026982	MET	416	CenterM-Z
2503	0.015102	MET	418	InertiaM-X
2505	0.016286	MET	418	InertiaM-Z
2520	0.018697	LEU	420	CenterM-Z
2529	0.024614	GLU	422	InertiaM-Z
2531	0.017402	GLU	422	CenterM-Y
2603	0.025589	GLU	434	CenterM-Y
2633	0.015029	ASP	439	CenterM-Y
2654	0.030855	ASP	443	InertiaM-Y
2699	0.025537	ILE	450	CenterM-Y
2706	0.016723	CYS	451	CenterM-Z
2710	0.016631	VAL	452	CenterM-X
2713	0.035574	ALA	453	InertiaM-X
2751	0.038688	ARG	459	InertiaM-Z
2819	0.026009	HIE	470	CenterM-Y
2875	0.017032	GLU	480	InertiaM-X
2883	0.018597	LEU	481	InertiaM-Z
2901	0.019823	GLN	484	InertiaM-Z
2930	**0.022772**	**PI**	**489**	CenterM-Y
2949	**0.025523**	**NA^+**	**496**	CenterM-Z
2953	**0.015137**	**NA^+**	**498**	CenterM-X
2999	**0.019929**	**Total**	**511**	T_M_ZY

Note: The legend is the same as that for Table 2.1. The most prominent key players are highlighted in red.

Table 2.3 also presents the results of the polynomial factor analysis. This is relevant to Section 2.2 in the main text.

Table 2.3 A list of dynamical variables acting as the mischievous monsters in Factor 1, for their negative loadings are equal to or noticeably lower than the limiting value of –0.010

Variable no.	Factor 1 load	Residue	Residue no.	Variable
2	-0.010248	MET	1	InertiaM-Y
22	-0.012011	TYR	4	CenterM-X
31	-0.010189	LEU	6	InertiaM-X
34	-0.012052	LEU	6	CenterM-X
59	-0.011767	GLY	10	CenterM-Y
83	-0.011448	VAL	14	CenterM-Y
86	-0.011013	GLY	15	InertiaM-Y
95	-0.010946	LYS	16	CenterM-Y
99	-0.01128	SER	17	InertiaM-Z
101	-0.011255	SER	17	CenterM-Y
124	-0.010637	ILE	21	CenterM-X
137	-0.011321	LEU	23	CenterM-Y
143	-0.011538	ILE	24	CenterM-Y
158	-0.011552	HIE	27	InertiaM-Y
167	-0.010907	PHE	28	CenterM-Y
190	-0.010021	TYR	32	CenterM-X
193	-0.010989	ASP	33	InertiaM-X
218	-0.010198	GLU	37	InertiaM-Y
237	-0.011441	TYR	40	InertiaM-Z
254	-0.011837	GLN	43	InertiaM-Y
255	-0.010711	GLN	43	InertiaM-Z
298	-0.011131	THR	50	CenterM-X
302	-0.010377	CYS	51	InertiaM-Y
314	-0.011272	LEU	53	InertiaM-Y
358	-0.011279	GLY	60	CenterM-X
374	-0.01087	GLU	63	InertiaM-Y
384	-0.010142	TYR	64	CenterM-Z
405	-0.01078	ARG	68	InertiaM-Z
407	-0.011039	ARG	68	CenterM-Y
411	-0.012041	ASP	69	InertiaM-Z

Variable no.	Factor 1 load	Residue	Residue no.	Variable
420	-0.011114	GLN	70	CenterM-Z
426	-0.011045	TYR	71	CenterM-Z
459	-0.011831	GLY	77	InertiaM-Z
463	-0.01186	PHE	78	InertiaM-X
476	-0.011509	CYS	80	InertiaM-Y
484	-0.011814	VAL	81	CenterM-X
489	-0.012072	PHE	82	InertiaM-Z
503	-0.010883	ILE	84	CenterM-Y
507	-0.011889	ASN	85	InertiaM-Z
515	-0.010909	ASN	86	CenterM-Y
520	-0.010821	THR	87	CenterM-X
526	-0.011813	LYS	88	CenterM-X
568	-0.012182	GLN	95	CenterM-X
584	-0.01145	GLU	98	InertiaM-Y
585	-0.011734	GLU	98	InertiaM-Z
586	-0.012095	GLU	98	CenterM-X
619	-0.010721	LYS	104	InertiaM-X
645	-0.012152	ASP	108	InertiaM-Z
648	-0.011959	ASP	108	CenterM-Z
650	-0.011928	VAL	109	InertiaM-Y
657	-0.010702	PRO	110	InertiaM-Z
666	-0.010089	MET	111	CenterM-Z
669	-0.011493	VAL	112	InertiaM-Z
678	-0.012043	LEU	113	CenterM-Z
681	-0.010528	VAL	114	InertiaM-Z
689	-0.01072	GLY	115	CenterM-Y
701	-0.011938	LYS	117	CenterM-Y
703	-0.011347	CYS	118	InertiaM-X
710	-0.011194	ASP	119	InertiaM-Y
715	-0.011236	LEU	120	InertiaM-X
716	-0.011098	LEU	120	InertiaM-Y
721	-0.01011	ALA	121	InertiaM-X

(*Continued*)

Table 2.3 (*Continued*)

Variable no.	Factor 1 load	Residue	Residue no.	Variable
722	−0.010118	ALA	121	InertiaM-Y
728	−0.011171	ALA	122	InertiaM-Y
737	−0.010519	ARG	123	CenterM-Y
746	−0.01061	VAL	125	InertiaM-Y
751	−0.010324	GLU	126	InertiaM-X
770	−0.011521	GLN	129	InertiaM-Y
775	−0.011588	ALA	130	InertiaM-X
781	−0.010744	GLN	131	InertiaM-X
787	−0.011611	ASP	132	InertiaM-X
788	−0.010339	ASP	132	InertiaM-Y
795	−0.01025	LEU	133	InertiaM-Z
798	−0.011227	LEU	133	CenterM-Z
807	−0.011565	ARG	135	InertiaM-Z
811	−0.010554	SER	136	InertiaM-X
812	−0.010914	SER	136	InertiaM-Y
817	−0.010622	TYR	137	InertiaM-X
836	−0.010552	PRO	140	InertiaM-Y
843	−0.010366	TYR	141	InertiaM-Z
845	−0.010844	TYR	141	CenterM-Y
846	−0.01047	TYR	141	CenterM-Z
849	−0.011758	ILE	142	InertiaM-Z
897	−0.011192	GLN	150	InertiaM-Z
905	−0.011645	GLY	151	CenterM-Y
908	−0.01199	VAL	152	InertiaM-Y
918	−0.010044	GLU	153	CenterM-Z
924	−0.011573	ASP	154	CenterM-Z
935	−0.011716	PHE	156	CenterM-Y
949	−0.010865	LEU	159	InertiaM-X
952	−0.011031	LEU	159	CenterM-X
961	−0.010183	ARG	161	InertiaM-X
982	−0.011169	ARG	164	CenterM-X

Variable no.	Factor 1 load	Residue	Residue no.	Variable
985	−0.010958	GLN	165	InertiaM-X
1007	−0.010689	PRO	168	CenterM-Y
1008	−0.010431	PRO	168	CenterM-Z
1011	−0.011699	GLU	169	InertiaM-Z
1013	−0.010427	GLU	169	CenterM-Y
1019	−0.011509	GLU	170	CenterM-Y
1030	−0.011912	TYR	172	CenterM-X
1054	−0.011183	LYS	176	CenterM-X
1076	−0.011259	LEU	180	InertiaM-Y
1088	−0.010243	LYS	182	InertiaM-Y
1111	−0.01057	VAL	186	InertiaM-X
1113	−0.011352	VAL	186	InertiaM-Z
1123	-0.0101	TYR	188	InertiaM-X
1125	−0.011072	TYR	188	InertiaM-Z
1134	−0.011496	ALA	189	CenterM-Z
1142	−0.010574	SER	191	InertiaM-Y
1149	−0.010952	HIE	192	InertiaM-Z
1157	−0.010839	VAL	193	CenterM-Y
1164	−0.011047	CYS	194	CenterM-Z
1183	−0.010207	ARG	198	InertiaM-X
1214	−0.012133	SER	203	InertiaM-Y
1238	−0.01079	ARG	207	InertiaM-Y
1246	−0.011889	ILE	208	CenterM-X
1268	−0.011601	GLU	212	InertiaM-Y
1270	−0.012145	GLU	212	CenterM-X
1278	−0.010364	LYS	213	CenterM-Z
1330	−0.011885	LEU	222	CenterM-X
1336	−0.012178	ASN	223	CenterM-X
1347	−0.011345	ARG	225	InertiaM-Z
1364	−0.010795	SER	228	InertiaM-Y
1371	−0.010702	MET	229	InertiaM-Z
1382	−0.011053	ASP	230	InertiaM-Y

(*Continued*)

Table 2.3 (*Continued*)

Variable no.	Factor 1 load	Residue	Residue no.	Variable
1384	-0.011539	ASP	231	CenterM-X
1406	-0.011277	THR	234	InertiaM-Y
1411	-0.01129	LEU	236	InertiaM-X
1415	-0.011168	LEU	236	CenterM-Y
1453	-0.011634	ALA	243	InertiaM-X
1454	-0.012106	ALA	243	InertiaM-Y
1457	-0.01048	ALA	243	CenterM-Y
1460	-0.010069	SER	244	InertiaM-Y
1461	-0.011919	SER	244	InertiaM-Z
1472	-0.010569	LEU	246	InertiaM-Y
1479	-0.011893	MET	247	InertiaM-Z
1489	-0.01168	GLN	249	InertiaM-X
1500	-0.01104	TYR	250	CenterM-Z
1521	-0.011109	THR	254	InertiaM-Z
1530	-0.010315	ALA	255	CenterM-Z
1535	-0.011769	THR	256	CenterM-Y
1539	-0.011164	GLN	257	InertiaM-Z
1551	-0.010091	VAL	259	InertiaM-Z
1575	-0.012002	LEU	263	InertiaM-Z
1581	-0.011973	LYS	264	InertiaM-Z
1584	-0.011809	LYS	264	CenterM-Z
1594	-0.010762	SER	266	CenterM-X
1605	-0.011545	LEU	268	InertiaM-Z
1610	-0.011705	LYS	269	InertiaM-Y
1613	-0.010024	LYS	269	CenterM-Y
1617	-0.010037	ILE	270	InertiaM-Z
1625	-0.011594	MET	271	CenterM-Y
1632	-0.011003	GLU	272	CenterM-Z
1647	-0.011834	GLN	275	InertiaM-Z
1648	-0.010917	GLN	275	CenterM-X
1657	-0.010137	CYS	277	InertiaM-X

Variable no.	Factor 1 load	Residue	Residue no.	Variable
1679	−0.010143	SER	280	CenterM-Y
1681	−0.010868	PRO	281	InertiaM-X
1686	−0.010487	PRO	281	CenterM-Z
1697	−0.010466	LYS	283	CenterM-Y
1702	−0.011352	LEU	284	CenterM-X
1715	−0.010747	LYS	286	CenterM-Y
1716	−0.011589	LYS	286	CenterM-Z
1731	−0.012006	ASP	289	InertiaM-Z
1752	−0.011046	THR	292	CenterM-Z
1767	−0.010723	THR	295	InertiaM-Z
1783	−0.010363	LEU	298	InertiaM-X
1787	−0.010654	LEU	298	CenterM-Y
1809	−0.011068	SER	302	InertiaM-Z
1812	−0.010927	SER	302	CenterM-Z
1814	−0.010919	GLU	303	InertiaM-Y
1822	−0.012124	LEU	304	CenterM-X
1823	−0.011426	LEU	304	CenterM-Y
1827	−0.010915	VAL	305	InertiaM-Z
1841	−0.010121	LYS	307	CenterM-Y
1852	−0.012074	PHE	309	CenterM-X
1869	−0.011546	SER	312	InertiaM-Z
1870	−0.010915	SER	312	CenterM-X
1880	−0.011042	ILE	314	InertiaM-Y
1886	−0.011023	LEU	315	InertiaM-Y
1890	−0.011373	LEU	315	CenterM-Z
1903	−0.01057	THR	318	InertiaM-X
1904	−0.011079	THR	318	InertiaM-Y
1906	−0.010594	THR	318	CenterM-X
1919	−0.01179	ARG	320	CenterM-Y
1922	−0.01166	TYR	321	InertiaM-Y
1934	−0.011652	TYR	323	InertiaM-Y
1936	−0.011297	TYR	323	CenterM-X

(*Continued*)

Table 2.3 (*Continued*)

Variable no.	Factor 1 load	Residue	Residue no.	Variable
1944	-0.01115	GLY	324	CenterM-Z
1945	-0.010842	CYX	325	InertiaM-X
1951	-0.01038	LEU	326	InertiaM-X
1954	-0.011658	LEU	326	CenterM-X
1955	-0.010736	LEU	326	CenterM-Y
1957	-0.010749	GLN	327	InertiaM-X
1966	-0.010775	LYS	328	CenterM-X
1967	-0.011973	LYS	328	CenterM-Y
1986	-0.010903	GLN	331	CenterM-Z
1990	-0.011146	HIE	332	CenterM-X
2003	-0.010444	TRP	334	CenterM-Y
2027	-0.010394	THR	338	CenterM-Y
2040	-0.010101	MET	340	CenterM-Z
2041	-0.011988	ARG	341	InertiaM-X
2053	-0.011512	ARG	343	InertiaM-X
2056	-0.010877	ARG	343	CenterM-X
2058	-0.011471	ARG	343	CenterM-Z
2066	-0.011112	VAL	345	InertiaM-Y
2080	-0.010915	GLY	347	CenterM-X
2091	-0.011647	VAL	349	InertiaM-Z
2101	-0.011649	LEU	351	InertiaM-X
2104	-0.012098	LEU	351	CenterM-X
2108	-0.010043	ARG	352	InertiaM-Y
2149	-0.011091	LEU	359	InertiaM-X
2150	-0.012113	LEU	359	InertiaM-Y
2160	-0.010136	ASN	360	CenterM-Z
2179	-0.011896	PHE	364	InertiaM-X
2182	-0.010534	PHE	364	CenterM-X
2195	-0.010274	ILE	366	CenterM-Y
2213	-0.010343	ASP	369	CenterM-Y
2232	-0.011774	SER	372	CenterM-Z

Variable no.	Factor 1 load	Residue	Residue no.	Variable
2233	−0.011306	PRO	373	InertiaM-X
2239	−0.01192	ILE	374	InertiaM-X
2255	−0.010335	ALA	376	CenterM-Y
2263	−0.01029	THR	378	InertiaM-X
2285	−0.010239	LEU	381	CenterM-Y
2314	−0.010239	VAL	386	CenterM-X
2321	−0.01147	GLN	387	CenterM-Y
2351	−0.010407	LEU	392	CenterM-Y
2352	−0.010009	LEU	392	CenterM-Z
2371	−0.011233	GLY	396	InertiaM-X
2395	−0.011274	PRO	400	InertiaM-X
2396	−0.011695	PRO	400	InertiaM-Y
2397	−0.01218	PRO	400	InertiaM-Z
2404	−0.010079	TYR	401	CenterM-X
2406	−0.011399	TYR	401	CenterM-Z
2409	−0.010277	MET	402	InertiaM-Z
2418	−0.012128	GLU	403	CenterM-Z
2467	−0.011192	ASN	412	InertiaM-X
2469	−0.010015	ASN	412	InertiaM-Z
2476	−0.011491	LYS	413	CenterM-X
2477	−0.010361	LYS	413	CenterM-Y
2480	−0.011524	HIE	414	InertiaM-Y
2490	−0.010406	ARG	415	CenterM-Z
2491	−0.010937	MET	416	InertiaM-X
2494	−0.011692	MET	416	CenterM-X
2509	−0.011375	PHE	419	InertiaM-X
2517	−0.012111	LEU	420	InertiaM-Z
2528	−0.011411	GLU	422	InertiaM-Y
2551	−0.010104	VAL	426	InertiaM-X
2553	−0.011437	VAL	426	InertiaM-Z
2571	−0.010805	LEU	429	InertiaM-Z
2587	−0.011393	THR	432	InertiaM-X

(*Continued*)

Table 2.3 (*Continued*)

Variable no.	Factor 1 load	Residue	Residue no.	Variable
2593	−0.010201	THR	433	InertiaM-X
2596	−0.011823	THR	433	CenterM-X
2599	−0.011232	GLU	434	InertiaM-X
2606	−0.01008	HIE	435	InertiaM-Y
2613	−0.011829	SER	436	InertiaM-Z
2622	−0.010364	ARG	437	CenterM-Z
2632	−0.010618	ASP	439	CenterM-X
2637	−0.010287	LEU	440	InertiaM-Z
2648	−0.010479	ARG	442	InertiaM-Y
2653	−0.010746	ASP	443	InertiaM-X
2669	−0.01024	ALA	445	CenterM-Y
2681	−0.01194	LEU	447	CenterM-Y
2694	−0.011708	GLU	449	CenterM-Z
2695	−0.011805	ILE	450	InertiaM-X
2697	−0.01148	ILE	450	InertiaM-Z
2703	−0.011549	CYS	451	InertiaM-Z
2712	−0.010892	VAL	452	CenterM-Z
2714	−0.011927	ALA	453	InertiaM-Y
2740	−0.01146	GLU	457	CenterM-X
2745	−0.010517	LEU	458	InertiaM-Z
2752	−0.011287	ARG	459	CenterM-X
2756	−0.010023	THR	460	InertiaM-Y
2773	−0.012032	ASN	463	InertiaM-X
2775	−0.011656	ASN	463	InertiaM-Z
2777	−0.012073	ASN	463	CenterM-Y
2785	−0.010392	ARG	465	InertiaM-X
2798	−0.011677	ALA	467	InertiaM-Y
2800	−0.011957	ALA	467	CenterM-X
2801	−0.012157	ALA	467	CenterM-Y
2814	−0.011306	GLN	469	CenterM-Z
2838	−0.011024	LYS	473	CenterM-Z

Variable no.	Factor 1 load	Residue	Residue no.	Variable
2843	-0.011161	LYS	474	CenterM-Y
2846	-0.011309	LEU	475	InertiaM-Y
2859	-0.01116	ALA	477	InertiaM-Z
2864	-0.010454	ILE	478	InertiaM-Y
2876	-0.011586	GLU	480	InertiaM-Y
2891	-0.01074	LEU	482	CenterM-Y
2897	-0.010681	GLN	483	CenterM-Y
2917	**-0.010146**	**MG^{++}**	**487**	CenterM-X
2936	**-0.011742**	**NA^{+}**	**491**	CenterM-Y
2976	**-0.01207**	**NA^{+}**	**505**	CenterM-Z
2983	**-0.01061**	**Total**	**507**	Total_CM
2992	**-0.010325**	**Total**	**509**	T_M_XZ
2998	**-0.01082**	**Total**	**511**	T_M_ZX
3000	**-0.010459**	**Total**	**511**	T_M_ZZ

Note: The legend is the same as that for Table 2.1. The most prominent mischievous monsters are highlighted in red.

Index

abscissa, 9–10, 178, 188
absolute temperature, 2, 15, 35, 50, 76, 83–84, 106, 125, 224
  negative, 53
  zero, 224
acceleration, 70, 72, 151–52, 183, 193
  zero, 153
action and reaction, 161–62, 248
actuarial usefulness, 215
acute psychoses, 202–3, 207
affection, 67
  peripheral neuronal, 67
affective disorders, 136, 143
  secondary, 129
affectivity, 135
aggregate states, 51
  realistic nonideal, 2
air, 95–96, 156, 180
  compressed, 95–96
  pneumatic, 180
alcohol toxicity, 63, 68
algebraic-logarithmic terms, 11
algorithm, 25–26, 245, 257
alienists, 135, 192, 209
Amber parameter database, 259
AmberTools, 259, 261
Ameline, Marius, 59, 63, 74–75, 77, 85, 91–94, 99, 101–2, 129, 132, 150–51, 153–56, 163, 166, 219–20
amino acids, 250, 258
amortization, 198, 230
amplitude, 140, 184, 192, 197
analogy, 108–10, 112, 115, 122, 142, 165, 168–71, 198
  acoustic, 192
  basic, 102
  clear, 124
  clear evolutionary, 115
  close, 103, 108
  fortuitous, 60
  intimate, 169
  phylogenetic, 110
  physiological, 108
  sensible, 169
  structural, 115
analysis
  detailed, 218
  detailed methodological, 240
  eigenvalue-eigenvector, 251
  final, 62
  logical, 217
  mathematical, 34, 37
  principal components, 251
  proper, 37
  quantitative, 214
  regression, 256
  statistical, 21, 218
  systematic statistical-mechanical, 21
Aneja–Johal approach, 25–26
animal-machine, 164
anomaly, observable, 93
antagonistic, 182–83, 201
antagonists, 182, 194
antithesis, 63, 65, 130, 140
apparatus, 115, 232
  nervous, 166
  schematic, 114
applied forces, 72, 151–52
approximation, 7–9, 11, 16, 50–51, 63–64, 68, 128
  analytical, 49

valid, 220
assimilation, 99, 110, 173
assimilation products, 173–74, 180
relevant, 175
astronomical phenomena, important, 160
asymptotes, 177
atomic structure, 228
atomic velocity, 260, 262
atomistic structure, 219
atomo-mechanical theories, 87, 92, 98–99, 101–2, 128–29, 143
contradictory, 101
universal, 91
atoms, 2, 20, 37–38, 89–90, 129–30, 214, 259
elastic, 90
hydrogen, 259

balance, 63, 97, 99, 134, 138, 183
perfect, 81
true dynamical, 153
basic law, 161
first, 102
second, 102, 221
third, 3
unique, 102
Bauer, Erwin, 156, 158–59
Bauer principle, 156
Bayesian, 23, 25, 29
Bayesian approach, 19–21, 28, 38
Bayesian representation, 38
Bayesian statistical thermodynamics of real gases, 25
Béclard, 60, 103, 105, 118–19
Becquerel, 104, 111–12
Berendsen thermostat, 260
Berthelot, 80, 93–94, 105, 143
Berthollet, 93, 143
beta functions, incomplete, 27–28, 30
biophysics, 155
macroscopic, 150, 159
biopolymers, 251
body balance, 151–53
Boltzmann, 2, 6, 19, 57, 217, 219
Boltzmann–Planck formula, 2, 6, 19, 57, 217, 219
Born, Max, 56
Bouty, Edmond, 60, 72, 143, 150
brain, 108, 110, 116–22, 127, 130, 136–37, 165, 167, 176, 179–81, 196, 199, 201–5, 207, 210
human, 171
impressed, 167
Brillouin, Léon Nicolas, 36
Buddhist methodology, 218
Buddhist theory, 216

Cajal, 109–10, 115
capsizability, 129, 131, 154
Carnot, 3, 14, 57, 60, 76, 80–82, 84–85, 91, 98–99, 101, 103, 119, 121, 141, 144
Carnot–Clausius principle, 98–99, 141
Carnot cycle, 153
Carnot principle, 60, 76, 80, 82, 84, 91, 101, 107, 111, 118, 120, 127, 130, 139, 154
Carroll, 253–57
Cartesian conception, 89
Cartesian impulses, 34
Cartesian theories, 90
CFA, *see* confirmatory factor analysis
chaos, 98–99
Chauveau, 103, 106–8, 146
chemical energy, 95, 111, 164–65
potential, 98
chemical poison, 66–67

chemical reactions, 5, 88, 93–94, 103–4, 112, 174, 244
 common, 170
 essential, 104
 pertinent, 106
chemical-theoretical assumption, 180
chemical thermodynamics, 41
chemical transformation, 171, 187
chemical work, 120–21
chemistry, 75–76, 153, 163, 169
 biophysical, 2
Clapeyron–Clausius equation, 41, 51, 53
classical-mechanical theory, single, 161, 216
classical mechanics, 161–62
 conventional, 160
 fundamental, 216
classical probability calculus, 213, 215
classical theories, relevant, 216
Clausius, Rudolf, 3, 15, 51, 76, 81–82, 86, 91, 97, 102, 106, 163, 166
coefficients, 7, 9, 11, 24, 71, 183, 190, 192, 194, 201–2, 206–8, 255
 collision, 260
 dementedness, 190
 vitality, 196
computer-aided simulation, 4, 240, 252
conception, 89, 92, 101–2, 112, 118
 energetic, 129, 134
 monistic, 105
 psychophysical, 141
 scientific, 65
 theoretical, 97
confirmatory factor analysis (CFA), 218, 250–54, 256–57
conflict, 249
 external, 249
 internal, 249
 noninstantaneous, 123
congenital intellectual weakness, 205
conjecture, 101
 philosophical, 129
conjugate gradient method, 259
consciousness, 100, 117–18, 121, 123, 128, 135, 137–38
 individual, 117
conscious process, 117
conservation, 70, 76, 87, 90, 135
continuity, 115, 123, 186
 histological, 115
contractility, 108–9
contributions, 12, 18, 60, 101, 146, 149, 159, 250–52
 explicit entropic, 13
 implicit entropic, 15
 random, 217
correlation coefficient, 243–44, 250, 253, 255
 conventional Pearsonian, 256
 pairwise, 252
 tetrachoric, 255
correlation, 3–4, 37, 40, 217–18, 244, 249, 251–56
 interparticle, 217
 maximum, 253
 observable, 218
 pairwise, 218, 239, 243, 249–51
 pertinent, 37
 tetrachoric, 255–56
cosmogony models, 98
counteractions, 3, 154
counterion-hydration sheath, 244, 258
coupling, interatomic, 2, 40

coupling constant, 260
current, 113, 169
  closed, 113
  electric, 169
  electro-capillary, 113
  galvanic, 111
  muscular, 113

D'Arsonval, 107–8, 113–14, 145
data
  artificial, 257
  binary, 255
data sets, 245, 250, 255
  binary, 255
  statistical, 241
De Boeck, 112, 114–15, 146
decomposition, 67, 115, 180
  chemical, 175
  double, 93
deflagration, 111
deformation, 113
degeneracy, 103, 133–34, 136
  mental, 135
degenerates, 129, 133–35, 138, 142
degeneration, 129, 133–34, 143, 207
  mental, 134, 139
degradability, successive, 120
degradation, 95, 127, 141
Delboeuf, 70, 125–26, 144, 146
delirium, 118, 135, 137, 191, 196, 206, 209
  acute, 203
  emotional, 135
  hallucinatory, 135, 137, 143
  polymorphous, 135
  sensory, 137
  systematized, 135
  systemized, 137
delusional interpretations, 135–36
delusionary interpretations, 135
delusions, 130–31, 134–35, 205
  primary, 129
  rudimentary, 205
  systematized, 137
dementia, 117–18, 191–92, 195–98, 203, 205, 207, 209
  accentuated, 205
  complete, 190
  early, 209
Descartes, 87, 89, 91, 101–2, 208
disassimilation, 130, 174, 180, 207
diseases, 66–67, 139, 192, 198
  mental, 59
disequilibrium, 133
disintegration, 117–19, 121
  functional, 117
  possible, 118
disorders, 135, 138, 252
  circulatory, 108
  intermittent, 205–6
  mental, 40, 136, 150
  myelitic, 67
  psychic, 68, 207
  psychosensory, 68
driving forces, 3–4, 6, 16–17, 25, 37, 234, 262
  realistic, 56
  relevant, 12, 21, 38
  zero, 6
dualism, 87, 90, 101
  particle wave, 92
duality, 216
  conceptual, 215
dynamical variables, 240, 243, 245–46, 249, 251, 262, 364, 368
  molecular, 250
  relevant, 258
dynamics, 43, 109, 244, 258
  functional, 241, 243
  global, 250

education, 90, 156, 226, 235, 262
first scientific, 225
physical, 233
EEC, *see* entropy-enthalpy compensation
basic, 38
perfect, 24–25
valid, 9, 15
valid linear, 10
EFA, *see* exploratory factor analysis
nonlinear, 245
polynomial, 245
EFAC, *see* exploratory factor analysis of correlations
effect and cause, 175
effects, 113, 182
acute, 203
allosteric, 43, 242–43
enthalpic, 32, 262
entropic, 16–17, 32, 253, 262
internal, 154
intrinsic common, 251
oxidation, 113
pathological, 204
periodic, 202
thermal, 104
eigenvalues, 251–52
eigenvectors, 244, 251–52
Einstein, Albert, 92, 223, 226–27, 230, 232, 234
Einstein's relativity theories, 102, 223
electrical conductivity, 114, 157
electricity, 62, 87, 91–92, 112, 114
electro-capillary concept, 103
energetics, 2–4, 39–41, 58–63, 73–75, 99–103, 107–9, 115–17, 121–23, 127, 129–31, 139, 141–42, 161–63, 165–67, 219–20
mechanical, 165
energy, 12–13, 62–63, 74–75, 77–80, 86–88, 90–99, 101–3, 105–7, 119–20, 127–28, 130–32, 140–43, 162–63, 165–66, 178–80
brain, 131
calorific, 63, 116
capillary, 113
cosmic, 122
electro-capillary, 122, 124
external, 123, 134
free, 8–10, 37
luminous, 95
manifestable, 92
mechanical, 104, 164
muscular, 124
nervous, 124, 130
observable, 92
potential, 3, 12, 37, 63, 92, 227, 262
psychic, 119, 127
usable, 97–98, 163
energy conservation, 13–14, 79, 86–87, 89–90, 92, 100–102, 127, 130, 141, 162–63, 165–66
energy conservation law, 92, 102
energy dissipation, 86–87, 92–93, 96–98, 101, 141, 180
energy/exergy, useful, 57
energy forms, 79, 93, 96, 103, 120, 123, 162–63, 165
physical, 127
psychic, 127
energy notion, 63, 75, 80, 87, 90–91, 102
energy transformation, 14, 62, 97, 103, 107, 118, 120, 122, 127, 165–66
Engelbrektsson, Nils, 2, 26, 33, 41–43, 51, 129, 132, 234
ensemble, 19, 33
conventional Gibbs, 38

grand-canonical, 33
microcanonical, 33
enthalpy, 3, 5–6, 9–10, 12–13, 15, 17–18, 37, 51, 63
entropy, 3, 5–6, 9–12, 33–37, 56–59, 61–63, 84–87, 91–92, 97–103, 127–34, 138–44, 154–55, 180–81, 219–20, 222–24
acquired, 62
differential, 36
low, 97
maximum, 17, 62
molar, 51
total, 97
zero, 6, 98, 154
entropy-enthalpy compensation (EEC), 1–2, 4, 9–10, 15–17, 24, 29, 32, 37, 42, 218, 234, 246, 248–49, 252, 262
entropy notion, 3, 19, 36, 40–41, 51, 62–63, 92, 97, 100, 102, 128–29, 139, 154–55, 215, 220–21
crucial, 41
fundamental, 36
environment, 12, 15, 84, 133–34, 138
current, 133
immediate, 95
professional, 230
enzymes, 240–41, 244, 251, 258, 261
Epstein, 226–27, 229–30
equilibrium, 15, 21, 39, 75, 82, 97–99, 107, 133, 138, 150–53, 161, 173, 180
actual, 17
broken, 138
dynamic, 152–53
entropy-enthalpy, 37
mystic, 154
stable, 75, 151
static, 152–53
universal, 98
unstable, 151–52
equilibrium thermodynamics, 3, 39, 41, 53, 150–51, 153, 213, 220–21
conventional, 1, 53, 85, 220
equivalence, 60, 79–82, 84, 87, 92, 98, 103–5, 118–19, 127, 129, 141, 161, 163, 165
reciprocal, 119
work-and-heat, 164
Euclid's postulate, 70
Euler angles, 250, 363
Euler's criterion, 41
evolution, 61–65, 87, 97–98, 100, 128–29, 133, 135, 137, 139–40, 181, 183, 186, 188, 191–92, 194–95
actual, 181
conceptual, 67
general, 67, 130, 137
progressive, 109
sensitive, 101
thermal, 101
excitation, 112, 115–16, 122, 124–27, 131–32, 142, 166, 174, 180, 186–87, 189–90, 209
cutaneous, 185
electrical, 111
external, 174–75, 184, 189
just-perceptible, 186
possible elementary, 29
prior, 142
relevant initial, 190
slightest, 111
excitement, 126, 129, 131–32
cerebral, 113
prevalent emotional, 36

exclusion, 71, 128
existence, 105, 113, 123–25, 127, 140, 161, 163, 165–66, 184, 189, 191, 209, 215, 232
  near-normal, 206
exploratory factor analysis (EFA), 4, 37, 43, 218, 220, 239, 241, 251–54, 256–57, 376
exploratory factor analysis of correlations (EFAC), 4, 40, 249
extensions
  arboreal, 110
  neuronal, 130
  spatial, 88

factor analysis, 218, 244, 249, 252–54, 256–57, 261
  applied, 248
  confirmatory, 218, 250–51
  conventional linear, 246
  linear, 243, 246
  nonlinear, 246, 256–57
  orthogonal, 256
  polynomial, 246, 257, 261, 363, 367
factor analysis of correlations, 218, 249, 253
factor loadings, 243, 246–48, 250
  pertinent, 247
  positive, 246
  relevant, 245
faculty, 192, 206, 229–30, 232
  intellectual, 207
  local, 230
  mental, 206
  pertinent, 196
  weakened, 191
Fechner, 121–22, 125–26, 179, 219–20
Fechner's law, 121, 175
Fechner's psychophysical theory, 219
first approximation, 4, 100, 102, 182, 210
  plausible, 246
forces, 2, 68, 71–75, 89–90, 101, 103–6, 147, 151–53, 161, 169, 182–83, 193–96, 199–204, 206–7, 209
  antagonistic, 182, 197, 199
  central, 79
  constant, 197, 199–200
  delayed, 201
  delaying, 206
  driving/livening, 75
  elastic, 107
  electromotive, 112
  external, 78, 81, 133
  favorable, 182, 194
  friction, 153
  gravitational, 75
  inciting, 184
  livening, 63, 91–92, 102
  livening/driving, 63, 73
  motor, 97
  nervous, 100, 124
  nonperiodic, 202
  psychic, 130
  realistic, 153
  single, 194
foreign energy, 94
foundation
  physical, 223
  statistical-mechanical, 2
  theoretical, 240
François-Franck, 112, 117, 146
Franzén, Karl, 2, 26, 33, 41–43, 51, 234
friction, 79, 153, 164
Frost–Kalkwarf approach, 41–42
functional dependence, 4, 50, 52
  actual, 3, 32
  approximate, 6
  general, 2

fundamental laws, 15, 73, 166
  first, 127
  second, 106

GAP, *see* GTPase-activating protein
gas, 13–14, 51, 78, 89, 126, 173, 176–80, 222, 224, 233
  ideal, 2, 4, 13–14, 26, 51–52, 217
  theory of, 179
Gauss hypergeometric, 27
Gaussian hypergeometric functions, 25, 28
GDP, *see* guanosine diphosphate 241–42, 244, 247, 257–60, 361
geometric progression (GP), 125, 171–73, 189
Gibbs, 2, 8–10, 15, 18–19, 26, 32–33, 37, 39, 56–57, 213–14, 217, 223–24
Gibbs and Helmholtz functions, 11, 18
Gibbs entropy, 33
Gibbs function, 2, 5, 7, 10, 12–15, 17–18, 38
Gibbs thermodynamic potential, 42
GP, *see* geometric progression
GTP, *see* guanosine triphosphate
GTPase-activating protein (GAP), 103, 241–42, 257–58, 260, 268
guanosine diphosphate (GDP), 241–42, 244, 247, 257–60, 361
guanosine triphosphate (GTP), 241–42, 244, 247, 258–60
Guldberg–Waage law, 173–74, 180–81

hallucinations, 68, 129, 132, 135, 137–38, 143, 191–92
  genital, 136
  injurious, 136
  terrifying, 136
Hamiltonian, 98–99, 163
Hamiltonian mechanics, 166
heat, 3, 12–14, 62, 76–83, 87–88, 92–99, 103–8, 112, 116–20, 126, 128, 141–42, 162–65, 170, 180
  absorbed, 83
  animal, 104, 106, 164
  constant, 82
  loss of, 82, 96, 164
  released, 62
  transformation of, 76, 80, 105
  uniformly distributed, 98
  useful, 162
heat absorption, 84–85, 94, 105
heat capacity, 14, 25, 28–29
  constant isochoric, 14
  isobaric, 14
  measurable, 29
heat energy, 62, 95–96, 104, 224
heat notion, 92, 128
heat release, 85, 93, 104, 108, 118, 128
heat sources, 83, 106
heat-work asymmetries, 213
Helmholtz, 2, 8, 10, 15, 18, 37, 86, 89–90, 104, 106, 123, 125
Helmholtz and Gibbs functions, 5, 14–15, 38
Helmholtz and Gibbs thermodynamic potentials, 42
Helmholtz energy, 9
Helmholtz function, 8, 11–14, 17–18, 30
Helmholtz hypothesis, 91
heredity, 129, 133–34, 139, 142–43, 191, 204
  healthy, 207
  normal, 133–34
  transformed, 204
hindrance, 17, 24–25, 32, 63, 154, 247, 253

hindrances/obstacles/resistances (HOR), 3, 6, 12, 16, 21, 32, 38, 63
HOR, *see* hindrances/obstacles/resistances
Horstmann–Liveing representation, 6, 63
Horstmann's energetics, 26
hydration-counterion sheaths, 241, 268
hydrolysis, 242, 258
hydrophobicity, 43

impressions, 72, 100, 103, 166, 179–80, 184, 186–87, 189, 229
 auditory, 184
 sensible, 117
 sensory, 123
 strong, 61
impulses, 135, 137–38, 153, 193, 205
 wild, 206
incubation, 206
 short, 207
indifferent balance, 134, 151–54
inertia, 183–84, 194, 201–2, 204–5, 207, 225, 258, 363
 cerebral, 134, 193, 202, 206
 progressive, 122
 total, 247
inertia/gyration tensors, 250
infection, 66–67, 203
 known, 201
instability, 62, 101, 133, 154
 excessive chemical, 111
intellectual work, 117–20
 static, 118
interconnections, 26–27, 218, 221
 mathematical/logical, 20
internal energy, 5, 8, 12–15, 25–26, 30, 76, 78–80, 84, 97, 123, 134
interrelationship, 19, 129
 close, 105
 dialectic, 19
 intrinsic, 18, 219
 time-entropy, 180
intervention, 88, 94, 204, 207
intoxication, 60, 66–68, 201
 acute, 203
 daily alcohol, 202
irreversibility, 87, 91, 137, 142, 166, 213
 practical, 99
irritation, 114, 124–25, 165–66, 174, 179–80, 184, 186
 external, 186
 increasing, 166
 relevant excitement, 174
 single tactile, 185
isothermal, 50, 82

Janet, Paul, 188–89, 195
Joffroy, 59, 61, 68, 149
Johal, 20, 25–26
Joule, 76–77, 79, 141

kinematics, 258
 complex, 69
kinetic energy, 3, 37, 63, 262
Kleiner, Alfred, 225, 230, 232
knowledge, 39, 61, 64–65, 155, 168
 acquired, 60
 exact, 184
 incomplete, 21
 limits of, 217
 scientific, 168

Laborde, 103, 108, 116–19, 147
Lambert function, 42
Landry disease, 67
Landry's paralysis, 63
Langevin thermostat, 260
language, 160

algebraic, 167
mechanical, 181, 195
ordinary, 195
ordinary everyday, 179
latent factors, 243, 245–46, 251–52
latent heat, 78
molar, 51
Lavoisier, 59, 88–89, 91, 104, 116
laws, 13, 19, 56, 58, 88–89, 91, 93–94, 111, 115, 121, 125–27, 160–63, 173–76, 179–81, 195–98
dementedness, 190
elasticity, 89
entropy conservation, 37
experimental, 108
insanity, 195
logarithmic, 175, 178, 180, 187, 189, 197
mechanical, 200
natural, 71
ordinary, 139
physico-chemical, 121
proportional, 175
simple dementia, 202
sinusoidal, 197
Léchalas, M., 91, 98
Lépine, 110, 147
Liebig, 59, 103–4, 106, 111, 147
Liebig's chart, 120
Liebig's theory, 107
linear factor analysis model, 245
Linhart, 2, 4, 6, 11, 15–16, 19–20, 24–26, 31–33, 35, 37–38, 57–58, 220–21, 227, 235
Linhart's approach, 2, 25, 29, 42, 57
Linhart's approximation, 25–26
Linhart's coefficient, 24
Linhart's expression for entropy, 220
Linhart's formula, 29
Linhart's function, 24
Linhart's parameter, 36
Linhart's process efficiency measure, 32
Linhart's theory, 25, 28, 32
Lippmann, 60, 108, 113–14, 145, 147
liquids, 50–51, 108, 113–14, 228
ingested, 68
Liveing, 32, 41, 63
Lombard, 116–18, 121, 128, 138
Lord Kelvin, 3, 144

machines, 79, 81, 83–84, 107, 144, 160, 165
chemical, 165
compressed air, 96
muscle-heat, 103, 106–7
perfect, 82–83
realistic, 81
macromolecules, 241
madness, 59, 135, 137, 139, 191–92, 196, 205
depressive, 209
intermittent, 191
interpretative, 205
lucid, 135
persecuted-persecutors, 138
systematic, 143
systematic hallucinatory, 135
Magnan, 59, 133–35, 147, 149
mania, 135–36, 143, 193, 205
delusional, 117
manifestation, 62, 89, 118, 120, 132, 138, 146
characteristic, 194
external, 118, 132
mental, 183
varied, 138
Margenau, 214
Mariotte, 126, 175, 177, 179

Mariotte's law, 176, 179
mathematical difficulties, 125, 195
mathematical expressions, 29, 82, 102, 179, 195–96, 219–20
mathematical form, 121, 221
  actual, 21
  approximate, 16
  conventional, 179
  elementary, 129
mathematical games, 57, 240
mathematical instruments, useful, 38
mathematical language, 80
mathematical model, 26
mathematical object, 176
mathematical relation, close, 178
mathematical relationship, simple, 186
mathematical statement, 84
mathematical technique, 214
mathematical tricks, skillful, 41–42, 53
mathematics, 65, 215, 233
  elementary, 71
matrix, 244, 251, 253, 257
matter, 2, 20, 39–40, 85, 88–90, 101, 151–52, 170, 215, 219, 228
  animate, 89
  inanimate, 62
  inert, 89
  living, 156
  white, 113
Maupertuis' principle, 63, 75, 138, 151
maximum work, principle of, 94, 96–97
Maxwell, 51, 90
Maxwellian distribution, 260
Maxwell relation, 52
Maxwell relationships, 5
Mayer, Robert Julius, 13, 60, 79, 103–5, 107, 147
Mayer principle, 76–77, 141
McDonald, 256–57
mechanical theory, 72, 75, 86, 91, 159, 163–64, 203–4, 210
mechanical work, 62, 75, 96, 104, 141
mechanics, 71–72, 86, 99, 162, 169, 181, 183, 200, 216, 231
  cerebral, 182, 193
  chemical, 173
  dethroned conventional, 74
  modern, 162, 167
  modern energy, 163
  molecular, 259
  nonrelativistic particle, 216
  nonrelativistic quantum particle, 216
  ordinary, 193
  valid, 210
mechanisms, 16–18, 64, 91, 107, 168–69, 171, 195, 197–98, 200, 204, 251–52
  actual, 252
  chemical, 180
  detailed molecular, 43
  enzymatic, 240
  general, 198
  guide, 169
  microscopic, 1, 37, 43
  state transition, 242
medicine, 59, 65, 150, 157
  clinical, 67
  experimental/statistical, 255
melancholy, 136, 143, 191
  distinctive, 136
  reputed, 136
memory, 1, 100, 129, 131, 137, 139, 143, 184, 234, 239
memory disorder, 139
mental capacity, 196

mental debility, 137
mental illnesses, 139
mental phenomena, 129–30, 142, 170, 200–201
  pathological, 103
mental states, 138, 141, 143, 206–7
  pathological, 192
metaphysical, 103, 128–30, 137, 150, 166, 168
metaphysical framework, 153
metaphysical Hamiltonian theory, 163
metaphysics, 151, 153–54
Meyer, Edgar, 226, 228–29, 231
Meyer's Judaism-anti-Semitism, 229
microbes, 66–67
microorganisms, 67
microparticle, 7, 19
modality, 17–18, 32, 68, 225, 233, 241, 249
  general, 80, 110
  microscopic, 37, 240, 262
  practical application, 256
  thermodynamic, 21
model, 35, 112, 115, 183, 241, 257
  cyclic-engine, 153
  mechanistic, 179
  polynomial, 257
  schematic, 115
  standard, 216
  stick, 242
molecular structure, 39, 100
molecular-targeted drugs, 242
Mosso, 116, 119, 128, 148
Mouret, 87, 91, 97, 99, 144
muscle fibers, 108–10, 121, 124
muscles, 60, 69, 103–5, 107–8, 110–11, 113–14, 117, 122, 126, 146–48, 164
  cardiac, 110
  long-time dead, 108
  thermodynamics of, 102, 118
  voluntary, 110
  working, 110
muscular contraction, 69, 107–8
  dynamic, 105
  static, 105
muscular work, 105–6, 118–20, 123, 164–65

nanoscience, 239, 261
nanotechnology, 239, 261
  modern bioinspired, 240
Napier, 172
nerves, 103, 108, 111–15, 122, 126, 130, 146
  artificial, 113–15
  healthy, 115
  living, 115
  motor, 109, 113
  natural, 113
  peripheral, 111
  schematic, 114–15
  sensory, 109
nervous centers, 111, 118, 128, 131
nervous system, 60–61, 80, 87, 103, 108–9, 115–16, 118–19, 122–24, 127, 130–35, 137–39, 142, 150, 165–66
  central, 122
  healthy, 142
neurons, 67, 108–12, 121–22, 124, 127, 130, 132, 138, 142
  elongated, 111
  isolated, 130
  motor, 109
  motor-sensory, 109
  peripheral, 127
  sensory, 109
Newton, Isaac, 3, 87, 91, 163, 208
Newtonian gravity, 62

Newtonian mechanics, 3, 249
Newtonian theories, 89
Newton's mechanics, 154
noise, 126, 184, 229
conventional thermal, 244–47, 261–62

objections, 94, 112, 116–17, 120, 162, 165, 227, 231
fundamental, 111
real, 94
opposition, 89–90, 92, 99, 101, 135, 162
radical, 91
sempiternal, 189
organism, 76, 132, 138, 170, 175, 180
organs, 88, 103, 111, 123, 142, 157, 165, 184–85, 194, 202
cerebral, 110, 188
pertinent, 164
relevant, 100
sensitive, 174
single, 88
oscillations, 66, 140, 149, 160, 184–86, 192, 197, 200, 202, 204, 206, 209
continual, 65
damped, 184–86, 188, 200
decreasing, 197
grandiose, 99
negative, 114
principal, 192
successive, 160
Ostwald, 70–71, 86, 88, 91, 145, 151

particles, 5, 8, 34, 51
cathode ray, 232
solid, 109
Pasteur, 67
pathogenesis, 137, 200
pathological inheritance, 134, 192–93
pathological states, 199
intermittent, 210
pathology, 132, 154
cerebral, 117
general, 201
infantile, 61
mental, 118, 149, 151
nervous system's, 142
patients, 68, 117, 135–39, 192
accentuated, 191
degenerate, 136
Patrizzi, 108, 148
Paulhan, 138, 148
PCA, *see* principal components analysis
perception, 69, 122, 127, 165, 185, 190, 195, 198
active, 103
duplicate, 185
periodicity, 202–3
PFA, *see* polynomial factor analysis
Pfaff transformations, 27
phase transition, 10, 38, 51
microscopic, 1, 42
philosophers, 175, 217, 219
phobias, 137, 193, 205
physical-chemical processes, 1, 18
realistic, 16–17, 38
physical-chemical systems, 17
realistic, 15–16, 18
realistic nonideal, 16
physical laws, 164, 175
ordinary, 126
physical sense, 3–4, 9–10, 12, 14, 32, 34, 38, 41, 63, 85, 150, 243–44, 247, 268
actual, 2, 15, 18, 32, 34, 40, 240, 252
clear, 3, 34, 252
distinct fundamental, 220

exact, 39
pertinent, 15
real, 102, 215
physical theories, 92, 216
modern, 216
productive, 213
useful, 154, 166
physicists, 50, 56, 58, 69, 107, 113, 128, 155, 161, 163, 165, 168, 181–82, 214–16, 220, 226
bachelor, 150–51
revolutionary, 102, 163
worldwide, 57
physics, 41, 57, 64, 75, 77, 113, 125, 150, 158, 162, 167–69, 179, 214, 216–17, 232–33
actual, 153
experimental, 141, 226, 233
high-frequency, 228
particle, 216
quantum, 39–41, 43, 223
realistic, 163
physiologists, 103–4, 107–8, 165, 187
physiology, 60, 103, 107, 109, 111, 113–14, 116, 121–22, 124, 127, 142, 147, 150, 179, 181
cerebral, 100
experimental, 107
general, 103
muscular, 108, 164
nutrition, 60
Planck, Max, 74, 217, 223–24, 227–29, 231, 233
Poincaré, H., 71, 87–88, 91, 145
polynomial exploratory factor analysis, 239
polynomial factor analysis (PFA), 246, 257, 261, 363, 367
posers, 10–11, 13–15, 18, 20, 22, 34, 119, 121, 131, 155–56, 214–16, 223–24, 245, 247–48, 252–53
basic, 13
crucial, 38
excellent, 155
general, 3
inconvenient, 91
philosophical, 129
tricky, 243
positivist, 137
positivistic methodologies, current, 216
Pouchet, 116, 120–21, 148
precision, 68, 76, 98, 166
current, 186
predictions, 167, 169, 193
verifiable, 159
preformed dipoles, 228
preformed electric dipoles, 232
pressure, 2, 12–14, 16, 23–24, 32, 36–37, 50–51, 78, 95, 126, 157, 175–79, 260
cohesion, 6
critical, 50
external, 179
higher, 6
internal, 14
lower, 6
negative, 52
reference, 9
relative, 11
vapor, 224
Prigogine's school, 102
principal components analysis (PCA), 251
principle of conservation, 87, 91–92
principle of Maupertuis, 63, 75, 138
principles, 40, 42, 60, 75–76, 79–80, 82, 86–90, 92, 94, 104–5, 141, 162–66, 186, 194–95, 216
consistent, 93

conventional, 162
core, 99
dialectic, 154
fundamental, 59, 94, 141, 193–94, 196
general, 60, 167
microscopical, 92
physical, 59
universal, 162
principles of energetics, 61, 107, 139
principles of mechanics, 162, 181
principles of thermodynamics, 56, 76, 81, 86, 92, 107–8, 130, 141, 163, 165–66
probabilistic interpretation, 3, 57
probability, 19, 21, 23, 29, 34, 213, 215, 217, 220, 240, 252, 254
magic, 35
probability theory, 4, 37, 40, 215, 224
standard, 29
problems, 3, 33, 35, 39, 51, 214, 218, 220, 223, 226, 231, 233, 250, 253–54, 256
anteriority, 161
biophysical, 59
complicated mathematical/numerical, 4
fundamental, 42
nontrivial, 35
pedagogical, 168
physical, 214
physical-chemical, 39–40
practical, 40
theoretical, 58
processes, 3, 15–17, 21–22, 24–25, 32, 37, 49, 62, 64–65, 95–96, 167, 183–84, 202, 244, 246–48
dynamical, 246
interatomic, 233
irreversible, 58
macroscopic, 213
melting, 233
mental, 181
oscillatory, 202
pathological, 61
practical, 58
psychic, 165, 201
simulated, 247
thinking, 120
working, 29
products, 13, 52, 72, 101, 110, 133, 142, 172, 174–75, 178–79, 187, 193, 195–96, 209, 241
atavistic, 133
carbonic acid, 95
decomposition, 67
excretion, 141
relevant excretion, 173
tried-and-true, 20
unstable, 112
progress, 17, 21, 43, 56, 60, 62, 100, 140, 163, 183, 253
actual, 65
insensitive, 99
scientific, 160
progression, 125, 171, 190
arithmetic, 125, 171
geometric, 125, 171
sensation's, 190
proofs, 103, 138, 142, 166, 226
direct, 123, 127
direct experimental, 127
direct legitimacy, 122
properties, 6, 14, 29, 69, 79–80, 89, 92, 104, 111, 128, 132, 180
chemical, 89
cyclic, 85
dominant, 69
electrical, 113
mathematical, 18
polarity, 86
proportionality, 175–76, 179

coefficient of, 174–75, 183, 186–87
law of, 180
psychic acts, 122–23, 138, 165, 182
simple, 174
psychic phenomena, 116, 119, 122, 127–29, 141, 165, 179, 181, 193, 195
psychological phenomena, 103, 165, 170, 174, 181, 183, 188, 193, 195, 210
psychological processes, 195, 200–201
conventional, 183
psychology, 124, 139, 143, 145, 160, 167, 219, 253
psychophysical law, 125–27, 165–66, 170, 175–76, 179–80
psychophysics, 129, 139, 166, 179–81, 219
true, 210
psychophysiology, 124, 165
Pugnat, 109, 148

quantum field theory, 216
quantum hypothesis, 227, 233
quantum mechanics, 33–34, 37, 57, 213–16, 220–21, 224, 252
quantum theory, 216, 224, 228–29, 231, 233

Rankine, 63, 74, 141
Ranvier, 114, 148
Ras-GDP, 241–42, 257–58
Ras-GTP, 241–42
Ras-GTP/GAP, 242, 257–58
Ratnowsky, Simon, 91, 221, 223–27, 229, 231–33
reactions, 94, 126, 136, 154, 161–63, 234, 243–44, 248
catalyzed, 241, 247
endothermic, 95
enzymatic, 43, 242
exothermic, 94
pertinent, 163
physiological, 174
reaction times, 123–24, 130, 142, 161, 165
muscular, 108
psychical, 127
real gases, 25–26, 29, 31, 40, 42
realistic processes, 3, 6, 16–17, 21, 25, 37, 56, 91, 224
reconstitution, 112, 115
reduction, 113, 205
reductionism, 215
Reichenbach, 217–18
relativity theory, 92, 234
representation, 170, 205, 257
all-atomic, 241
atomistic, 20, 40, 151
continuum, 34
definitive, 182
energetic, 166, 246
exact mathematical, 35
standard, 255
reversibility
perfect, 82
right, 131
reversible modifications, 132
Riemannian rectangles, 35
Riemannian sum, 34–35
Rosecrance, 249
Rummel, Rudolph Joseph, 249

sadness, 135–36
Scherrer, Paul, 226–27
Schiff, 103, 116–18, 123, 128, 138, 149
scholastics, 153
Schrödinger, Erwin, 33–34, 92, 225, 228, 230–31, 233

science, 63, 68–69, 71, 103, 107, 143–46, 159, 170, 181, 213, 217–18, 222–23, 232–33, 262
biological, 60, 157
end of, 216
materials, 243
modern, 68
natural, 101, 218, 251
observational, 60
physical, 59–60, 101
political, 248
psycho-chemical, 122
psychological, 122
relevant applied, 163
scientific character, 159, 193, 210
scientific methods, 61, 89
scientific research, 34, 68, 158, 168, 235, 262
scientific theory, 159–60, 167–69, 186
scientists, 159, 164, 167–68, 170
modern, 90
natural, 218
theoretical, 215
sensations, 69, 76, 88, 90, 117–18, 122, 124–27, 138, 143–44, 174, 179–80, 187–88, 190, 192, 197–98
fresh, 131
magnitude of, 175
negative, 125
new, 166
ordering, 69
perceived, 122
persistent, 192
psychological, 219
subjective, 72
simulations, 241–42, 251–52, 257–61
relevant molecular dynamical, 43
sleep, 110, 130
artificial, 130
induced, 110
natural, 110
neuronal, 111
solenoids, 168–69
solids, 224–25, 228, 233
Sommerfeld, Arnold, 226, 228, 230–31
stability, 61–63, 75, 96, 100, 120, 122, 127, 131–34, 136, 138, 140, 142, 151, 154, 249
acquired, 131
diminished, 205
extreme, 143
greatest, 96
mental, 142
Stallo, 71, 90–91, 151
statistical-mechanical approach, 3, 29, 37, 39
statistical mechanics, 2–3, 18, 20, 25, 32–34, 39–41, 43, 213–15, 220, 224
conventional, 38, 217
statistical thermodynamics, 39, 150
conventional, 38
statistics, 3–4, 28
comorbidity, 255
conventional correlational, 254
stress, 14, 92, 250
stretch, 64, 188, 218, 248
structural changes, 241–42
structure-function relationships, 241
subsystems, 21, 23, 25, 63
independent, 217
supramolecular, 239, 246–47, 261
surroundings, 12
natural, 244
system's, 12
unmediated, 21

symptoms, 66, 68, 131, 135, 139, 192, 194
  dominant, 135, 170, 194
syndromes, principal, 134
synthesis, 63, 65, 130, 140, 168
  general, 170

temperature, 6, 9, 13–14, 16–17, 21–26, 36–37, 50–51, 76–79, 81–84, 94–96, 111–12, 114–15, 117–19, 125, 260
  absolute critical, 50
  cerebral, 117–18
  critical, 50
  epicranial, 117
  function of, 16, 25, 36
  initial, 22, 26
  negative, 125
  notions of, 76, 128
  pertinent, 81
  reference, 6, 9, 260
  relative, 10
  total, 21
  zero, 9, 17
tension, 108, 111–12, 223
  high capillary, 111
  high chemical, 100–101
theoretical physicists, 33, 57–58, 227
theoretical physics, 38, 214, 223, 225–29, 231–33
theory, 36, 98–99, 103–4, 106–7, 109–12, 159–60, 164, 167, 175, 183–86, 191–93, 210, 215–16, 221–22, 233
  atomic-mechanical, 92
  atomo-physical, 91
  chemical, 104
  electro-capillary, 108, 112
  emission, 90
  energetic, 165
  general, 93
  grand, 64
  kinetic, 90
  metaphysical, 163
  molecular, 90, 228
  muscle-heat-machine, 105
  psychological, 179
  thermodynamic, 58
  wave, 90
theory of light, 90, 228
thermal machines, 81, 164–65
  maximum-yielding, 82
thermal motion, 3, 12, 15, 244
  chaotic, 15, 43, 241
  stationary, 16
thermodynamic entropy, 240, 253
thermodynamic equation, 2, 4, 40
  universal, 2, 26, 41–42, 51
thermodynamic functions, 10
  conventional, 16
  pertinent, 16
thermodynamics, 12–13, 56, 58–59, 61, 91–92, 102–3, 106–8, 118, 120–21, 141, 150, 213, 220–21, 246, 248–49
  applied, 42
  different, 58, 63, 155, 219
  engineering, 40–41
  first basic, 19
  macroscopic, 4, 20, 39–41
  second basic law of, 6, 51
thermodynamic system, 21
thermodynamic variables, 14
Thompson galvanometer, 114
time and space, 63, 69, 103, 127
Tolman, 57–58, 155
total energy, 17, 92, 94, 141
transformation, 76, 80–82, 84–87, 91–92, 95, 98, 100, 102–4, 107, 122–24, 127–28, 134, 141, 163–65, 173–74
  current, 91
  heat-to-work, 118

inverse, 104, 162, 164
irreversible, 84–85, 91
nontrivial mathematical, 16
reversible, 76, 84, 86
uncompensated, 86
work-to-heat, 118
transitions, 60, 86, 134
cyclic, 241
phase, 53
Trincher, Karl, 13–14, 156–57, 159
Trinczer, Karl, 156–58

universe, 84, 87–92, 98–99, 101, 105, 141, 163, 166, 180
limited, 94
utility, 96, 130, 159, 163, 165
lost, 96
system's, 96

validity, 39, 58, 126, 129, 166, 182, 185, 207, 214, 223
empirical, 53
general, 26, 37
restricted, 35
van der Waals approximation, 2, 41, 52
van der Waals equation, 4–5, 50
van Gehuchten, 115, 146
variable, 9, 53, 162, 170, 179, 182, 215, 217, 219, 243, 246–48, 250–52, 257–58, 268–377
dimensionless, 24
dynamic, 258
hidden/latent, 218
intensive, 2, 14
internal domestic, 249
kinematic, 268
local, 250
pertinent, 217
polar, 98
random, 28, 213
relative energy, 11
time-dependent, 241, 250
variation, 77–78, 84–85, 108–9, 111–14, 126–27, 131, 133, 147, 149, 182, 193, 195, 197, 203–4, 221–22
appreciable, 115, 126
concomitant, 128
correlative, 115
electrical, 111, 115
elementary, 87, 100
excitability, 111
logarithmic, 203
minimum, 125
small, 182
thermal, 112, 121
total, 78
zero, 133
velocity, 70, 73, 101–2, 193, 232
vessels, 95–96, 113
empty, 95
filled, 95
solid, 95

Ward, 87, 100, 145
water, 93, 95, 105, 114, 258–60
acidulated, 113–14
alcoholic, 114
Watt, James, 77
weakening, 192, 196, 204
intellectual, 190, 195–96, 198
wear
cerebral, 198
organic, 100
wear and tear, 180
Weber, 124–25, 127, 142, 174–75, 179, 186, 219–20
Weber–Fechner law, 122, 124, 142, 174, 181, 220
Weber's law, 124–25, 189
Weyssenhoff, Jan von, 233–34
Wiener, Norbert, 36

work amount, 12, 25–26, 29–30, 78
work and heat, 13, 162, 164–65
work
  intrinsic, 105
  preliminary, 119
  useful, 105
work-to-heat relation, 164
Wundt, 69, 100, 111, 123, 145, 149
Wundt's hypothesis, 112

X-rays, 187, 257

Young, Sydney, 50

zero entropy change, 154